DEVELOPMENTS IN QUATERNARY SCIENCE 5
SERIES EDITOR : JIM ROSE

ICELAND –
MODERN PROCESSES AND
PAST ENVIRONMENTS

Developments in Quaternary Science
(Series editor: Jim Rose)

For further information as well as other related products, please visit the Elsevier homepage
(http://www.elsevier.com)

Cover photograph by courtesy of Oddur Sigurðsson.

Developments in Quaternary Science, 5
Series editor: Jim Rose

ICELAND – MODERN PROCESSES AND PAST ENVIRONMENTS

by

C. Caseldine
University of Exeter,
Exeter, U.K.

A. Russell
Univ. of Newcastle-upon-Tyne,
Newcastle-upon-Tyne, U.K.

J. Harðardóttir
Hydrological Service,
National Energy Authority,
Reykjavik,
Iceland

Ó. Knudsen
Jarðfræðistofan ehf.
Reykjavik,
Iceland

ELSEVIER

2005

Amsterdam – Boston – Heidelberg – London – New York – Oxford – Paris
San Diego – San Francisco – Singapore – Sydney – Tokyo

ELSEVIER B.V.
Radarweg 29
P.O. Box 211, 1000 AE Amsterdam
The Netherlands

ELSEVIER Inc.
525 B Street, Suite 1900
San Diego, CA 92101-4495
USA

ELSEVIER Ltd
The Boulevard, Langford Lane
Kidlington, Oxford OX5 1GB
UK

ELSEVIER Ltd
84 Theobalds Road
London WC1X 8RR
UK

First edition 2005

Library of Congress Cataloging in Publication Data
A catalog record is available from the Library of Congress.

British Library Cataloguing in Publication Data
A catalogue record is available from the British Library.

ISBN: 0-444-50652-7
ISSN: 1571-0866 (series)

♾ The paper used in this publication meets the requirements of ANSI/NISO Z39.48-1992 (Permanence of Paper).
Printed in The Netherlands.

1460 5518

feo

ICELAND – MODERN PROCESSES AND PAST ENVIRONMENTS

ICELAND – MODERN PROCESSES AND PAST ENVIRONMENTS

Foreword

"Iceland – Modern processes and past environments", edited by Chris Caseldine, Andrew Russell, Jórunn Harðardóttir and Óskar Knudsen is the fifth volume in the *Elsevier Book Series* on *"Developments in Quaternary Science"*.

This book series, *Developments in Quaternary Science* is designed to provide an outlet for topics that require substantial space, are linked to major scientific events, require special production facilities (i.e. use of interactive electronic methods), or are archival in character. The subjects covered by the series will consider Quaternary science across different parts of the Earth and with respect to the diverse range of Quaternary processes. The texts will cover the response to processes within the fields of geology, biology, geography, climatology archaeology and geochronology. Particular consideration will be given to issues such as the Quaternary development of specific regions, comprehensive treatments of specific topics and compendia on timely topics such as dating methodologies, environmental hazards and rapid climate changes. This series will provide an outlet for scientists who wish to achieve a substantial treatment of major scientific concerns and a venue for those seeking the authority provided by such substantial treatments. This series is also linked to the Quaternary journals: *Quaternary Science Reviews, Quaternary International* and *Quaternary Research*, and should be seen as an outlet for publications that cover research that is complementary to that published in these journals.

Iceland – Modern process and past environments contains 13 research papers on topics that reflect the ways that a highly sensitive part of the Earth may respond to both large and small changes of climate and crustal activity. Thus, attention is given to the behaviour of the adjacent ocean and offshore processes, the response of glacier margins and the development of glacier processes, the nature and scale of chemical weathering processes and the development of soils and vegetation to Holocene-scale climate variation, particularly since the full expression of the Little Ice Age. Likewise a number of papers consider the effects of crustal processes with significant new analyses or reviews of glacio-isostatically forced shorelines, jökulhlaup processes, subglacial volcanic processes and the climatic effects of atmospheric SO_2 from flood lava eruptions. Amongst these straight scientific investigations is the welcome presence of a paper that evaluates the way that environmental and climatic changes were perceived and reported in written records, and so provides that critical insight into the way that humans respond to external stresses; something that is surely about to become more important for society as our population spreads to more sensitive areas of the globe and the effects of global warming become more apparent.

In addition to the research papers, the editors have provided an incisive brief introduction to the volume highlighting the main scientific and societal issues considered. In this introduction all of the papers in the book are identified by author and is available via the Elsevier web site at: http://www.elsevier.com/locate/series/dqs

where full details of the book and the contents are readily available. This is a new innovation for the *Developments in Quaternary Science* Book Series and we hope that it will make the contents of this important publication more accessible.

The publication of *"Iceland – Modern process and past environments"*, edited by Chris Caseldine, Andrew Russell, Jórunn Harðardóttir and Óskar Knudsen continues a tradition, begun in 1989 of bringing together important work carried out in a region that provides crucial insight into the way that sensitive parts of the Earth may respond to climatic and internal forces. This tradition is reflected in two previous publications that have been benchmarks in the study of Icelandic science. The publication of this volume brings together the Elsevier Quaternary book series and the Icelandic environmental research series, and it gives me great pleasure to welcome what is undoubtedly an important set of high quality research findings. These papers will provide a new benchmark for the study of maritime sub-arctic processes and climate changes and their expression in an area of active crustal spreading.

Jim Rose,
Series Editor

Preface

The idea for this volume came from the meeting *Iceland 2000* held at the University of Keele in April 2000. This meeting was the third in an occasional series where researchers interested in a range of environmental themes in Iceland had assembled and discussed current results and ideas in a very supportive and constructive atmosphere. The first of these was held in the University of Aberdeen in April 1989 with papers published as *Environmental Change in Iceland: Past and Present*, edited by Judith K. Maizels and Chris Caseldine (1991). The second took place in the University of Munich in December 1991 published as *Environmental Change in Iceland*, edited by Johann Stötter and Friedrich Wilhelm (*Münchener Geographische Abhandlungen Reihe B, Band 12*, 1994).

Rather than publish the research papers arising out of the Keele meeting it was felt that the time was right for a series of papers reviewing current research over a range of themes, incorporating very recent research and also pointing to future problems that require addressing. Thus following the meeting a number of scientists were approached to contribute, many of whom were at the meeting. Although it has not proved possible to cover every aspect of relevance to the theme given the incredible diversity of the Icelandic landscape and the need to publish within a reasonable timescale, the collection does span a wide range of some of the most important aspects of research in Iceland. Furthermore, it is the first volume to marry together research on current processes and past environments over such a range of topics, and as such we hope it will not prove the last such venture along this theme, and that areas not covered may be covered in any future revisiting of the themes discussed here. With the accelerating pace of research activity and the constantly evolving nature of ideas about environmental issues one of the main benefits of producing such a volume is to provide a benchmark against which future research can be evaluated.

Putting together the volume has required help from a number of people and we would like to thank in particular Sue Rouillard from the Cartography Unit in the Department of Geography, University of Exeter, who dealt so patiently with preparing all the figures and text for submission, and Professor Jim Rose of the Department of Geography, Royal Holloway, University of London for his constant encouragement and assistance.

With so much Icelandic material there is always the question of the extent to which Icelandic characters are used as against the 'internationalisation' of characters that often occurs in journals and material written in English. We have decided to use the correct Icelandic letters throughout but in the references have ordered papers according to the shorter English alphabet. Thus the Th sound, Þ, is used for all names whether originally printed in Icelandic or English, but is placed in the reference list where Th would appear.

A remarkable amount has been achieved but there is still much to be done, and increasingly there is a danger that as science becomes more and more compartmentalised crucial links between those reconstructing the past and those observing the present will be lost. A cursory inspection of the bibliography to this volume shows this trend and also emphasises the debt owed to early workers, particularly in the 20th century, whose visions were relatively broad; none more so than Sigurður Þórarinsson, to whom there are citations in virtually every contribution.

Whilst the overarching theme of the volume is to bring together the past and present there are a number of other issues that recur throughout.

1.1 The unique qualities of the Icelandic landscape

The juxtaposition of a heavily glaciated landscape with one of the most active volcanological environments on the globe, at a latitude at times marginal for human occupation, means that both today and in the past there has been a unique opportunity for the interplay of processes creating an extremely distinctive landscape, one that presents significant present and future challenges to humans. The Icelandic landscape thus offers the opportunity to view not only the results of these interacting processes, as shown for example in terms of landform assemblages by **Evans**, but also to view them in action, whether they be glacial or volcanic, an opportunity to view the 'consequences' whilst enjoying the 'essence' of scientific enquiry *sensu* Gould. Offshore the environment is equally significant with major boundaries and rapid changes in the character of the water masses, as demonstrated by **Jónsson and Valdimarsson**. Occasional rare opportunities have allowed observation of current interactions as seen in the case of the 1996 Gjálp eruption by **Guðmundsson**, and also by **Russell, Fay, Marren, Tweed and Knudsen**.

1.2 Scale and magnitude of processes and events

Issues of scale and magnitude have long been at the heart of geological and geomorphological enquiry, especially as researchers have become increasingly aware of the inability of current process studies to explain the nature of the events that occurred in the past to produce observed landscape features. As a 'gradualist' view of the physical landscape was found wanting it has been studies in environments such as Iceland that have revealed the likely magnitude of previously unobserved and unmeasured activity. Studies of sandar evolution in Iceland reviewed by **Russell, Fay, Marren, Tweed and Knudsen** have made clear the significance of jökulhlaups in building up extensive thicknesses of deposits with discharges clearly well in excess of anything yet measured directly, reminding us that the future may well offer challenges requiring a planning envelope far wider than perhaps believed likely. Similarly **Þórðarson** outlines how the impacts of flood lava eruptions have been underestimated, with previous views of them as being short-lived with little wider impact changing to a realisation that they are capable of releasing huge amounts of SO_2 into the atmosphere. Because of the lack of detailed observations of oceanographic change through time **Jónsson and Valdimarsson** also demonstrate a significant underestimation of the degree of variability in both the character and scale of water movement around Iceland.

1.3 Significance of the areas of limited or poor understanding

As should be the case in a volume of this nature most contributions highlight areas of limited understanding emphasising the need for future research to be concentrated on certain key topics. The oceanographic environment is one in which interpretation of past changes has relied on an all too simple view of the structure of the ocean floor around Iceland. **Jónsson and Valdimarsson** point out the complexity of the ocean structure around Iceland in an area crucial to the return of NADW as part of the thermohaline circulation.

On land **Hallsdóttir and Caseldine** emphasize the relatively weak chronological basis for understanding ecosystem change, despite the opportunities for using tephrochronology as an additional and potentially very precise chronostratigraphic tool, and similarly reveal the limited extent of understanding of just how terrestrial ecosystems responded to forcing over the Holocene in this sub-arctic region. The recent glacial record as summarised by **Sigurðsson**, based predominantly on detailed and consistent amateur observations, is still relatively poor in resolution before the latter part of the 20th century in comparison with more populated regions. Interpreting the record is complicated by the widespread, and often recurrent, occurrence of glacial surges, yet still reveals underlying similarities across a country which experiences very steep climatic gradients. For the earlier part of the Holocene **Wastl and Stötter**, whilst showing what can be achieved from extensive and detailed stratigraphic and morphological studies, similarly underline complexities in interpreting chronostratigraphic data.

1.4 Paradigms and perspectives

In a recent paper Ogilvie and Jónsson (2001) presented a thought-provoking analysis of how perspectives on the nature of 'natural' climate variability changed through time in Iceland depending on the climate experienced by the researchers concerned, thus creating significant paradigm shifts within which research was carried out. This approach is perhaps one that could be applied to other areas and which is in evidence here, where recent research based in Iceland has significantly altered current understanding. For understanding the origins of hyaloclastites the Gjálp eruption and the emergence, albeit briefly, of a newly-formed predominantly hyaloclastite ridge has allowed both confirmation of current thinking over processes and provided major new insights, as well as allowing estimation of extremely high rates of heat transfer between magma and ice **(Guðmundsson)**. **Evans** shows how the diversity of glacial environments in Iceland can contribute to a landsystems approach to landform development emphasising the value of seeing landscapes as a complex but coherent unit rather than as a combination of unique landforms.

Within the field of Quaternary studies reconstruction of former glacial limits and associated sea-level elevations has engendered an ever-changing set of paradigms and these are summarised and placed in context by **Norðdahl and Pétursson,** and also by **Andrews**. Relatively small changes in either the extent or age of various limits can have major impacts on modelling and understanding the wider implications of the assumed distributions of ice and sea through time. As future research concentrates increasingly on a model-based approach it is essential not only that the data are precise and accurate but that the way such data are interpreted is understood and, if not necessarily agreed with by all researchers, is still clearly embedded in a comprehensible methodology. Interpretations will continue to change through time, as will methodologies, but enduring logical structures are essential to any mode of scientific enquiry.

Given the theme of the volume and the sparsely populated nature of much of Iceland it is not surprising that the human perspective is not a prominent feature but a lot of the work discussed does have importance for communities. Hazards in the form both of volcanic threats and flooding risks from jökulhlaups are considered by **Guðmundsson, Þórðarson** and **Russell, Fay, Marren, Tweed and Knudsen.** Even though the threats in Iceland may appear to be well contained much can be learned from Iceland of value to communities elsewhere where there is a much closer relationship between human settlement and environmental hazard. Detailed analysis of climatic records derived from a range of human sources by **Ogilvie** reveals as much about how humans observe and react to climate, as it does about the climate itself.

1.5 Lessons for the future

There are many lessons for the future, both in specific scientific terms, and in the implications of the reviewed research for landscape conservation and human responses, both in Iceland and more widely. Lying on the fringes of the Arctic that is predicted to respond most quickly and noticeably to future predicted trace-gas induced warming, it is essential that we comprehend as fully as possible current processes and how their present operation compares to that in the recent past. Without this understanding any identification of the fingerprints of future change will only be possible within wide error ranges. If the predicted changes do operate over the next century then the impacts on Iceland could be severe, especially in the biotic environment affecting soils and weathering rates, as discussed by **Arnalds** and **Gíslason** respectively. After a century of relatively quiescent volcanism threats of major eruptions appear more likely, especially from Katla, thus improving our understanding of how to predict both the timing and character of such events is important.

The need for continued support for research in a wide range of aspects of the Icelandic environment discussed here is highlighted by all contributors, requiring increased levels of co-operation and certainly increased funding; the responsibility not just of Icelandic researchers but of the wider academic community. More can be learned of a diverse range of topics of wide significance in the small area of Iceland and its immediate marine environment than in most other areas of the globe.

2. Late Quaternary marine sediment studies of the Iceland shelf – palaeoceanography, land/ice sheet/ocean interactions, and deglaciation: a review

John T. Andrews

INSTAAR and Department of Geological Sciences, University of Colorado, Box 450, Boulder, CO 80309, USA

The Iceland margin is crossed by a series of broad, relatively shallow troughs which contain various thicknesses of late Quaternary sediments. The bounding slopes have a variety of morphologies and sediment types. Dates from Marine Oxygen Isotope Stage (MIS) 2 and 3 have been obtained from sediments recovered from Húnaflóaáll (N. Iceland), Djúpáll (NW Iceland), and the Látra Bank (W. Iceland). Sediments of this age have also been recovered from the slope of W and NW Iceland above the Denmark Strait sill, and from the base of the slope below Djúpáll. However, it is still not clear where the margins of the Iceland Ice Cap were located during the Last Glacial Maximum (LGM). It appears probable that the extent of ice on the North West Peninsula was restricted during the LGM and did not extend any distance across Djúpáll (the trough which extends from Ísafjarðardjúp to the shelf break). In places on the shelf, cores have penetrated through Holocene and late glacial marine sediments into underlying diamictons. Radiocarbon dates from sediments immediately above diamictons from troughs off N and SW Iceland give dates on deglaciation of ca 13.4k ± ^{14}C yr BP (with a 400 yr ocean reservoir correction). It appears that deglaciation of the shelf occurred rapidly after ca 13.4k ± ^{14}C yr BP and the ice was grounded near the present coast by 12.5 k ^{14}C yr BP. On the NW and N Iceland shelves, 3.5 kHz seismic surveys reveal regional reflectors which can be tracked over 10s to 100s of kilometers. Recovery of sediment cores which penetrate these reflectors indicate that they mark the Saksunarvatn and Vedde ashes with ages of 9k ± and 10.3k ± ^{14}C yr BP. A variety of evidence, including foraminifera and carbonate variations, suggests that the Holocene of N Iceland in particular was marked by rapid changes in oceanography. In the last 5k yr there has been a noticeable cooling of the climate, which led to the reoccurrence of species of cold water benthic foraminifera and a reduction in net carbonate production. The Little Ice Age is a marked feature in the sediment record from Reykjarfjörður, a small fjord on the Strandir coast of N Iceland.

2.1 INTRODUCTION

The climate of Iceland is strongly influenced by changes in ocean surface temperatures and changes in sea-ice extent and duration (Björnsson, 1969; Ogilvie, 1991, 1997; Ólafsson, 1999; Sigurðsson and Jónsson, 1995; Stötter *et al.*, 1999). Hence it seems obvious that significant information about the climate history of Iceland can be gained by an examination of offshore marine sediments. Furthermore, some of the large questions which overlie our knowledge of Iceland during the Quaternary are potentially answerable only from an examination of the records from the Iceland margin (i.e. the fjords, shelf, troughs, and slope). These larger questions must also be addressed by a study of the land records, but issues such as: the extent of ice around Iceland during the Last Glacial Maximum (LGM), the history of glaciation of the shelf, and high-resolution changes in palaeoceanography, must invariably come from studies of the offshore marine sediment sequences (Andrews *et al.*, 2000; Eiríksson *et al.*, 2000a, b; Haflidason *et al.*, 2000; Jennings *et al.*, 2000).

It is worth noting that the Icelandic offshore marine record has two elements that are highly unusual in the marine realm. The first is a richness of locally derived volcanic tephras which form critical isochrons in the shelf records and across the North Atlantic (Kvamme *et al.*, 1989; Sejrup *et al.*, 1989; Lacasse *et al.*, 1998; Haflidason *et al.*, 2000). The second is the high meltwater and sediment events associated with jökulhlaups, especially in SW and S Iceland which are today, and were probably major sediment transfer processes during deglaciation (Björnsson, 1992; Geirsdóttir *et al.*, 2000; Jennings *et al.*, 2000; Maria *et al.*, 2000).

2.1.1 Background

Iceland is well suited for late Quaternary marine investigations because a series of fjords and troughs provide excellent sediment traps and preservation potential. Seismic surveys have been undertaken by the Marine Research Institute over several years (Thors and Helgadóttir, 1991; Thors and Boulton, 1991; Thors and Helgadóttir, 1999), and seismic profiles have been obtained on portions of the W, NW, and N Iceland shelf in 1988, 1993, 1996, 1997 and 1999 (Helgadóttir, 1997; Syvitski *et al.*, 1999; Andrews *et al.*, 2000).

The Icelandic shelf is crossed by a series of broad and relatively deep troughs that extend from the coastline seaward to the shelf break. In places the troughs are structurally controlled but some element of glacial erosion is commonly held to be responsible for their broad geometries and bathymetry. The troughs commonly slope seaward and reach depths of between 200 and 650 m.

The present oceanographic and hydrographic setting of Iceland suggests that records from the Iceland margin should typify changes within the broader North Atlantic region. This is because the Iceland shelf represents the interplay between warm, salty water moving northward in the Irminger Current, and cold, fresher Polar Water entrained in the East Iceland Current (Fig. 2.1) (Stefánsson, 1962, 1969; Malmberg, 1969, 1985; Malmberg and Magnússon, 1982). In recent decades water column temperatures have shown average changes of over 5°C due to changes in atmospheric surface forcing (Ólafsson, 1999), and to changes in surface salinities (Dickson *et al.*, 1996, 1988; Belkin *et al.*, 1998).

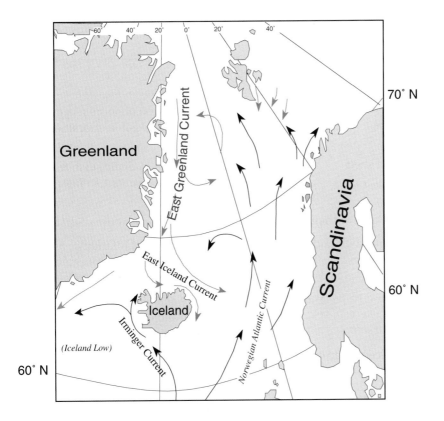

Fig. 2.1 Surface currents in the vicinity of Iceland and the northern North Atlantic. The Irminger Current is a branch of the Norwegian Atlantic Current, also referred to as the North Atlantic Drift. The East Iceland Current branches toward northern Iceland from the cold Polar East Greenland Current which represents surface outflow from the Arctic Ocean.

2.1.2 History of Quaternary marine studies

Although Quaternary studies, especially glacial and palaeoclimatic investigations, on Iceland have a long history, the importance of studying the Quaternary offshore sediment sequences along the Iceland margin has just started to be realized (Table 2.1). Prior to the late 1990s the few papers that had been published dealt with evidence for changes in relative sea level (Thors, 1978; Thors and Helgadóttir, 1991; Thors and Boulton, 1991), nature of shelf and fjord sediments (Thors, 1974; Haflidason, 1983), and some seismic interpretation of shelf stratigraphy and morphology (Egloff and Johnson, 1978). A significant start had been made by Helgadóttir (1984) on faunal and isotopic changes in Jökuldjúp, off SW Iceland, whereas Ólafsdóttir (1975) had described a massive ridge at the shelf/slope break and suggested it formed a major terminal moraine of the Iceland Ice Cap. In addition, a large and significant data base was built up, largely unpublished, from the Marine Research Institute (Iceland), consisting of detailed high-resolution seismic profiles, a large number of large diameter, 1-2 m long, gravity cores and some radiocarbon dates (Thors and Helgadóttir, 1999). Research on cores from the Denmark Strait, immediately west of Iceland (Fig. 2.2) had been very influential in defining late Quaternary oscillations of the marine Polar front (Ruddiman and McIntyre, 1981; Ruddiman *et al.*, 1994), and in indicating significant Holocene palaeoceanographic variability (Kellogg, 1984).

Table 2.1 List of pertinent papers on the Quaternary marine geology and paleoceanography of the Iceland margin (see references for full citation) .

Authors & year	Area	Remarks
Egloff and Johnson, 1978	Western Iceland shelf and slope	First major publication on offshore seismics. Identified a number of possible moraines and noted the Snorri Drift
Thors and Boulton, 1991	N Iceland shelf	Use of acoustic surveys to distinguish features on the seafloor which document an interval(s)? of lower Holocene sea level.
Ólafsdóttir, 1975	Western Iceland shelf	Documented large moraine at the shelf break
Syvitski et al., 1999	SW Iceland shelf and western slope	High-resolution seismics of Jökuldjúp, SW Iceland. Study of the Latra End Moraine and seismic survey of the west-central shelf
Eiríksson et al., 2000a	East-central N. Iceland	Study of downcore changes in foraminifera and sedimentology; chronology based on AMS ^{14}C dates and tephrochronology
Andrews et al., 2000	SW to N-central Iceland	Survey of basal AMS ^{14}C dates from cores collected in 1997. Discussion of seismic stratigraphy and magnetic susceptibility
Jennings et al., 2000	Jökuldjúp, SW Iceland	Discussion of site seismic and lithostratigraphy for HU93030-006 and environmental reconstruction based on benthic foraminifera for the period ca 10-13 k ^{14}C yr BP
Eiríksson et al., 2000b	N-E Iceland shelf	Stress on the importance of distinguishing tephras in marine cores as an aid to linking land and marine records. Documentation of late Holocene changes including an overall cooling in the last 5 k ^{14}C yr BP
Andrews et al., 2001a	N W Iceland shelf	Multiproxy study of a marine core from the head of Reykjarfjörður, a small fjord on the north coast of the NW Peninsula. An overall deterioration of climate is noted over the last 4 k ^{14}C yr BP with a major change ca 1 k ^{14}C yr BP. A clear Little Ice Age event is seen in the foraminifera

A major change occurred in the mid 1990s with cruises by CSS *Hudson* (cruise HU93030) (Hagen, 1995; Syvitski *et al.*, 1999), R/V *Jan Mayen* (JM96-) (Cartee-Schoolfield, 2000; Jennings *et al.* subm.), the PANIS 1995 cruise HM107 (Eiríksson *et al.*, 2000a,b), and the extensive Icelandic/USA cruise in 1997 on the *Bjarni Sæmundsson* (Helgadóttir, 1997; Andrews *et al.*, 2000). German cruises had also recovered cores from the Blosseville Basin, north of the Denmark Strait (site PS2644, Fig. 2.2) (Voelker, 1999). These efforts culminated in 1999 with the International IMAGES V cruise on the *Marion Dufresne* where around 10 giant piston cores were retrieved with lengths between 18 and 40 m. The IMAGES cores on the Iceland margin were taken by four groups, consisting of two Icelandic teams, a USA (Colorado) effort, and cores retrieved by the French.

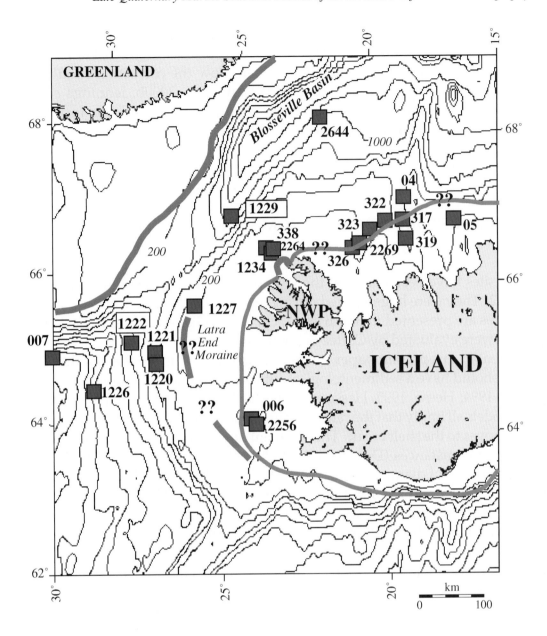

Fig. 2.2 Ice limits and marine cores discussed in this chapter. The grey line shows the postulated limit of ice during the Last Glacial Maximum (see text for discussion). Notice the ?? around large sections of the Iceland Ice Cap. The squares show the core sites. Sites 007 and 006 are from cruise HU93030. Sites 1222, 1229 etc are from cruise JM96. Sites 338, 322 etc are from the 1997 cruise B997. Sites 04 and 05 are from cruise HM107 taken in 1995. Locations 2256, 2269 are from the MD99cruise.

2.1.3 Review objectives

This review will concentrate on materials collected prior to 1999 and will include a survey of published results plus the presentation of some previously unpublished data. The review will concentrate on 1) evidence for the extent of glacial ice during the LGM ca 22k cal yr; 2) the timing of deglaciation of the Iceland shelf; and 3) changes of oceanographic conditions during the Holocene.

It is important to note that the issue of the correct ocean reservoir correction is exceedingly significant for the correlation of events around Iceland. It is known that it varies both temporally and spatially (Bard *et al.*, 1994; Voelker *et al.*, 1998; Haflidason *et al.*, 2000) but there is no common agreement as to how the correction should be applied. For this reason, most papers (e.g. Eiríksson *et al.*, 2000b; Jennings *et al.*, 2000) apply a 400 yr correction to the laboratory reported dates, and this format is used in the data on the figures and the date list (Table 2.2). However, later discussions will further highlight the problem.

2.2 GLACIATION HISTORY DEDUCED FROM SLOPE AND PROXIMAL DEEP-SEA CORES

A number of sites have been cored on the W and N Iceland slopes. These sites are shown on Fig 2.2. They were largely cored during the 1996 *Jan Mayen* cruise. Available radiocarbon dates are presented in Smith and Licht (2000) (Table 2.2) and JM96-1222, -1225, and -1229 were evaluated by Cartee-Schoolfield (2000).

A key argument in this present paper is that the inference of an ice sheet at the shelf break (Fig. 2.3) should have a sedimentological/lithofacies signature (Yoon *et al.*, 1991; Andrews *et al.*, 1994; Hesse, 1995; Hesse and Khodabakhsh, 1998; Vorren and Labert, 1997). These models all imply that the rate of sediment accumulation (SAR) will increase as an ice sheet moves to the shelf break. The sediment delivered to the slope will depend on glaciological circumstances (Dowdeswell *et al.*, 1996; Syvitski *et al.*, 1996a,b), but will include the supply of diamictons (as debris flows), turbidites, settling of sediment from meltwater plumes, and the input of iceberg rafted detritus (IRD).

A number of sites have also been studied from more distant localities, including our work at HU93030-007 on the western side of Denmark Strait (Andrews *et al.*, 1998a), and cores to the north and south of Denmark Strait from the Blosseville and Irminger basins respectively (Elliott *et al.*, 1998; Hagen, 1999; Voelker *et al.*, 1998; Voelker, 1999; Cartee Schoolfield, 2000; van Kreveld *et al.*, 2000). These cores should contain some information on the glacial history of Iceland in terms of changes in sedimentation rates, which will increase during periods of glacial advance and retreat (Haflidason *et al.*, 1998; Andrews, 2000), and changes in the contribution of glacially eroded Icelandic sediment in the form of ice rafted (IRD) basalts (Bond and Lotti, 1995; Elliott *et al.*, 1998). The pervasive presence of basaltic and rhyolitic angular glass shards is difficult to interpret in terms of glacial history because such sediments can be derived directly from explosive volcanism on Iceland (e.g. North Atlantic ash zones I and II (Ruddiman and Glover, 1972; Grönvold *et al.*, 1995; Lacasse *et al.*, 1996)) or they can be reworked and re-distributed as they are transported seaward on and within the Iceland ice cap(s) (Bond *et al.*, 2001). This glacial transport process can lead to significant delays in large volumes of ash being deposited (Jennings *et al.*, 2002). The role of sea-ice is probably also an important consideration in the transport of tephras around the North Atlantic.

The data from cores on the NW and W Iceland slopes (Cartee-Schoolfield, 2000) do not unambiguously negate or confirm ice at the shelf break during the LGM. If the Iceland Ice Cap stood at the shelf break during the LGM, as several reconstructions

Table 2.2 Radiocarbon dates (see Fig. 2.6 for location) from marine cores associated with delimitation of the LGM and deglaciation. Dates have a 400 yr ocean reservoir correction applied.

Core #	Date and error	Reference	Comments
Dates within diamictons			
B997-322	>37, 800	Andrews and Helgadóttir, 1999	136 cm in core
B997-323	>25,900	Andrews and Helgadóttir, 1999	90 cm in core
B997-326/1	23,170 ± 340	Andrews and Helgadóttir, 1999	at 190 cm
JM96-1227	35,650 ± 560	Syvitski *et al.*, 1999	base of core at 56 cm
Dates pertaining to deglaciation on the N and W Iceland shelf			
HM107-05	13,580 ± 90	Eiríksson *et al.*, 2000a	basal dates from 393-394.3
HM107-05	13,290 ± 100	Eiríksson *et al.*, 2000a	—
HM107-05	13,700 ± 140	Eiríksson *et al.*, 2000a	—
HM107-04	11,640 ± 80	Eiríksson *et al.*, 2000a	basal date 392 cm
B997-317PC1	11,870 ± 100	Kristjánsdóttir, 1999	basal date 245 cm
B997-319GGC	11,700 ± 110	Kristjánsdóttir, 1999	basal date 173 cm
B997-322	11,570 ± 130	Andrews *et al.*, 2000	date from above basal diamicton
B997-323	13,040 ± 190	Andrews *et al.*, 2000	date from above basal diamicton
B997-326/1	13,435 ± 215	this paper	date immediately above diamicton
B997-326/2	12,755 ± 95	this paper	date from base of core
B997-336	13,280 ± 70	this paper	date from base of core
HU93-006	12,705 ± 85	Jennings *et al.*, 2000	date from base of core
MD99-2256	13,390 ± 80	this paper	date above basal diamicton
Basal dates on continuous(?) sequences			
B997-338	34,200 ± 640	Harðardóttir *et al.*, 2002	basal date
JM96-1229	30,870 ± 380	Cartee-Schoolfield, 2000	basal date
JM96-1221	19,350 ± 180	Smith and Licht, 2000	benthic foraminifera
JM96-1221	17,840 ± 80	Smith and Licht, 2000	planktonic foraminifera
JM96-1222	23,185 ± 180	Smith and Licht, 2000	—
JM96-1225	39,750 ± 2600	Smith and Licht, 2000	—
JM96-1220	17,690 ± 80	Smith and Licht, 2000	—

suggest (Webb *et al.*, 1999; Bourgeois *et al.*, 2000), we might *a priori* predict high rates of sediment accumulation (Andrews and Syvitski, 1994) from a combination of glacial processes (Fig. 2.3). Off the NE Canadian margin, adjacent to the Laurentide Ice Sheet, average sedimentation rates (Andrews *et al.*, 1998b) during the last glaciation were typically 50cm/kyr but increased to over 100cm/kyr during Heinrich events (Jennings *et al.*, 1996). On the adjacent E. Greenland slope the sediment accumulation rate in HU93-007 (Fig. 2.2) between 14 and 28 k ^{14}C yr BP was over 30 cm/ky (Andrews *et al.*, 1998a, 1999).

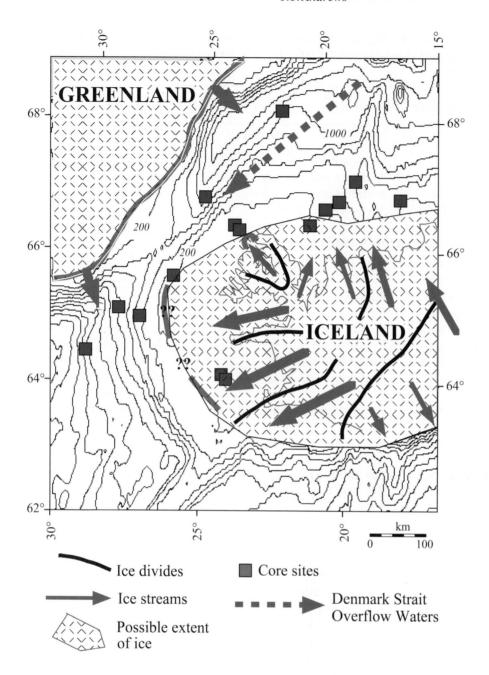

Fig. 2.3 Cartoon showing possible ice extent across Denmark Strait during the LGM and transport paths for icebergs and sediment plumes. Location of possible ice streams within the Iceland Ice Cap shown after Bourgeois et al. (2000) and Stokes and Clark (2001). Volcanic ash plumes also moved sediment onto the ice sheets and onto the ocean surface (sometimes onto sea ice).

Radiocarbon dates from the short gravity cores from the JM96- cruise off W and NW Iceland (Hagen, 1999; Smith and Licht, 2000; Cartee-Schoolfield, 2000) indicate generally slow rates of sediment accumulation (SAR) (Fig. 2.4), even though some sediment may have been deposited instantaneously as sediment gravity flows. SAR is quite low (10-12 cm/kyr) on the slope sites, seaward of the Latra End Moraine (Ólafsdóttir, 1975;

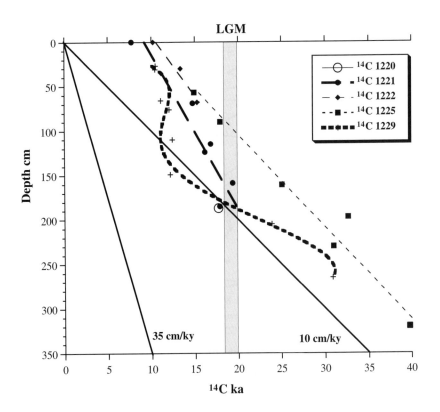

Fig. 2.4 Depth versus radiocarbon dates from five sites on the slope and basin margin off W and NW Iceland (Cartee-Schoolfield, 2000; Smith and Licht, 2000). The solid lines inclined lines give rates of sediment accumulation (10 and 35 cm/kyr). For all cores but JM96-1229 the sediment accumulation rate is ca 10-12 cm/kyr, but there is a noticeable interval of rapid accumulation in JM96-1229 between 10 and 12 ka. The shaded grey area indicates the span of the LGM.

Syvitski *et al.*, 1999), and it appears that deposition may have effectively ceased at many sites around 10 k ^{14}C yr BP (Fig. 2.4). At JM96-1221 on the upper slope, the average SAR is 14 cm/kyr and on the lower slope, at JM96-1222 (Fig. 2.2), the SAR is similar. The only site with an interval of rapid sedimentation is JM96-1229 from the southern limit of the Blosseville Basin, below Djúpáll (Fig. 2.2) (Cartee-Schoolfield, 2000). In this core, the slow SAR between 30 and sometime after 20 k ^{14}C yr BP was interrupted by an interval of rapid SAR between 10 and 12 k ^{14}C yr BP (Fig. 2.4). This interval of rapid sediment accumulation may reflect the rapid deglaciation of the N. Iceland shelf shortly after ~13 k ^{14}C yr BP (Andrews and Helgadóttir, 1999; Eiríksson *et al.*, 2000a).

Magnetic susceptibility (MS) of sediment offers the prospects of correlating events between land and adjacent marine depo-centers (Robinson *et al.*, 1995; Stoner and Andrews, 1999). It is an estimate of the concentration of magnetite in a unit volume (e.g. JM96-1220) or normalized for the mass per unit volume (Walden *et al.*, 1999). In Figure 2.5 the MS is shown for cores along the upper slope ca 500 m water depth, and immediately west of the Latra End Moraine (Fig. 2.2), and a site at 1040 m water depth toward the base of the slope. Correlations are based on major features of the records

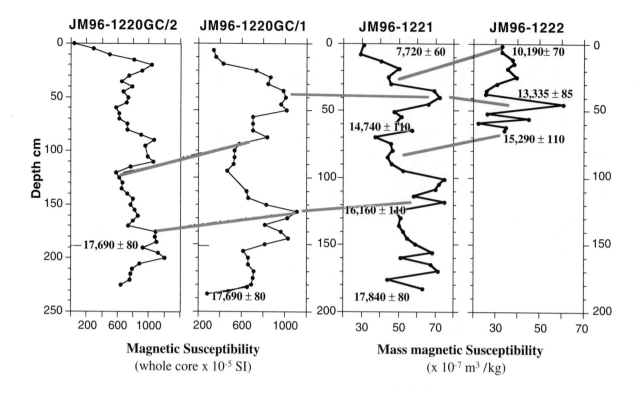

Fig. 2.5 Example of magnetic susceptibility logs from the west Iceland margin (see Fig. 2.2 for location) showing radiocarbon dates and possible correlative events (grey lines).

and are limited by a few radiocarbon dates on planktonic foraminifera (Smith and Licht, 2000). The base of the cores date from immediately post-LGM time and show two major fluctuations in this parameter. The peak in MS sometime post-14,740 and 15,290 and pre-13,335 ^{14}C yr BP might be correlative of the North Atlantic Heinrich event 1, which has been recorded on the adjacent East Greenland slope (Stein *et al.*, 1996; Andrews *et al.*, 1998a). Better dating is required to accept or reject this thesis, but Fig 2.5 indicates that changes in MS are occurring along and down the slope. It will be important to distinguish the origins of lithofacies on these slopes in order to distinguish between intervals of downslope gravity flows versus emplacement of iceberg rafting detritus (IRD) into hemipelagic sediments (Cartee-Schoolfield, 2000).

In conclusion, although definitive studies have yet to be made, the relatively low SARs on the slope off W and NW Iceland are not easily reconciled with an active ice cap located at the shelf break during the LGM. Cores are available for faunal and isotopic studies and progress on the regional ice sheet/ocean interactions will be presented in the next few years. These studies will complement available research on sites well north and south of the W/NW Iceland margin (Elliott *et al.*, 1998; van Kreveld *et al.*, 2000).

2.3 MARINE EVIDENCE FOR ICE EXTENT ON THE ICELAND SHELF DURING LGM AND THE TIMING OF DEGLACIATION

Bourgeois *et al.* (2000) and Stokes and Clark (2001) present geomorphological data and arguments for an ice sheet over Iceland, largely extending to the shelf break, with a series of ice streams embedded within it (Fig. 2.2). There are, however, some who infer that the extent of glacial ice on the Northwest Peninsula (NWP) during the Late Weichselian was restricted (Hjort *et al.*, 1985; Norðdahl, 1990; Voelker, 1999). The age of the massive Látra end moraine (Ólafsdóttir, 1975; Syvitski *et al.*, 1999) is unclear given the evidence for a restricted ice cap on the NWP. A series of moraines have been identified from bathymetric and seismic surveys along the Iceland margin (Ólafsdóttir, 1975; Egloff and Johnson, 1978), but their precise age(s) remain elusive. The key papers for this interval (Table 2.1) pertaining to the Iceland shelf are those of Syvitski *et al.* (1999), Eiríksson *et al.* (2000a,b), Andrews *et al.* (2000), Jennings *et al.* (2000), and Andrews *et al.* (2002). A date from the base of a 56 cm stiff diamicton just north of the moraine gave an age of 35,560 ± 560 (Table 2.2) (Syvitski *et al.*, 1999), but the stratigraphic relations between the core site and the moraine have not been explicitly investigated.

Site B997-338 in Djúpáll (Fig. 2.2) shows continuous sedimentation through late oxygen stage (OS) 3 and 2 (Andrews *et al.*, 2002b). The basal date of 34,200 ± 640 ^{14}C yr BP (Table 2.2) is followed by increasingly younger dates (Smith and Licht, 2000) and there is no evidence for reworking. This site indicates that the ice on the NWP did not extend any great distance out onto the shelf from Ísafjarðardjúp. A large ridge which loops around the end of this fjord system may be the LGM limit (Thors and Helgadóttir, 1999; unpubl.). These data appear to require a limited ice cap over the NWP (Rundgren and Ingólfsson, 1999). It is unclear whether a restricted ice extent on the NWP is glaciologically compatible with the formation of the Látra End Moraine (Fig. 2.2) during the LGM. To my mind it is difficult to generate sufficient ice to flow seaward to the shelf break off NW Iceland and produce this massive ridge (Ólafsdóttir, 1975; Syvitski *et al.*, 1999) and at the same time have a restricted local ice cap over the NWP.

2.3.1 Deglaciation

Dates from the base of marine cores on the Iceland shelf provide, in most cases, a minimum date on deglaciation (Andrews *et al.*, 2000; Eiríksson *et al.*, 2000a; Jennings *et al.*, 2000). This is because most of these cores did not penetrate the full glacial/deglacial sediment sequence, and did terminate in a stiff, compact diamicton which probably represents a subglacial facies (till). Cores from Jökuldjúp (MD99-2256) and Húnaflóaáll (B997-322, -323, and -326PC1) (Andrews and Helgadóttir, 1999; Helgadóttir and Andrews, 1999; James, 1999) do penetrate till-like sediments. However, only the Jökuldjúp site is clearly interpreted as a till/marine sequence (Jennings *et al.*, 2000). Here the date at the contact is 13,390 ± 80 ^{14}C yr BP (Table 2.2) (Fig. 2.6). In the area of Húnaflóadjúp (sites 322, 323, 326, Table 2.2, Fig. 2.2) only thin (0.5 to 1.5 m thick) postglacial sediments cover the diamictons. It is presently uncertain (Andrews and Helgadóttir, 1999; Helgadóttir and Andrews, 1999; James, 1999; Principato, 2000) whether the ice extended seaward of these sites, but in 323 the date near the contact is 13,040 ±

190 ^{14}C yr BP, and at 326PC1 a date of 13,435 ± 215 ^{14}C yr BP (Fig. 2.6) has been obtained immediately above a date near the top of the diamicton of ca 25 k ^{14}C yr BP. Where diamictons have been cored the resulting dates on marine carbonates (foraminifera and molluscs) range in age from ca 23 to >38 k ^{14}C yr BP (Table 2.2). The oldest deglacial date from basal core sediments off Iceland is that of ca 13,580 ± 90 ^{14}C yr BP, the median date from three separate assays from the base of HM107-05 (Eiríksson *et al.*, 2000a, p. 27) (Table 2.2; Fig. 2.6). In Djúpáll, NW Iceland, B997-336 has a basal date of 13,280 ± 70 ^{14}C yr BP (Andrews *et al.*, 2000; Smith and Licht, 2000). Neither this core site, nor HM107-05, specifically delimit the LGM nor deglaciation for their respective area (Fig. 2.6).

It is certainly true that the range of till contact and basal dates from W and N Iceland marine cores are only a few hundred years older than the first dates on deglaciation of Iceland within the present coastline (Ingólfsson, 1988; Ingólfsson and Norðdahl, 1994; Ashwell, 1996; Eiríksson *et al.*, 1997). These dates, of which there are relatively few, are frequently in the range of between 12 and 13 k ^{14}C yr BP. Overall the combination of the marine and terrestrial dates on deglaciation indicate that the Iceland Ice Cap retreated rapidly in probably less than 1 k yr (possibly catastrophically?) from the mid-shelf to the land and this was coincident with the formation of the marine limit (Ingólfsson *et al.*, 1995, 1997; Rundgren *et al.*, 1997).

Rundgren, Björck and others (Björck *et al.*, 1992; Rundgren, 1998) cored Torfadalsvatn on the Skagi Peninsula and obtained basal radiocarbon dates of 11,300 ± ^{14}C yr BP. The crude outline of the Iceland Ice Cap on Fig. 2.6 (in keeping with our lack of detailed evidence) outlines the approximate limits of the ice cap about 13 ± k ^{14}C yr BP. This can be compared to our data from the East Greenland margin (Andrews *et al.*, 1996; Jennings *et al.*, 2002) where deglaciation also appears to have commenced by approximately 14 k ^{14}C yr BP but was probably still offshore in Kangerlussuaq Trough at 13 k ^{14}C yr BP (Fig. 2.6).

2.3.2 Discussion

The dates on deglaciation noted above are all based on applying a 400 yr reservoir correction. However, it is known that off N Iceland at the Younger Dryas time that the correction was larger in the range of 800 to 1100 yrs (Bard *et al.*, 1994; Haflidason *et al.*, 2000). The age of the Vedde ash in the Greenland ice core record is close to 10.3 k ^{14}C yr BP (11.9k cal yr BP) (Grönvold *et al.*, 1995). The research of Bond *et al.* (2001) suggests that there have been more than one major eruption during the "Vedde interval", and Jennings *et al.* (2002) have shown that late deglacial sediments in the Kangerlussuaq Trough, East Greenland (Fig. 2.2) contain tephra shards with a geochemical signature of Ash Zone II, an ash erupted around 50,000 years or so ago! It is assumed that the ash was erupted on the Greenland Ice Sheet and calved off some 36,000 ± yrs later! Bond *et al.* (2001) also issue a similar warning about processes which cause a delay in delivery of an ash to the sea-floor.

To the north of Iceland in core PS2644 (Fig. 2.2), Voelker *et al.* (1998) and Voelker (1999) noted even larger correction errors, and off N Iceland Eiríksson *et al.* (2000a) identified the Borrobol ash in core HM107-05 (Fig. 2.2) with a corrected (400 yr) basal age of ca 13,700 ^{14}C yr BP (Table 2.2), whereas the estimate on the terrestrial equivalent

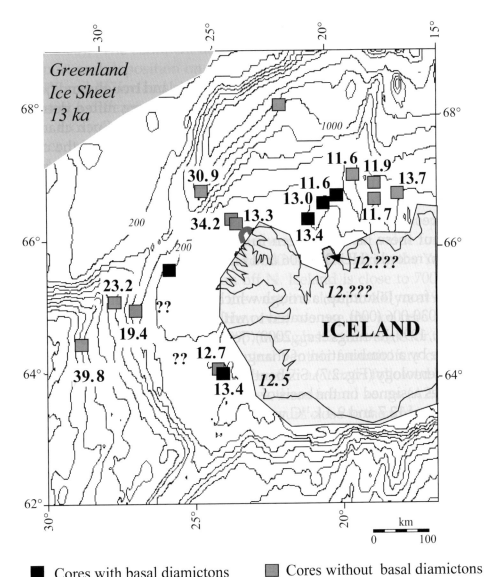

■ Cores with basal diamictons   **■** Cores without basal diamictons

12.7   Radiocarbon dates ka from the shelf (see Table 2)

12.7   Radiocarbon dates ka from land

Fig. 2.6 Basal radiocarbon dates on the deglaciation of the Iceland shelf. Dates are corrected by subtracting 400 yrs for the ocean reservoir. This may be too small, hence dates may well be maximum age estimates for deglaciation. The outline of the ice cap is intentionally crude to indicate our lack of detailed information about its marginal configuration.

in Scotland (Turney *et al.*, 1997) is closer to 12,500 ^{14}C yr BP. This might imply a reservoir correction of ~1600 yrs! Thus it is probable that the deglaciation dates on Fig 2.6 should be considered maximum age estimates.

2.5. CHANGES DURING THE HOLOCENE

Visible (> 2 mm) ice-rafted sediments are present in cores off N. Iceland until about 9 k ^{14}C yr BP (Castaneda, 2001; Smith, 2001). Ice retreated from the north coast toward the central mountains leaving a rich variety of ice marginal landforms and sediments (Kaldal and Víkingsson, 1990). After ca 9 k ^{14}C yr BP the sediments are characterized by a decrease in the WCMS reflecting the combination of a decrease in sediment density, and an increase in the diamagnetic components of the sediments, specifically the carbonate and carbon content resulting in dilution of the WCMS signal. Andrews *et al.* (2002c) compared aspects of the sediment from the Iceland margin with the adjacent East Greenland shelf and Harðardóttir and Andrews (in prep) have discussed the sediment magnetic records from cores off NW Iceland and E Greenland.

Papers dealing with the overall oceanographic variability for the Holocene include studies off N Iceland by Eiríksson *et al.* (2000b) and off SW Iceland by Hagen (1995). Hagen's study (1995) showed significant cooling during the Preboreal cold event (Hald and Hagen, 1998) but little subsequent change. Off N Iceland Eiríksson *et al.* (2000a) note an increase in arctic benthic foraminifera with a date of 7.3 k ^{14}C yr BP, which is associated "...presumably" (p. 39) with the 8.2k cal yr event in the Greenland ice core records and triggered by meltwater from the Laurentide Ice Sheet (Alley *et al.*, 1997; Barber *et al.*, 1999). However, the temporal resolution in existing cores around Iceland is relatively coarse because of a combination of modest rates of sediment accumulation and sampling at 5-10 cm intervals. Hence, early Holocene variations (6 to 10 k ^{14}C yr BP) are presently not resolved at decadal to centennial resolution. Eiríksson *et al.* (2000a,b) note that marine conditions became increasingly arctic or polar over the last 5000 yrs as shown by the decrease in certain planktonic foraminifera and an increase in typical arctic species such as *Elphidium excavatum* f. *clavata*. Similarly, Andrews *et al.* (2001a) showed that this latter species reach maximum values during the height of the Little Ice Age in north Iceland fjord waters.

Specific attention has been paid to the records from the last 5 ka (Andrews *et al.*, 2001b; Eiríksson *et al.*, 2000a,b; Jennings *et al.*, 2001), especially in terms of changes in the faunas, sedimentological parameters, and fluxes of carbonate and total organic carbon (TOC) (Andrews *et al.*, 2001b,c). These records have the resolution to resolve multicentury oscillations. Changes in the marine environment and comparison with the adjacent land records have also been a major concern of these papers.

Fig 2.9 illustrates some of the details of the marine record for the last 4-5 cal k yr BP where the variations in carbonate content of the sediment is used as a proxy for marine productivity (Andrews *et al.*, 2001a,b; Castaneda, 2001), based on the work of marine biological research of Þórðardóttir (1977, 1984) in north Iceland waters, and specifically the important effect on productivity of the very cold period of the late 1960s associated with the North Atlantic Great Salinity Anomaly (Malmberg, 1985; Dickson *et al.*, 1988; Belkin *et al.*, 1998; Ólafsson, 1999). The data on Fig. 2.9 are from site 328 at the head of Reykjarfjörður (Andrews *et al.*, 2001b). The chronology is based on 11 radiocarbon dates and the age of the data series is estimated from a linear regression to the calibrated dates. The record is marked by a series of oscillations (twelve in all) labelled a to l on Fig. 2.9. These data indicate that the marine environment has oscillated at multicentury

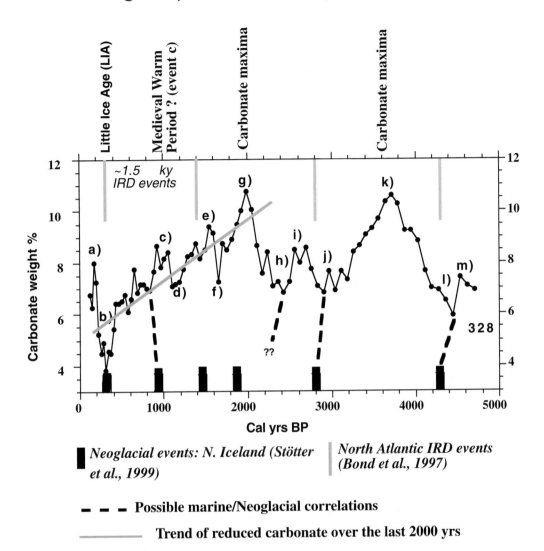

Fig. 2.9 Plot of the carbonate weight % values from B997-328 showing pronounced oscillations over the last 4-5k cal yr (lettered a to l) and showing the possible dates of neoglacial events (Stötter et al., 1999) and IRD events in the North Atlantic (Bond et al., 1997) (after Andrews et al., 2001b). The inclined grey line indicates a steady trend toward lower carbonate values over the last 2000 yrs.

time-scales. The deepest minima in the carbonate weight percent data indicates reduction in carbonate production of between 3 and 4% and occurred 200-400 yrs ago, hence is coeval with the Little Ice Age (Fig. 2.9, event b) (cf. Þórarinsson, 1953a).

Andrews *et al.* (2001b,c) showed that the record from Reykjarfjörður correlated extremely well with other records from the inner and mid-shelf of north Iceland. Furthermore, it had a strong parallel with the reconstructed temperature at Summit, Greenland, derived from borehole temperatures and isotopic data (Cuffey and Clow, 1997; Alley *et al.*, 1999a).

The moraine record of Neoglacial events on N Iceland (Guðmundsson, 1997; Stötter *et al.*, 1999) is sporadic and dating is not always easy. Estimated ages of Neoglacial events are shown on Fig. 2.9, as are the intervals of the rafting of sediments into the

North Atlantic (Bond *et al.*, 1997, 1999). Refinement of the chronology of the carbonate events illustrated on Fig. 2.9 will require the identification of some of the younger tephras on sediments from the inner and mid-shelf. This is required to ascertain with certainty any temporal and spatial changes in the ocean reservoir effect which might lead to differences of 100-300 yrs in the corrected dates (Eiríksson *et al.*, 2000b).

2.6 CONCLUSIONS

In the last decade the pace of research on the Late Quaternary sedimentary and micropalaeontological records from the western and northern Iceland margin has increased measurably. Coring expeditions in 1993, 1995, 1996, 1997 and 1999 have retrieved a number of piston and gravity cores which contain high-resolution records of changes during the interval of deglaciation as well as during the Holocene (Table 2.1). Because research on the Iceland margin is so new, it will require come time before all the relevant questions connected with ice extent, date of deglaciation, and subsequent palaeoceanographic changes, have had a chance to be answered. However, this review has indicated the extent of the available data and it clearly indicates that studies of the sediments of the Iceland fjords, troughs, and slope will provide the critical evidence to answer some of the persistent questions that arise in considering the Late Quaternary history of Iceland and its continental margins. Furthermore, a knowledge of ice sheet-ocean interactions around the Iceland margin are vital to add verification and insights into the role of the Iceland ice cap(s) and glaciers in North Atlantic Late Quaternary events (cf. Bond and Lotti, 1995). Of particular importance is to understand the glacial/volcanic sediment transfer processes which probably consist of a mixture of processes including airfall of tephras (onto glacial and sea ice, and into the marine environment directly), iceberg rafting, meltwater plumes, rafting on sea-ice, large glacial lake outbursts, and gravity flows.

ACKNOWLEDGEMENTS

I wish to acknowledge the National Science Foundation of the USA who have supported the research based at the University of Colorado (ATM-9531397, OPP-972510, OCE98-09001). I wish to thank Drs Anne Jennings, Guðrun Helgadóttir, Jorunn Harðardóttir, and Áslaug Geirsdóttir, and Stephanie Cartee-Schoolfield, Mikie Smith, Isla Castaneda, and Greta Krisjansdóttir, for their research contributions, many of which are noted in this review. I am grateful to the Universities of Exeter and Keele, UK, for their assistance in supporting me at the Iceland 2000 meeting at the University of Keele, April 2000. I am especially grateful to Áslaug Geirsdóttir for her detailed, painstaking, and critical review of this paper.

3. Relative sea-level changes in Iceland: new aspects of the Weichselian deglaciation of Iceland

Hreggviður Norðdahl[1] and Halldór G. Pétursson[2]

[1] University of Iceland, Science Institute, Dunhagi 3, IS-107 Reykjavík, Iceland
[2] Icelandic Institute of Natural History, IS-600 Akureyri, Iceland

3.1 INTRODUCTION

During the latter part of the 20th century research effort in studies of the Weichselian history of Iceland was intensified and about a decade ago three papers appeared, providing a good review of the status of Weichselian studies at that time (Einarsson and Albertsson, 1988; Ingólfsson, 1991; Norðdahl, 1990a). Since then much new data on the Weichselian development of Iceland have accumulated – not only from sections on land but also from sediment cores retrieved from the shelves around Iceland (see e.g. Andrews et al., 2000; Eiríksson et al., 2000a). It is, therefore, considered timely to present another review paper, this time with the main emphasis on relative sea-level (RSL) changes, a new approach to the history of the last deglaciation of Iceland.

In this paper we follow the chronostratigraphical terminology for Norden as proposed by Mangerud et al. (1974). All radiometric dates are ^{14}C dates except otherwise indicated and all ^{14}C ages of marine organisms mentioned in the text have been corrected with respect to the ^{13}C/^{12}C ratio, and in accordance with the apparent sea-water reservoir age of 365 ± 20 ^{14}C years for living marine organisms around Iceland (Håkansson, 1983). Weighted mean age and standard deviation of dated samples are calculated with the Radiocarbon Calibration Program (CALIB) Rev 4.1.2 (Stuiver and Reimer, 1993; Stuiver et al., 1998). For clarification, directions of glacial striae are given in true degrees along the direction of glacier flow.

3.2 WEICHSELIAN RECORDS IN ICELAND

Situated on the Mid-Atlantic Ocean Ridge on top of the Kverkfjöll hot spot (Fig. 3.1), Weichselian strata in Iceland contain subaerial and subglacial volcanic formations meaning that records of former climatic variations and climatic changes are not only preserved in glacigenic and non-glacigenic sedimentary formations, but also in subaerial

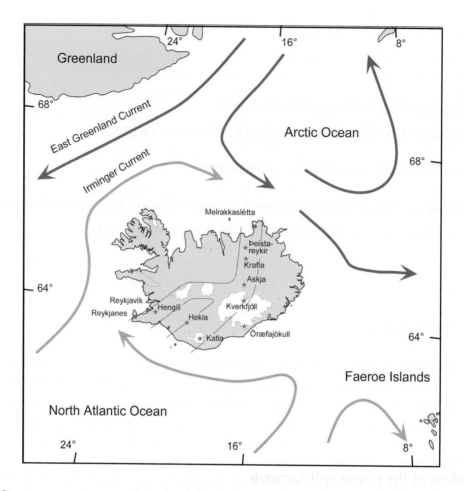

Fig. 3.1 Ocean currents around Iceland (Valdimarsson and Malmberg, 1999) with Early and Middle Weichselian sediments (dots) and a few central volcanoes (stars) mentioned in the text.

and subglacial bedrock formations. The Weichselian history of Iceland is, therefore, obtained from data collected from sedimentary sequences on land, from offshore sediment cores, as well as from the lava-pile of Iceland.

3.2.1 Early Weichselian Strata

The earliest known Weichselian strata in Iceland are lava horizons within the bedrock of the Krafla central volcano in northern Iceland (Fig. 3.1), indicating that at least parts of the northern volcanic zone were ice-free at the time of eruption (Sæmundsson, 1991, 1992). During an early Weichselian stadial Iceland was most likely covered by a continuous ice sheet reaching out on the shelves around Iceland (Fig. 3.2). Some ^{40}Ar/^{39}Ar dates of subglacial silicic hyaloclastic breccias in the Krafla area have yielded ages between 85 and 90 k yr BP for this stadial (Sæmundsson, 1991, Sæmundsson *et al.*, 2000). In southern Iceland the bedrock in the Hengill central volcano is considered to be of the same age and reflects a pattern of stadial – interstadial conditions there (Sæmundsson, 1992).

3.2.2 Middle Weichselian Formations

Newly discovered marine sediments and bedrock formations on the Reykjanes peninsula and in the Reykjavík area in southwestern Iceland have been dated to the Middle Weichselian Substage. Bedrock and sediment formations of a suggested Middle Weichselian age have also been described on the Melrakkaslétta peninsula in northeastern Iceland (Pétursson, 1986, 1988, 1991) (Fig. 3.1).

The Rauðamelur sedimentary sequence (Sæmundsson, 1988) comprises two stadial phases with expanding glaciers and deposition of tills, and two interstadial phases with raised RSL and accumulation of littoral and sublittoral lee-side spit formations (Fig. 3.3). These sediments rest on a striated bedrock suggesting that the outermost part of the peninsula was overridden by glaciers flowing towards the northwest (305°). A dated sample of a whalebone collected from the lower spit formation has yielded a finite age of 34,735 ± 1400 ^{14}C yr BP (Table 3.1; 1) (Norðdahl and Sæmundsson, 1999). These sediments are partly capped by a lava flow – the Rauðamelshraun lava – erupted when RSL had regressed to a position below the sediments. Later, the area was again overridden by glaciers advancing in a northwesterly direction (320°) and depositing till on top of the lava flow and the lower spit formation. The till is in turn discordantly overlain by the upper spit formation representing a general ice retreat and transgression of RSL (Sæmundsson and Norðdahl, 2002). The sediments of Rauðamelur have subsequently neither been overridden by glaciers nor have they been inundated by the sea. Three dates from marine shells yielded a mean age of 12,325 ± 85 ^{14}C yr BP (Norðdahl and Sæmundsson, 1999) and a single date of 12,635 ± 130 ^{14}C yr BP (Jóhannesson *et al.*, 1997) have produced a weighted mean age of 12,355 ± 80 ^{14}C yr BP (Table 3.1; 2-5) for the formation of the upper spit formation. At that time RSL was approximately on a level with the spit formation now situated at 20-25 m asl, some 40-45 m below the 70 m marine limit (ML) shoreline just north of the Rauðamelur site (Sæmundsson and Norðdahl, 2002). According to Guðmundsson (1981) this area has tectonically subsided by some 50 m in Postglacial times, explaining the different height of Rauðamelur and an assumed synchronous Bölling ML shoreline at about 70 m asl.

Skálamælifell and other tuyas about 15 km east of Rauðamelur (Fig. 3.3) have been ^{40}Ar-^{39}Ar dated to 42.9 ± 7.8 k yr and correlated with the Laschamp geomagnetic event (Levi *et al.*, 1990). The altitude of the tuyas, reaching as high as 250 m asl, is an approximation to the thickness of a glacier covering the area and at that time most likely extending beyond the present coastline. Stratigraphical data from this area show a subsequent retreat of the glaciers and subaerial eruption of lava as the volcanic zone became ice-free. The Rauðamelshraun lava might possibly be correlated with these lava flows.

Fossiliferous and partly lithified and stratified silty fine sand, resting on glacially striated (9°) bedrock a few km north of Rauðamelur (Fig. 3.3), was deposited at the Njarðvíkurheiði locality in relatively shallow and quiet marine environment (Fig. 3.3). Six samples of marine shells from these sediments have yielded a weighted mean age of 21,525 ± 105 ^{14}C yr BP (Table 3.1; 6-11). These sediments are discordantly overlain by coarse grained littoral sediments (Jóhannesson and Sæmundsson, 1995; Jóhannesson *et al.*, 1997). At Suðurnes in Reykjavík (Fig. 3.3), Eiríksson *et al.* (1997) have described fossiliferous marine sediments subsequently overridden by glaciers. Two samples have

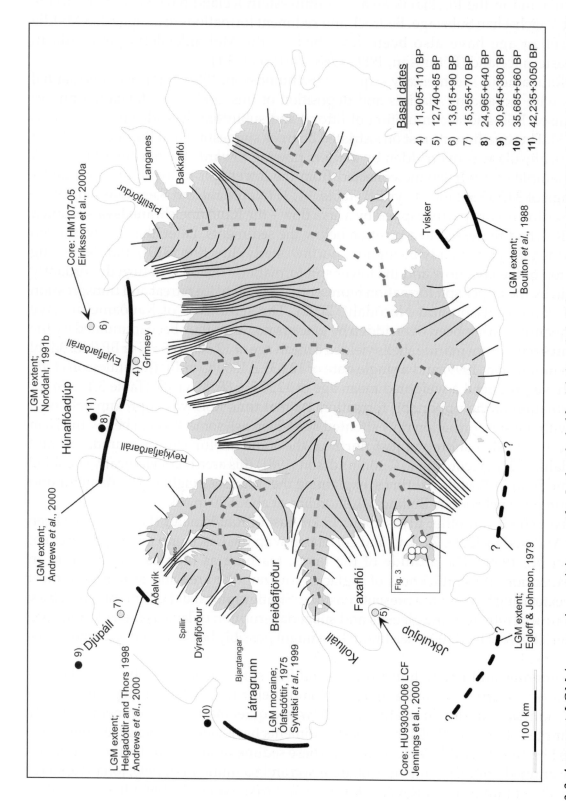

Fig. 3.2 Apparent LGM ice margin position on the Iceland shelf with basal dated cores pre-dating (black dots) and post-dating (grey dots) the LGM. The basal dates are listed in Table 3.1, 15–22. Sediments pre-dating the LGM on land (open circles) are reviewed in Fig. 3.3. The ice flow pattern is based on Bourgeois et al., 1998.

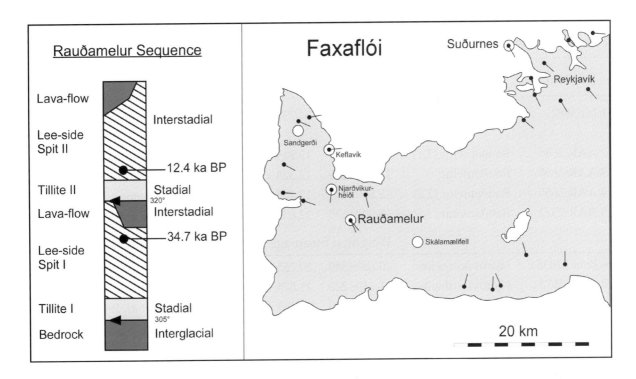

Fig. 3.3 The Rauðamelur sedimentary sequence on the Reykjanes Peninsula in southwestern Iceland with location of other pre-LGM Middle Weichselian Interstadial sediment and bedrock formations. Glacial striae are partly after Kjartansson (1960).

yielded an age of 27,735 ± 410 and 28,385 ± 480 ^{14}C yr BP with a weighted mean age of 28,010 ± 310 ^{14}C yr BP (Table 3.1; 13-14).

All these marine sediments show that in Middle Weichselian time coastal areas in Iceland were ice-free and inundated by the sea. Finite datings yielding ages between 34.7 and 20.3 k ^{14}C yr BP allows a tentative correlation with the Ålesund Interstadial in Western Norway (Larsen *et al.*, 1987).

3.2.3 Last Glacial Maximum (LGM)

Until recently, the only available information on the Weichselian maximum extent of the Icelandic inland ice sheet were glacial striae in coastal regions and on islands reflecting ice flowing out onto the shelves around Iceland. Till-like deposits and glacial striae on the island of Grímsey (Keith and Jones, 1935; Einarsson, 1967; Hoppe, 1968) and a conspicuous moraine-like ridge on the shelf off Breiðafjörður (Ólafsdóttir, 1975) are the earliest evidence for an extended Icelandic ice sheet (Fig. 3.2). The moraine-like ridge is more than 100 km long and is found some 130 km off land at depths between 200-350 m. The ridge is usually 20-30 m in relief but in places it is more than 100 m high (Ólafsdóttir, 1975).

Based on geomorphological evidence such as glacial striae and periglacial landforms, Sigurvinsson (1982) postulated a LGM ice sheet reaching at least 25-35 km off the Dýrafjörður fjord in Vestfirðir, the Northwestern Peninsula of Iceland (Fig. 3.2), about 10-15 km inside Spillir, a rugged area on the sea floor. In much the same way Hjort *et al.*

Table 3.1. Radiocarbon dated shells of marine organisms of Bölling and pre-LGM age collected on present-day dry land and from the Iceland shelf.

Lab Num	Sample	^{14}C age BP	Res. age BP	Reference
1 Lu-3682	Rauðamelur	35,100±1400	34,735 ±1400	Norðdahl & Sæmundsson 1999
2 AAR-3655	Rauðamelur (2,1)	12,610±90	12,245 ± 90	Norðdahl & Sæmundsson 1999
3 AAR-3655	Rauðamelur	12,690±100	12,325 ±100	Norðdahl & Sæmundsson 1999
4 AAR-3655	Rauðamelur (2,2)	12,850±90	12,485 ± 90	Norðdahl & Sæmundsson 1999
5 AAR-2572	Rauðamelur	13,000±130	12,635 ±130	Jóhannesson *et al.* 1997
		Weighted mean age: 2-5	12,355 ± 80	^{14}C yr BP
6 Beta-82638	Njarðvíkurheiði	20,630±340	20,265 ±340	Jóhannesson *et al.* 1997
7 AAR-2573	Njarðvíkurheiði	21,440±220	21,075 ±220	Jóhannesson *et al.* 1997
8 AAR-2574	Njarðvíkurheiði	20,750±220	20,385 ±220	Jóhannesson *et al.* 1997
9 AAR-2575	Njarðvíkurheiði	21,500±290	21,135 ±290	Jóhannesson *et al.* 1997
10 AAR-2576	Njarðvíkurheiði	22,060±230	21,695 ±230	Jóhannesson *et al.* 1997
11 AAR-2577	Njarðvíkurheiði	23,980±230	23,615 ±230	Jóhannesson *et al.* 1997
12 AAR-2803	Sandgerði	24,510±200	24,145 ±200	Jóhannesson *et al.* 1997
		Weighted mean age: 6-11	21,525 ± 105	^{14}C yr BP
		Weighted mean age: 6-12	22,070 ± 90	^{14}C yr BP
13 AAR-1900	1639	28,100±410	27,735 ±410	Eiríksson *et al.*, 1997
14 AAR-1901	1640	28,750±480	28,385 ±480	Eiríksson *et al.*, 1997
		Weighted mean age: 13-14	28,010 ± 310	^{14}C yr BP
15 AAR-4209	97-317PC, 245cm	12,270±100	11,905 ±100	Andrews *et al.* 2000
16 AA-12896	193030-006LCF 1235cm	13,105±85	12,740 ± 85	Andrews *et al.* 2000
17 AAR-3383	HM107-05, 393.0-394.3cm	13,980±90	13,615 ± 90	Eiríksson *et al.* 2000
18 CAMS-46527	96-1234GCC, 259cm	15,720±70	15,355 ± 70	Andrews *et al.* 2000
19 CAMS-27763	97-323PC1, 292cm	25,330±640	24,965 ±640	Andrews *et al.* 2000
20 CAMS-44859	96-1229GGC, 265cm	31,310±380	30,945 ±380	Andrews *et al.* 2000
21 CAMS-42012	96-1227GGC, 56cm	36,050±560	35,685 ±560	Andrews *et al.* 2000
22 AAR-3887	97-322PC, 161cm	42,600±3050	42,235 ±3050	Andrews *et al.* 2000

(1985) have proposed a LGM position of the ice-edge some 10-20 km beyond the Aðalvík bay in northwestern Vestfirðir (Fig. 3.2). On the basis of calculations of glacier gradients, based on glacial and periglacial landforms in northern Iceland, Norðdahl (1983, 1991) suggested glaciers reaching onto the shelf and to a position just north of Grímsey (Fig. 3.2).

As more and more data have been collected from the shelf around Iceland through geophysical survey and sediment coring and re-evaluation of morpho-sedimentological evidence (e.g. Andrews *et al.*, 2000; Eiríksson *et al.*, 2000a) a more accurate picture of the LGM glacier extent has emerged. Off southeastern Iceland Boulton *et al.* (1988) described major till masses – up to 100 m thick – at the shelf edge (Fig. 3.2), presumably representing a LGM glacier extent there. A moraine-complex, the Tvísker-moraine about 20 km off the present coast is thought to reflect a major standstill during the Late Weichselian deglaciation of southeastern Iceland. On the basis of geophysical data, Egloff and Johnson (1978) found that an acoustically transparent, 100-350 m thick formation on the shelf-edge off southwestern Iceland (Fig. 3.3), probably represents till deposits of Late Pleistocene age. Though these deposits have not yet been dated they may very well represent the LGM extent as well as being the result of repeated ice advances reaching the edge of the shelf. Geophysical and core data gathered from the shelf west off Vestfírðir indicate that the Icelandic ice sheet episodically covered the southwestern part of the shelf, at least three times during the Quaternary (Syvitski *et al.* 1999). Furthermore, Syvitski *et al.* (1999) are of the opinion that the moraine-like ridge off Breiðafjörður was formed by a grounded glacier marking the LGM extent of the Icelandic inland ice sheet. The proposed LGM age of the moraine is based on a single date from a 50 cm gravity core retrieved from the seaward side of the moraine providing a maximum age of about 36 k ^{14}C yr BP (Fig. 3.2, Table 3.1; 21). Andrews *et al.* (2000) have provided three additional basal dates from the sea-ward side of the moraine yielding an age of 15.7, 18.0 and 18.2 k ^{14}C yr BP, respectively but they also contemplated the possibility that the moraine could in fact be older than the LGM.

Just north of the town of Keflavík in southwestern Iceland (Fig. 3.3) an about 5 m thick sequence of sediments – resting on glacially striated (265°) bedrock – is made up of ca 3 m thick basal till overlain by ca 2 m of coarse grained gravel and sand sediments (Kjartansson, 1955; Matthíasdóttir, 1997). The basal till encloses sedimentary clasts containing fragments of *Hyatella arctica* obviously carried by a glacier from the bottom of Faxaflói and onto land at Keflavík. Similar sedimentary clasts have been found in wave-washed basal till on top of striated (215°) bedrock about 6 km west of Keflavík (Fig. 3.3). A radiocarbon date of these shell fragments has yielded an age of 24,145 ± 200 ^{14}C yr BP (Table 3.1; 12) (Jóhannesson *et al.*, 1997). This date and the other six dates from the Njarðvíkurheiði locality, yielding a weighted mean age of 22,070 ± 90 ^{14}C yr BP (Table 3.1; 6-12), pre-date the LGM glacier advance across and beyond the present coastline of the outer Reykjanes peninsula.

Helgadóttir and Thors (1998) and Andrews *et al.* (2000) have identified and located possible end-moraines at different locations along the Djúpáll trough on the shelf off the Ísafjarðardjúp fjord in northwestern Vestfirðir (Fig. 3.2). There, a large moraine-like ridge, some 30 km off land and post-dated by basal core-dates between 9.0 and 15.4 k ^{14}C yr BP, may mark the LGM position of the glaciers. If the main Djúpáll ridge is in fact a LGM moraine then its position is in a good agreement with the postulated LGM glacier

Fig. 3.4 Intensive periglacial frost shattering of basaltic bedrock on the outermost part of the Langanes Peninsula in northeastern Iceland.

extent (Sigurvinsson, 1982; Hjort *et al.*, 1985) south and north of the Djúpáll trough (Fig. 3.2). In the Reykjafjarðaráll trough northeast off Vestfirðir and north of the Húnaflói bay Andrews *et al.* (2000) have suggested a LGM position of the ice-edge some 70-80 km off land (Fig. 3.2). This assumption is based on basal core-dates in the area, where the LGM position is pre-dated by two samples yielding an age of 25.0 and 42.2 k ^{14}C yr BP (Table 3.1; 19,22), and post-dated by samples yielding ages between 9.4 and 9.7 k ^{14}C yr BP, north and south of the proposed LGM limit (Fig. 3.2).

At present nothing is yet known about the LGM extent of the Icelandic inland ice sheet on the shelf off northeastern and eastern Iceland except that glacial striae show that the glaciers reached beyond the present coastline. The outermost part of the

Langanes peninsula in northeastern Iceland is characterized by intensive periglacial frost shattering of the bedrock (Fig. 3.4) possibly indicating a lengthy ice-free situation there. The radial pattern of glacial striae and other geomorphological features on the inner parts of Langanes show that the peninsula was not overridden by glaciers and ice streams flowing along the Þistilfjörður and Bakkaflói fjords (Fig. 3.2) (Pétursson and Norðdahl, 1994).

3.2.4 Ice Divides and Ice Free Areas

For quite some time we have visualized a Weichselian ice sheet in Iceland with a major ice-divide over the south-central parts of the island (Einarsson, 1961, 1968, 1994) and ice streams extending radially across the present coastline. A later modification of the LGM ice sheet configuration accounts for a 1500-2000 m thick ice-dome above the central Icelandic highlands (Norðdahl, 1990) with topographically constrained flow of ice streams and outlet glaciers (Fig. 3.2) (Bourgeois *et al.*, 1998; Norðdahl, 1983, 1991). At the LGM, when the ice sheet reached far out onto the shelves around Iceland, a major ice divide is considered to have been extended westwards towards northwestern Iceland and towards the northeast and north to the northern part of the neo-volcanic zone of Iceland. Thus, instead of a radially flowing ice sheet we now visualize an ice sheet with ice flowing away from an elongated ice divide above the centre of the island with more topographically constrained ice flow in areas closer to the present coasts. A modelling experiment reveals the fundamental importance of the role of geothermal heat flux across the neovolcanic zone on the basal thermodynamics of the Icelandic inland ice sheet, leading to a considerably thinner ice sheet and rapid mass turnover of the ice sheet (Hubbard *et al.*, in press). The pattern of glacial striae in coastal parts of Iceland was formed in deglacial times when the edge of the Late Weichselian ice sheet retreated to positions inside the present coastline (Kjartansson, 1955). Therefore, the flow of the much thinner and peripheral parts of the ice sheet was much more dependent on local topographical control which led to formation of temporary and local ice divides e.g. on Melrakkaslétta (Kjartansson, 1955; Pétursson, 1986) and in northwestern Iceland (Kjartansson, 1969).

In spite of the fact that in LGM times an ice sheet extended onto the shelves around Iceland, ice-free enclaves existed on steep mountain sides above ice streams in northwestern, central northern and eastern Iceland. Furthermore, remote and peripheral parts of Iceland, such as Langanes and the Bjargtangar headland north of Breiðafjörður (Fig. 3.2), may have remained more or less ice-free in LGM times (Pétursson and Norðdahl, 1994; Norðdahl and Pétursson, 2002). This, however, is contrary to a general view of an extended ice sheet in northwestern Iceland. A pronounced alpine landscape with well developed corries and signs of extensive and prolonged periglacial activity in the coastal mountains of eastern Iceland (Kjartansson, 1962a; Sigbjarnarson, 1983; Norðdahl and Einarsson, 1988, 2001), could also point to moderate LGM glacier extent there. Assumed LGM ice-free areas – nunataks in a huge ice sheet far inside the glacier margin – were probably not particularly viable as a refugium. There only the most tolerant species could have survived the Weichselian glaciation and that is, according to Buckland and Dugmore (1991) and Rundgren and Ingólfsson (1999), very uncertain.

3.3 INITIAL DEGLACIATION

Having defined the approximate timing of the Weichselian maximum glacier extent in Iceland to have most likely occurred at, or just after, about 20 k ^{14}C yr BP, the initial deglaciation of the shelves around Iceland occurred subsequent to this, and was well under way at about 12.7, 15.4 and 13.6 k ^{14}C yr BP, according to basal dates of sediment cores off western, northwestern and northern Iceland respectively (Andrews *et al.*, 2000; Eiríksson *et al.*, 2000a). The earliest Lateglacial dated marine shells found on present-day dry land in the coastal regions of Iceland were collected in western Iceland (12,800 ± 200 ^{14}C yr BP; Ashwell, 1975) and northeastern Iceland (12,655 ± 90 ^{14}C yr BP; Pétursson, 1986, 1991) (Table 3.2; 1-2). It therefore seems obvious that the shelf off western and northern Iceland was deglaciated approximately at the same time as the coastal parts of Iceland.

3.3.1 Deglaciation of the Iceland Shelf

As mentioned above the timing of the maximum glacier extent onto the Iceland shelf is based on data from Reykjanes and several sediment cores (Norðdahl and Sæmundsson, 1999; Andrews *et al.*, 2000; Eiríksson *et al.*, 2000a) collected at various locations on the shelf around Iceland (Fig. 3.2). Basal dates from these sediment cores show ages between 3,515 ± 65 and 42,235 ± 3050 ^{14}C yr BP (Andrews *et al.*, 2000). When examining these cores in the literature we see that none of them is supposed to have penetrated what is called 'glaciomarine mud' and into the underlying 'ice-contact sediments' (Jennings *et al.*, 2000). Nevertheless, the basal parts of four of these cores are alleged to pre-date the LGM situation (Fig. 3.2, Table 3.1; 19-22), while the basal parts of the rest of the cores are assumed to post-date the LGM situation on the shelf around Iceland. Sediment cores retrieved in Eyjafjarðaráll off northern Iceland and in Djúpáll off northwestern Iceland have yielded Lateglacial basal dates of 11,905 ± 100 and 15,355 ± 70 ^{14}C yr BP, respectively (Fig. 3.2, Table 3.1; 15,18) (Andrews *et al.*, 2000), providing a minimum estimate for the age of the deglaciation of these parts of the Iceland shelf.

Two sediment cores – 09030-006 LCF (Jennings *et al.*, 2000) and HM107-05 (Eiríksson *et al.*, 2000a) – have produced ages of approximately equal age or greater than the oldest Lateglacial dated material collected on present-day dry land in Iceland. In Jökuldjúp a ca 17 m long sediment core (93030-006 LCF) reached to within a metre above the ice-contact sediments. Foraminifera in the lowest part of the glaciomarine mud indicate ice-distal conditions and immigration of slope species onto the shelf in association with relatively warm Atlantic water (Jennings *et al.*, 2000). A series of ^{14}C dates has produced an age-depth model with a basal date of about 12,740 ± 85 ^{14}C yr BP (Table 3.1; 16) (Syvitski *et al.*, 1999). Extrapolation of the age-depth curve down to the boundary between the massive glaciomarine mud with drop stones and the ice-contact sediment suggests an age of about 13.0 k ^{14}C yr BP for the onset of marine conditions in Jökuldjúp.

Off northern Iceland, Eiríksson *et al.* (2000a) have dated a ca 4 m long core (HM107-05), which reached into a relatively massive marine mud deposit, with a basal date of 13,615 ± 90 ^{14}C yr BP (Table 3.1; 17). Foraminiferal studies of the core show that the environment was similar to the present one, characterized by relatively warm water showing that the Irminger Current must have reached the area at that time. Furthermore,

10). The existence of shells in these sediments shows that the southern coast of Hvalfjörður had been deglaciated and submerged by the sea at that time. These terraces have subsequently not been overridden by glaciers.

Marine sediments of Bölling age have been most thoroughly studied in the Melabakkar-Ásbakkar Cliffs in the Melasveit area in western Iceland (Ingólfsson, 1987, 1988) and in coastal cliffs along the western coast of the Melrakkaslétta peninsula in northeastern Iceland (Fig. 3.6) (Pétursson, 1986, 1991). The age of these sediments has been determined with 14 and 5 dated samples of marine shells, respectively.

3.4.1.1 The Melasveit – Andakíll Sediment Terraces

The Melabakkar-Ásbakkar Cliffs in the Melasveit area in western Iceland form an almost continuous 5 km long and 15-30 m high constantly changing vertical section in consolidated and slightly lithified fossiliferous sediments (Fig. 3.7). These exposures have been described and studied by a number of scientists, such as Thoroddsen (1892) and Bárðarson (1923). A comprehensive review and additional studies were carried out by Ingólfsson (1984, 1985, 1987, 1988), who dated the sediment sequence and outlined an interpretation of the glacial history of the lower Borgarfjörður region.

At the base of the Melabakkar-Ásbakkar cliffs and overlying glacially striated basalt bedrock is a fossiliferous unit, the Ásbakkar diamicton, which in the lower part is mainly made of stratified silty diamicton with occasional gravel clasts (Ingólfsson, 1987). The fauna of the Ásbakkar diamicton resembles the recent *Macoma calcarea* community which, according to Thorson (1957), occurs widely in the Arctic Seas and prefers silty and sandy silt fjord bottoms with relatively low salinity, water temperatures below +5 °C, and water-depths less than 45-50 m (Ingólfsson, 1988). At present the Annual Sea Surface Temperature (ASST) around Faxaflói is about +6 °C (Jónsson, 1999). Six samples of marine shells have yielded ages between 11,945 ± 110 and 12,505 ± 110 ^{14}C yr BP (Table 3.2; 7,11-15) with a weighted mean age of 12,130 ± 70 ^{14}C yr BP. Upwards the unit

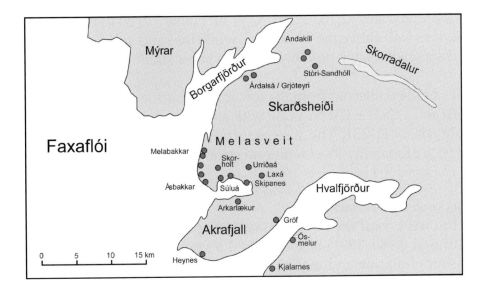

Fig. 3.7 Radiocarbon dated marine shells (dots) of Bölling – Alleröd age in the Lower Borgarfjörður and Hvalfjörður area in Western Iceland.

grades into sandy-silty massive diamicton with an increased number of gravel and boulder clasts, and also with fewer marine shells, probably indicating increased proximity to the source of the sediments (Ingólfsson, 1987) or decreased water-depth due to regression of RSL.

The area east of the coastal cliffs and south of the Skarðsheiði massif (Fig. 3.7) is covered with sediments deposited and moulded by glaciers and the sea in Lateglacial times (Ingólfsson, 1988; Magnúsdóttir and Norðdahl, 2000). At Laxá and Gröf (Fig. 3.7) in the eastern part of the area fossiliferous diamictons contain a mollusc fauna very similar to the *Macoma calcarea* fauna of the Ásbakkar diamicton (Ingólfsson, 1988). Samples retrieved from the deposits at Laxá and Gröf have yielded age of 12,105 ± 110 and 12,475 ± 110 ^{14}C yr BP, respectively (Table 3.2; 16-17).

Sediment terraces at Árdalsá north of Skarðsheiði (Fig. 3.7) are made of about 25 m thick massive to crudely stratified silty-sandy diamicton with subfossil marine shells in its lower and upper parts. At about 50 m asl the diamicton is truncated and overlain by 17 m thick unit of beach gravel and sand sediments which constitute the upper part of a terrace reaching 60-70 m asl (Ingólfsson, 1988). Two samples of marine shells, one from the lower part and another from the upper part of the diamicton, have been dated yielding an age of 12,100 ± 150 ^{14}C yr BP (Ashwell, 1975) and 12,145 ± 140 ^{14}C yr BP (Ingólfsson, 1988), respectively (Table 3.2; 19,18). Furthermore, the marine fauna at Árdalsá is similar to the *Macoma calcarea* fauna of the Ásbakkar diamicton (Ingólfsson, 1988). The altitude of the fossiliferous diamicton and the beach deposits indicates that in late Bölling times RSL was there at, or above, the 50 m level.

In Andakíll north of Skarðsheiði (Fig. 3.7) sediment terraces reaching 40 and 70 m asl have been studied by Ashwell (1967) and Torfason (1974). About 10 m thick fossiliferous silty-sandy diamicton with pebble and cobble clasts is there overlain with a 20 m thick unit of foreset bedded sand and gravel sediments, which in turn are discordantly overlain by about a 2 m thick unit of stratified sand and gravel sediments reaching about 40 m asl (Torfason, 1974). Shells collected at about 20 m asl in the basal part of the terrace have been dated to 12,240 ± 200 and 12,100 ± 250 ^{14}C yr BP (Table 3.2; 20-21) (Ashwell, 1975). At that time RSL was well above the 20 m level, while the sand and gravel units may have been formed somewhat later, possibly during a subsequent transgression of RSL.

The dates of 14 shell samples collected from the lowermost marine units in this area lie between 11,905 ± 160 and 12,505 ± 110 ^{14}C yr BP with a weighted mean age of 12,155 ± 45 ^{14}C yr BP (Table 3.2; 7,11-23). The age results of all the 14 samples are indistinguishable at a 95% probability level with one standard deviation (Stuiver and Reimer, 1993).

3.4.1.2 Western Melrakkaslétta Sediments

In coastal cliffs on the west coast of the Melrakkaslétta peninsula in northeastern Iceland (Fig. 3.6), Pétursson (1986, 1991) has described a very detailed and complex stratigraphical succession of marine and glacial sediments and a subaerial lava flow reaching well beyond the onset of the Weichselian glaciation in Iceland. These sediments rest on a bedrock that is either made of striated lava flows or hyaloclastite.

The Lateglacial history of the coastal cliffs is based on dated sediment sequences at

Rauðinúpur, Hvalvík, Kópasker (Röndin), and Auðbjargarstaðir (Fig. 3.6). Since the formation of the early Bölling (ca 12.7 k ^{14}C yr BP) shoreline in Hvalvík, the marine environment was gradually changed from glaciomarine to a pronounced marine environment with deposition of a 4-5m thick fossiliferous sand unit containing shells of *Hiatella arctica* and *Mya truncata* (Pétursson, 1986). The lower part of this sand unit is made of stratified sand of different grain sizes while the upper part of the unit is made of somewhat coarser sand and gravel sediments, deposited when RSL was most likely situated above the present 30 m level. At Hvalvík moulds of *Mya truncata*, with especially well developed siphon, display the fight of the molluscs to survive against a growing rate of sediment influx when RSL was lowered. At Kópasker (Röndin) this unit is overlain by a unit consisting of well rounded gravel sediments, containing a few ice-rafted stones of foreign lithology, deposited in a sublittoral coastal environment when RSL had regressed below the 10 m level (Pétursson, 1991). The Hvalvík and Kópasker localities (Fig. 3.6), were subsequently overridden by glaciers flowing across the present coast.

In this area, five dated samples of marine shells collected from the sandy sediments at Rauðinúpur, Hvalvík, Kópasker (Röndin), and Auðbjargarstaðir, have yielded Bölling ages between 11,955 ± 110 and 12,475 ± 140 ^{14}C yr BP with a weighted mean age of 12,135 ± 85 ^{14}C yr BP (Table 3.2; 24-28). The ages of these samples are not significantly different at one standard deviation with a 95% probability level.

3.4.2 Alleröd Marine Sediments

Fossiliferous sediments of Alleröd age (proper) have been dated in much the same places in southwestern, western, and northeastern Iceland as sediments of Bölling age (Figs. 3.5, 3.6). In a few places in the Reykjavík area fossiliferous sediments are found on top of glacial diamictites overlying glacially striated bedrock. On the island of Viðey and in Kópavogur samples of marine shells have yielded ages of 11,335 ± 120 ^{14}C yr BP and 11,225 ± 100 ^{14}C yr BP, respectively (Table 3.3; 1-2) (Hjartarson, 1993). In the University Campus area in Reykjavík two samples of marine shells from sediments overlain by glaciomarine deposits, yielded Alleröd ages of 11,255 ± 255 ^{14}C yr BP and 11,390 ± 115 ^{14}C yr BP (Table 3; 3-4), indicating that the area may subsequently have been overridden by glaciers (Norðdahl, 1991a).

As the Bölling–Alleröd deglaciation proceeded at least parts of the northern volcanic zone became ice free, and a subaerial eruption occurred in the vicinity of the Þeistareykir volcanic centre, some 40 km inside the present coastline of northern Iceland (Fig. 3.1). Lava flows from this eruption reached the Kelduhverfi area (Sæmundsson, 1973) and Stöplar in the Reykjahverfi area (Norðdahl and Pétursson, 1994) north and west of the erupting volcano (Fig. 3.6). Subsequently, in Younger Dryas times these areas were overridden by glaciers, as indicated by glacial striae and till on top of the lava flow (Elíasson, 1977; Norðdahl and Pétursson, 1994). The same pattern of glacier retreat and advance has been suggested for parts of the volcanic zone on the Reykjanes peninsula in southwestern Iceland (Sæmundsson, 1995). Considering how far RSL had regressed in late Bölling and early Alleröd times, and how far glacio-isostatic recovery had progressed, it is tempting to conclude that considerable areas of the country had become ice-free before the onset of the Younger Dryas ice advance.

Table 3.3 Radiocarbon dated shells of marine organisms of Alleröd age collected on present day dry land in Iceland.

LabNum	Sample	^{14}C age BP	Res. age BP	References
1 AAR-0120	Viðey-glacial till	11,700 ± 120	11,335±120	Sveinbjörnsdóttir & Johnsen 1991
2 AAR-0125	Kópavogur marine sediment	11,590 ± 100	11,225±100	Hjartarson 1993
3 U-2596	Félagsstofnun studenta	11,620 ± 255	11,255±255	Norðdahl 1990
4 Ua-2724	Tjarnargata 30 (Nr 1371/ BAL 2444)	11,755 ± 115	11,390± 115	*New date*
5 Lu-2372	Ásbakkar - Ásgil 2	12,080 ± 120	11,715±120	Ingólfsson 1988
6 Lu-2373	Ásbakkar - Ásgil 3	11,910 ± 140	11,545±140	Ingólfsson 1988
7 Lu-2376	Ásbakkar 2	11,830 ± 100	11,465±100	Ingólfsson 1988
8 Lu-2196	Ásbakkar - Ásgil 1	11,980 ± 130	11,615±130	Ingólfsson 1985
		Weighted mean age: 5-8	11,530 ± 110	^{14}C yr BP
9 Lu-2056	Súluá	11,330 ± 80	10,965± 80	Ingólfsson 1985
10 Lu-2340	Arkarlækur 1	11,350 ± 100	10,985±100	Ingólfsson 1988
11 Lu-2338	Heynes	11,430 ± 140	11,065±140	Ingólfsson 1985
12 Lu-2524	Urridaá	11,520 ± 100	11,155±100	Ingólfsson 1988
		Weighted mean age: 9-12	11,050 ± 80	^{14}C yr BP
13 Lu-3342	Melar	11,530 ± 100	11,165±100	Ásbjörnsdóttir & Norðdahl 1995
14 U-2019	Ekruhorn	11,620 ± 240	11,255±240	Kjartansson 1966
15 Lu-2778	Tjaldanes 1	11,640 ± 110	11,275± 110	Ásbjörnsdóttir & Norðdahl 1995
16 Lu-2776	Ballará 1	11,720 ± 80	11,355± 80	Ásbjörnsdóttir & Norðdahl 1995
		Weighted mean age: 13-16	11,275 ± 80	^{14}C yr BP

Marine sediments in Iceland attributed to the Alleröd Chronozone have been most carefully studied in the Melabakkar-Ásbakkar Cliffs in western Iceland, in sediment terraces at Melar and Saurbær in eastern Breiðafjörður (Fig. 3.6), and in the coastal cliffs of Fossvogur in Reykjavík.

3.4.2.1 The Melabakkar and Ásbakkar Cliffs

Stratigraphically overlying the Ásbakkar diamicton and Ás beds in the Melabakkar-Ásbakkar Cliffs in Western Iceland are the fossiliferous Látrar beds glaciomarine sediment facies (Ingólfsson, 1988). These sediments are mainly composed of gravelly sand deposits that are succeeded by fossiliferous laminated sand and silt sediments with numerous gravel and boulder clasts. The marine molluscs have been dated to 11,530 ± 110 ^{14}C yr BP, a weighted mean of four samples (Table 3.3; 5-8) collected from the cliffs. The allogenic fauna of the Látrar beds resembles the *Macoma calcarea* faunal assemblage in the Ásbakkar diamicton but with fewer species and individuals, reflecting a marine environment with relatively rapid sedimentation or brackish water (Ingólfsson, 1987). During the Látrar marine phase fine grained fossiliferous silt and silty-sand sediments also accumulated in places outside the Melabakkar-Ásbakkar Cliffs, namely at the Súluá, Urriðaá, Arkarlækur, and Heynes localities (Fig. 3.7) (Ingólfsson, 1988).

The interstratified silt and sand sediments at the Súluá locality contain high arctic species such as *Buccinum groenlandicum* and *Portlandia arctica* indicating very cold marine conditions in the area at that time. This faunal assemblage relates very well to a transitional mollusc community between the boreal-arctic *Macoma calcarea* community of the Látrar beds and a high arctic *Portlandia arctica* community (Ingólfsson, 1988). A 10-15 m thick sequence of stratified silt with intrabeds of silty sand with spread granule and pebble clasts is located at Heynes south of Akrafjall (Fig 3.7). The silty sediments are fossiliferous with similar faunal composition as the Súluá deposits, containing *Portlandia arctica*, an indicator of high arctic marine environment at this location (Ingólfsson, 1988).

The silty-sandy fossiliferous sediments at Heynes, and Urriðaá, and Arkarlækur north of Akrafjall (Fig. 3.7) have been lithostratigraphically correlated with the Súluá interstratified silt and sand facies of the Látrar bed unit (Ingólfsson, 1988). Four mollusc samples collected at Súluá, Heynes, Arkarlækur, and Urriðaá, respectively (Table 3.3; 9-12) have yielded a weighted mean age of 11,050 ± 80 ^{14}C yr BP, for the upper part of the Látrar beds which places its formation to the transition between the Alleröd and Younger Dryas Chronozones.

3.4.2.2 The Holtaland – Tjaldanes Terraces

The Holtaland – Tjaldanes terraces comprise a series of fossiliferous strata exposed in about 5 km long and up to 40 m high cliffs facing the Gilsfjörður fjord (Fig. 3.8). A comprehensive description of the sediments and mollusc fauna was given by Bárðarson (1921). At the base of the terraces at Kaldrani is a 10-15 m thick dark grey massive silty deposit – the *Portlandia* diamicton, with numerous up to 0.5 m large, rounded and sometimes striated clasts in the lowermost part of the unit. Between 2 and 15 m up in the basal diamicton, Bárðarson (1921) collected shells of *Portlandia arctica* and a few other arctic species. Above the *Portlandia* diamicton is a 2-6 m thick unit – the *Pecten* deposits, made of stratified silt and sand deposits with increasing sand content upwards. A few pebble and cobbled sized clasts are found dispersed in the unit. The *Pecten* sediments carry mollusc shells in great quantity especially in the sandy upper part of the unit with *Pecten islandicus* as the most prominent species. The third and topmost unit – the *Mytilus* deposits is about 2-6 m thick and is made of stratified sand and gravel deposits with a few thin sandy silt layers in the basal part of the unit. According to Bárðarson (1921) this unit is poor in fossils with very few individuals of a small number of species. The edge of the terrace at Kaldrani reaches about 25 to 30 m asl.

Kjartansson (1966) studied the Holtaland – Tjaldanes terraces and collected mollusc shells at Ekruhorn (Fig. 3.8) for dating of the basal high-arctic *Portlandia* diamicton. A sample of *Mya truncata* yielded an age of 11,255 ± 240 ^{14}C yr BP (Table 3.3; 14). Later, Andrésdóttir (1987) collected shells of *Mya truncata*, *Hiatella arctica*, and *Balanus* sp. at Tjaldanes, which have been dated to 11,275 ± 110 ^{14}C yr BP (Table 3.3; 15) (Ásbjörnsdóttir and Norðdahl, 1995). The age of the samples is not significantly different at 95% probability level.

Bárðarson (1921) concluded that a basal diamicton containing a high-arctic *Portlandia arctica* faunal assemblage indicates decreasing glacier proximity in a marine environment less than 15 m deep but with continuously rising RSL. A mollusc fauna characterized by *Pecten islandicus* represents a marine environment somewhat warmer than during formation of the *Portlandia* diamicton, but harsher than the present-day marine environment in the inner parts of Breiðafjörður. Again, according to Bárðarson (1921), the *Pecten* unit was formed when the water-depth was somewhere between 20 and 50 m. The uppermost *Mytilus* unit was formed when RSL was regressing from the ML shorelines. The *Mytilus edulis* faunal assemblage of the uppermost unit conforms to a present-day littoral environment (Bárðarson, 1921).

3.4.2.3 The Melar Terrace

At Melar, on the outer parts of Skarðsströnd (Fig. 3.8), a coastal terrace displays about a 15 m thick sequence of marine and littoral sediments (Andrésdóttir, 1987; Ásbjörnsdóttir and Norðdahl, 1995). Studies of the foraminifera fauna in the marine part of the sediments, show that they can be correlated with the sediments of the Holtaland – Tjaldanes terraces, displaying a comparable history of environmental changes. At Melar the lowermost 1 m of the marine sediments, resting on about a 2 m thick till unit which in turn overlays glacially striated basalt bedrock, is barren of foraminifera. A 9 m thick sequence of stratified silt and sandy-silt sediments, with occasional clay layers and ice-rafted pebble sized clasts, contain tests of foraminifera especially in the lowermost 1.5 m and the uppermost 4 m of the sediments. The marine sediments of the Melar terrace are discordantly overlain by a 3 m thick unit made of stratified sand and gravel deposits. This unit is barren of both foraminifera tests and mollusc shells.

The lowermost 1.5 m of the Melar sequence contains an arctic *Elphidium excavatum* (15-70%) - *Casidulina reniforma* (15-20%) faunal assemblage, which is correlated with the *Portlandia* diamicton in the Holtaland – Tjaldanes terraces. The uppermost 4 m of the fine grained marine sediments at Melar contain an *Elphidium excavatum* (76-92%) – *Casidulina reniforma* (2-14%) faunal assemblage indicating an environment with decreasing water-depth and temperature. This part of the Melar sediments is correlated with the *Pecten* beds in the Holtaland – Tjaldanes terraces (Ásbjörnsdóttir and Norðdahl, 1995). The topmost gravel unit at Melar is correlated with the *Mytilus* unit of the Holtaland – Tjaldanes terraces representing the time when RSL was situated at the 10-15 m level at Melar and Holtaland – Tjaldanes terraces, respectively (Ásbjörnsdóttir and Norðdahl, 1995).

Two samples of barnacle shells, collected at about 2 m, and between 9 and 11 m in the Melar sequence, have been dated to 11,165 ± 100 and 11,355 ± 80 ^{14}C yr BP, respectively (Table 3.3; 13,16) (Ásbjörnsdóttir and Norðdahl, 1995). The succession of the age of these samples is reversed but their age is statistically the same at the 95% probability level with a weighted mean age of 11,280 ± 100 ^{14}C yr BP. Therefore, we consider the formation of the Melar sequence to have occurred in late Alleröd times. A weighted mean age of all dated samples from the Holtaland – Tjaldanes and the Melar terraces is 11,275 ± 80 ^{14}C yr BP (Table 3.3; 13-16) and their age is statistically the same on a 95% probability level.

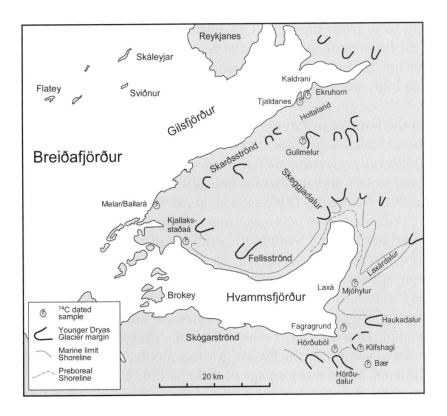

Fig. 3.8 Radiocarbon-dated marine samples in the Dalir (Hvammsfjörður and Gilsfjörður) area in western Iceland with the approximate position of Younger Dryas glacial margin and raised shorelines and place names mentioned in the text.

3.4.2.4 The Fossvogur Layers

The Fossvogur marine sediments have been known to the geological community for more than a century and many geologists have contributed to our present understanding of the geological history contained in these sediments (Geirsdóttir and Eiríksson, 1994). The sediments form about 3-5 m high sea cliffs on the southern coast of Reykjavík (Figs. 3.6, 3.13) and rest on the Reykjavík dolerite lavas, a glacially sculptured basalt bedrock with distinct glacial striae that are mainly orientated towards the west-northwest (Hjartarson, 1999).

The Fossvogur sedimentary sequence reveals a history of an advancing glacier in pre-Alleröd times and a subsequent retreat that was followed by a transgression of the sea during the Alleröd with formation of fossiliferous near-shore sediments. Furthermore, the deposition of the marine sequence on top of lithified lodgement till was frequently interrupted by debris flows originating from coastal erosion of basal till higher up on the Öskjuhlíð hill above the Fossvogur sea cliffs. In the upper part of the sedimentary sequence about 1-2 m thick clast-supported diamictite has been interpreted as a subaqueously formed debris-flow associated with increased calving at an advancing tidewater glacier margin. Succeeding this debris flow in the sedimentary sequence is a unit made of laminated and plane-bedded fine grained silt and sand sediments with convoluted bedding and distorted lamination suggesting a high rate of deposition, while the glacier successively withdrew from the area. Crudely to plane-bedded gravel sediments on top of parts of the Fossvogur sequence have been interpreted as a channel fill from a stream cutting down into the underlying sediments when RSL finally regressed to a position below the 3-5 m high sea cliffs at Fossvogur (Geirsdóttir and Eiríksson, 1994).

The marine macrofauna of the Fossvogur Layers contain species that are all known to live in low arctic – high boreal Icelandic waters today. Two faunal assemblages have been distinguished; one sublittoral preferring mixed muddy bottom with scattered harder parts such as pebbles and stones, and another littoral-sublittoral preferring mixed bottom where harder parts occurs frequently (Eiríksson *et al.*, 1991).

The age of the Fossvogur sediments has been securely determined by dating 30 samples of marine shells (Hjartarson, 1989; Sveinbjörnsdóttir *et al.*, 1993). Although these samples were taken from four different sections in the Fossvogur sediments their ages show a relatively narrow spread at the transition between the Alleröd and the Younger Dryas Chronozones (Sveinbjörnsdóttir *et al.*, 1993). The age of all the samples within each section is the same at a 95% probability level, except in one section containing the uppermost unit where a single sample from 9 makes their age different at the 95% probability level. The oldest dated sample in the Fossvogur sediments, taken from the lower marine part of the sediments, has yielded an age of 11,435 ± 170 ^{14}C yr BP and the youngest sample, taken from the upper most marine part of the sediments, has been dated to 10,585 ± 170 ^{14}C yr BP. A weighted mean age of some 30 dated samples is 11,070 ± 35 ^{14}C yr BP putting the formation of the sediments at the Alleröd – Younger Dryas transition and, furthermore, places a climatic deterioration and expansion of the tidewater glacier to the end of the Alleröd or the very beginning of the Younger Dryas (Sveinbjörnsdóttir *et al.*, 1993; Geirsdóttir and Eiríksson, 1994).

3.5 YOUNGER DRYAS AND PREBOREAL MARINE SEDIMENTS

Due to the apparent late Alleröd and early Younger Dryas rise of RSL in Iceland the extent of shallow coastal marine environments must have increased. The exact extent of the Younger Dryas and subsequent Preboreal marine environment in Iceland is not known but the distribution of fossiliferous sediments dated to the period between 11.0 and 9.0 k ^{14}C yr BP provides a minimum estimate of the extent of the marine environment (Fig. 3.9). At present, some 101 samples of marine mollusc and barnacle shells, collected throughout the country, have been dated to the Younger Dryas and Preboreal Chronozones. Out of the 86 formally published dates 39 samples have yielded Younger Dryas ages between 10,985 ± 100 and 10,005 ± 90 ^{14}C yr BP, and 47 have yielded Preboreal ages between 9,995 ± 90 and 9,055 ± 80 ^{14}C yr BP. The majority (74) of these samples were retrieved from fine-grained sediments at altitudes between present sea-level and the 35 m level, while 11 samples were collected from altitudes between 55 and 85 m asl in southern Iceland and 1 in northeastern Iceland. The somewhat limited spatial distribution of Younger Dryas – Preboreal marine shells in Iceland is most likely due to the presence of harsh near-terminus marine environment and expanding glaciers.

A sudden termination of the Younger Dryas Stadial conditions, supported e.g. by

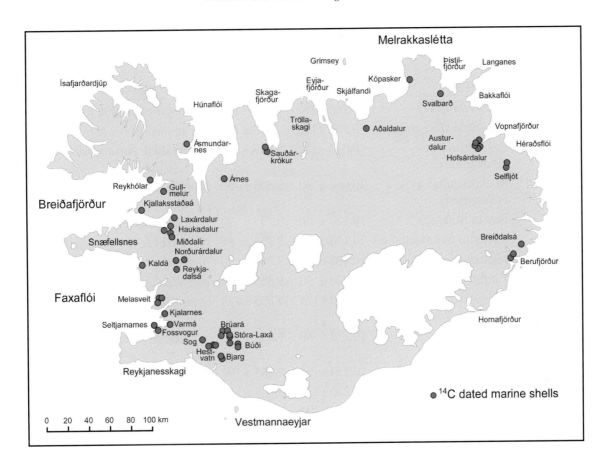

Fig. 3.9 Location of ¹⁴*C dated marine sediments (dots) of Younger Dryas – Preboreal age. The radiocarbon dates are listed in Table 3.4.*

studies of Greenland ice cores (Dansgaard *et al.*, 1989; Alley *et al.*, 1993) and high resolution North Atlantic deep-sea cores (Lehman and Keigwin, 1992), caused a considerable retreat of the Icelandic inland ice sheet, glacio-isostatic uplift of coastal areas and regression of RSL. Studies of ice-core and deep-sea data have also revealed a distinct but short-lived deterioration of environmental conditions in Preboreal times with a positive alteration in the mass-balance of the Icelandic ice sheet leading to a glacier advance and rise of RSL (Norðdahl and Ásbjörnsdóttir, 1995; Rundgren *et al.*, 1997; Norðdahl and Einarsson, 2001). The youngest reliable date of marine shells have yielded ages greater than 9.4 k ¹⁴C yr BP, which indicates that at about that time or little later RSL must have regressed below present sea-level.

Mapping features such as raised shorelines, and termino-glacial marine and terrestrial deposits has enabled us to distinguish between two successive glacier advances and to determine their extent in Younger Dryas and Preboreal times, respectively (Norðdahl, 1990). The spatial distribution of radiocarbon dated samples of marine shells yielding ages between 11.0 and 9.0 k ¹⁴C yr BP along with dated tephra beds again provides a minimum estimate of the extent of the marine environment, RSL changes, and location of presumably ice-free areas.

Table 3.4. Radiocarbon dated shells of marine organisms of Younger Dryas and Preboreal age collected on present day dry land in Iceland.

Lab. Num.	Sample	¹⁴C age BP	Res. age BP	Reference
1 AAR-0156	Dalasýsla (Bær)	9,780 ±130	9,415 ± 130	Sveinbjörnsdóttir & Johnsen 1991
2 AAR-0158	Dalasýsla (Kvisthagi)	9,780 ±110	9,415 ± 110	Sveinbjörnsdóttir & Johnsen 1991
3 Lu-3110	Fagragrund in Haukadalur	10,030 ±120	9,665 ± 120	Norðdahl & Ásbjörnsdóttir 1995
4 Lu-3112	Kjallaksstaðaá at Fellsströnd	10,060 ± 90	9,695 ± 90	Norðdahl & Ásbjörnsdóttir 1995
5 Lu-3109	Bær in Miðdalir	10,120 ± 90	9,755 ± 90	Norðdahl & Ásbjörnsdóttir 1995
6 Lu-3115	Mjóhylur in Laxárdalur2	10,120 ± 90	9,755 ± 90	Norðdahl & Ásbjörnsdóttir 1995
7 Lu-3114	Mjóhylur in Laxárdalur 1	10,130 ± 90	9,765 ± 90	Norðdahl & Ásbjörnsdóttir 1995
8 Lu-3111	Hörðuból in Hörðudalur	10,220 ± 90	9,855 ± 90	Norðdahl & Ásbjörnsdóttir 1995
9 AAR-0155	Dalasýsla (Kjallaksstaðaá)	10,220 ±125	9,855 ± 125	Sveinbjörnsdóttir & Johnsen 1991
10 AAR-0157	Dalasýsla (Laxárdalur 1)	10,220 ± 60	9.855 ± 60	Sveinbjörnsdóttir & Johnsen 1991
11 Lu-3113	Kvisthagi in Miðdalir	10,230 ± 90	9,865 ± 90	Norðdahl & Ásbjörnsdóttir 1995
12 AAR-0085	Dalasýsla (Fagragrund)	10,230 ±170	9,865 ± 170	Sveinbjörnsdóttir & Johnsen 1991
13 AAR-0154	Dalasýsla (Hörðuból)	10,270 ±100	9,905 ± 100	Sveinbjörnsdóttir & Johnsen 1991
14 AAR-0133	Dalasýsla (Laxárdalur 2)	10,280 ±110	9,915 ± 110	Sveinbjörnsdóttir & Johnsen 1991
15 Lu-2777	Gullmelur	10,580 ±100	10,215 ±100	*New date*

Weighted mean age: 1-14   9,735 ± 50   ¹⁴C yr BP

| 16 Lu-3118 | Helgafellsmelar | 10,580 ± 90 | 10,215 ± 90 | Ingólfsson *et al.* 1995 |
| 17 U-2898 | Varmá, Helgafellsmelar | 10,780 ±110 | 10,415 ± 110 | Norddahl 1991a |

Weighted mean age: 16-17   10,315 ± 105   ¹⁴C yr BP

18 U-2817	Varmá, Helgafellsmelar	10,180 ±150	9,815 ±150	Norðdahl 1990
19 AAR-0003B	Austurströnd	10,180 ±150	9,815 ±150	Hjartarson 1989
20 AAR-0003A	Austurströnd	10,290 ±210	9,925 ±210	Hjartarson 1989
21 U-0412	Reykjavíkurflugvöllur	10,310 ±260	9,945 ±260	Einarsson 1964
22 U-0415	Reykjavíkurflugvöllur	10,230 ±190	9,865 ±190	Einarsson 1964
23 U-0413	Reykjavíkurflugvöllur	9,940 ±260	9,575 ±260	Einarsson 1964
24 U-0414	Reykjavíkurflugvöllur	10,450 ±160	10,085 ±160	Einarsson 1965

Weighted mean age: 18-24   9,875 ± 90   ¹⁴C yr BP

25 Lu-3802	Breiðdalsá	9,980 ±110	9,615 ±110	Norðdahl & Einarsson, 2001
26 Lu-3800	Fagrihvammur	10,170 ±110	9,805 ±110	Norðdahl & Einarsson, 2001
27 Lu-3668	Teigarhorn	10,480 ±120	10,115 ±120	Norðdahl & Einarsson, 2001

Weighted mean age: 25-27   9,880 ± 100   ¹⁴C yr BP

| 28 Lu-2401 | Stóra Laxá at Hrepphólar 1 | 10,110 ±140 | 9,745 ±140 | Hjartarson & Ingólfsson 1988 |
| 29 Lu-2402 | Stóra Laxá at Hrepphólar 2 | 9,960 ±160 | 9,595 ±160 | Hjartarson & Ingólfsson 1988 |

Table 3.4 continued.

Lab. Num.	Sample	^{14}C age BP	Res. age BP	Reference
30 Lu-2403	Þjórsá 1	10,360 ± 90	9,995 ± 90	Hjartarson & Ingólfsson 1988
31 Lu-2404	Þjórsá 2, near Minnahof	10,220 ± 90	9,855 ± 90	Hjartarson & Ingólfsson 1988
32 AAR-1241	Búdafoss	10,290 ±140	9,925 ±140	Geirsdóttir *et al.* 1997
33 AAR-1242	Búdafoss	10,120 ±150	9,755 ±150	Geirsdóttir *et al.* 1997
34 AAR-1243	Búdafoss	10,130 ±170	9,765 ±170	Geirsdóttir *et al.* 1997
35 Lu-2596	Oddgeirshólar	10,440 ± 90	10,075 ± 90	Hjartarson & Ingólfsson 1988

Weighted mean age: 28-34   9,845 ± 75 ^{14}C yr BP

36 Lu-3119	Kjölur, Aðaldalur	10,530 ± 80	10,165 ± 80	*New date*
37 Lu-2673	Fell 2	9,980 ± 70	9,615 ± 70	Norðdahl & Hjort 1987
38 Lu-2675	Skógaeyri 1	10,050 ± 90	9,685 ± 90	Norðdahl & Hjort 1987
39 Lu-3289	Vs-1	10,140 ±130	9,775 ±130	Sæmundsson 1995
40 Lu-2674	Fell 3a	10,230 ± 90	9,865 ± 90	Norðdahl & Hjort 1987
41 AAR-0886	VN-3	10,270 ±150	9,905 ±150	Sæmundsson 1995
42 AA-21465	Sval. S3	10,400 ± 75	10,035 ± 75	Richardson 1997
43 T-4467	Kópasker	10,570 ± 80	10,205 ± 80	Pétursson 1986

Weighted mean age: 37-41   9,655 ± 70 ^{14}C yr BP

3.5.1 The Dalir Area, Western Iceland

Raised fossiliferous marine sediments, mainly composed of stratified clay, silt and sand sediments, are found at altitudes between present sea-level and about 50 m around the Hvammsfjörður fjord in Western Iceland (Fig. 3.8), showing that the area has been inundated by the sea. Marine limit shorelines are situated at 65-70 m asl from the mouth of Hvammsfjörður and eastwards along the entire fjord with conspicuous lower shorelines at 50-55 m asl and 30-35 m asl (Norðdahl and Ásbjörnsdóttir, 1995). The marine limit shorelines are truncated at the mouth of all of the tributary valleys between Skeggjadalur and Hörðudalur, while the lower shorelines and the 30-35 m shorelines in particular, can be followed across the tributary valleys (Fig. 3.8). The different extent of raised shorelines was most likely controlled by more extended glaciers during the formation of the marine limit shorelines at 65-70 asl. Later, especially during the formation of the 30-35 m raised shorelines, the glaciers had retreated well into the tributary valleys (Norðdahl and Ásbjörnsdóttir, 1995).

Seven samples of marine shells have been collected from the fine-grained sediments around Hvammsfjörður at elevations between 6 m and 27 m asl (Fig. 3.8). All these samples were dated at the ^{14}C Laboratory at the University of Lund, Sweden and at the AMS Dating Laboratory at the University of Aarhus, Denmark, yielding sea corrected ages between 9,415 ± 110 and 9,915 ± 110 ^{14}C yr BP (Table 3.4; 1-14). A weighted mean of all these dates is 9,735 ± 50 BP where the age of individual samples is statistically inseparable at a 95% probability level. All these samples have been related to raised shorelines at or somewhat below the 30-35 m level and, therefore, they date an early Preboreal position of RSL in the Hvammsfjörður area and consequently they postdate the formation of the marine limit shorelines at 65-70 m asl.

The course of RSL changes in the Hvammsfjörður area has been determined from sedimentary sequences in the Laxárdalur valley (Fig. 3.8) where glacially striated bedrock at the mouth of river Laxá is overlain with till. The till is in turn succeeded by stratified sand and gravel sediments forming irregular kames and eskers along the valley. The kame and esker formations are partly and in places totally covered by thinly bedded and laminated clay, silt and sand sediments reaching as high as 50 m asl in Laxárdalur. A few whole shells of *Mya truncata* and fragments of other mollusc shells have been retrieved from these sediments along with foraminifera tests clearly showing that these sediments were accumulated from suspension in relatively calm marine environment. Some 15 km up into the Laxárdalur valley these fine grained sediments are covered with coarse grained sand and gravel sediments forming a dissected outwash formation graded to RSL situated as high as 70 m asl (Norðdahl and Ásbjörnsdóttir, 1995).

In the lower part of the Laxárdalur valley the fine grained marine sediments are discordantly overlain by fossiliferous clayey-silty sand sediments of sublittoral origin which in turn are succeeded by re-worked sediments. This arrangement of the sediments in Laxárdalur has earlier been described e.g. by Thoroddsen (1905-06), Bárðarson (1921), Kjartansson and Arnórsson (1972), Norðdahl and Ásbjörnsdóttir (1995) and Jónsdóttir and Björnsdóttir (1995). At an exposure at Mjóhylur the following succession of sediments has been observed in about a 15 m high terrace situated between 15 and 30 m asl on the south side of river Laxá (Figs. 3.8, 3.4, 3.10):

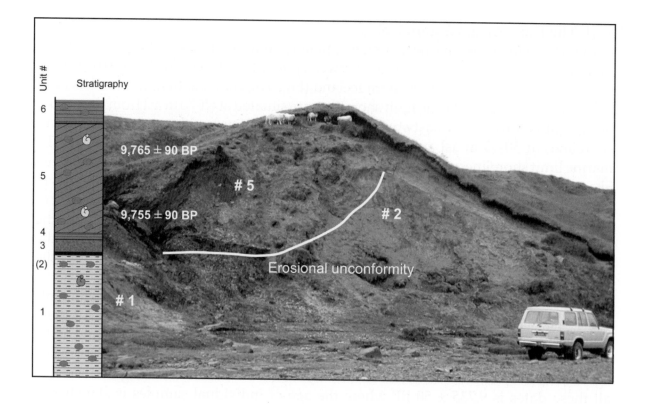

Fig. 3.10 The Mjóhylur locality in the Laxárdalur valley in the Dalir district in western Iceland (based on: Norðdahl and Ásbjörnsdóttir, 1995).

Unit 1 at the base of the terrace is made up of a crudely stratified or massive greyish silty-sandy clay deposit with gravel, pebble, and cobble sized clasts some of them with distinct glacial striae. Occasional mollusc shells or shell fragments have been found in this part of the terrace. The lower boundary of unit 1 is hidden at the logging site but the upper boundary is diffuse and the sediments are successively changed into unit 2, which is about 4 m thick and made of laminated and thinly bedded brownish clay and clayey silt sediments with much fewer gravel and pebble sized clasts. About 500 m downstream from the Mjóhylur site units 1 and 2 are partly covering an esker or a kame, which in turn rests on till or glacially striated bedrock. Units 1 and 2 are interpreted as a glaciomarine deposits with ice-rafted clasts where the increased stratification and reduced number of ice-rafted clasts most likely reflects decreased glacier proximity during deglaciation of Laxárdalur.

Unit 2 has a sharp erosional contact to the overlying units which occupy a channel about 5 m deep eroded into the sediments of units 1 and 2 (Fig. 3.10). The erosion of this channel must have occurred when base level (and RSL) was lowered from its maximum position close to the 65-70 m level and down to, or below, the base of the erosional channel at about 20 m asl.

Unit 3, about 0.1-0.3 m thick, is made of sandy gravel containing numerous shell fragments with a sharp and apparently horizontal contact to unit 4, which is about 0.1 m thick made of greyish clay with mollusc shells and shell fragments. Unit 4 has a sharp to graded contact to the overlying unit 5.

Unit 5 is about 10 m thick and made of laminated clayey-sandy silt and sand. From bottom upwards the laminated sediments are successively replaced by thinly bedded medium and coarse sand sediments with occasional sand layers. The stratification of unit 5 has a distinct but apparent 30° dip reflecting transport of the sediments towards the northeast. Whole mollusc shells and shell fragments are found in the sandy sediments in the lower half of unit 5 with *Mya truncata in situ* (Fig. 3.11) showing that at least some of the shells have not been moved or otherwise transported into the erosional channel. Two mollusc samples collected at about 21 and 25 m asl respectively have yielded a weighted mean age of 9,700 ± 45 ^{14}C yr BP (Table 3.4; 6,7,10,14). The sediments of units 3-5 were accumulated in an erosional channel when RSL rose to an altitude above the base of the channel. The marine environment in and around the channel is reflected in the composition of the mollusc fauna living there at the time, a faunal assemblage characterized by an almost equal mixture of Arctic (58%) and Boreal (42%) species (Bárðarson, 1921). Furthermore, this faunal assemblage indicates a relatively shallow marine environment characterized by freshwater inflow from melting and retreating glacier.

Unit 6, the uppermost part of the Mjóhylur-terrace is made of apparently horizontally stratified sand and gravelly sand and massive greyish clay with sand lenses. The top of unit 6 is situated at about 33 m asl and very close to marine shorelines at 30-35 m asl, explaining the textural contrasts within unit 6.

On the basis of studies of the sediments in the Laxárdalur valley it is possible to reconstruct the course of RSL changes in the Hvammsfjörður area (Fig. 3.12). Subsequent to deglaciation and formation of kames and eskers in Laxárdalur the area was inundated by the sea forming the marine limit shorelines at 65-70 m asl. At that time glacier termini

Fig. 3.11 An in situ bivalve of Mya truncata in unit 5 of the Mjóhylur section (Photo: Norðdahl, H.).

were situated in all the tributary valleys around Hvammsfjörður (Fig. 3.8). Following the formation of the ML shorelines the glaciers retreated and because of subsequent glacio-isostatic readjustment of the crust RSL was lowered to a position below the 20 m level (Fig. 3.12). Later, RSL was raised to a temporary position at 30-35 m asl and shallow-water/sublittoral sediments filled the channel previously eroded into the marine sediments at the Mjóhylur locality (Fig. 3.10). This transgression of RSL, which was most likely caused by a glacier re-advance in the area (Norðdahl and Ásbjörnsdóttir, 1995), has been dated to 9,735 ± 50 ^{14}C yr BP, a weighted mean of 14 ^{14}C dates from the area (Table 3.4; 1-14). Since the formation of the 30-35 m shorelines RSL was apparently lowered to a position below present sea-level (Fig. 12) during a general Preboreal retreat of Icelandic glaciers (Norðdahl, 1990; Ingólfsson and Norðdahl, 1994; Ingólfsson *et al.*, 1995).

The age of the ML shorelines in the Hvammsfjörður area is still unknown otherwise than they are older than the ^{14}C dated (9,735 ± 50 ^{14}C yr BP) 30-35 m shorelines. A single mollusc sample has been collected at about 50 m asl from the Gullmelur terrace in Saurbær (Fig. 3.8), a terrace that has been related to a base-level situated close to the 80-90 m level there (Andrésdóttir, 1987; Bárðarson, 1921). The mollusc sample has yielded a ^{14}C age of 10,215 ± 100 ^{14}C yr BP (Table 3.4; 15). This altitude of a base-level in Saurbær might possibly be compared with the height of the ML shorelines at 65-70m asl in the Hvammsfjörður (Fig. 3.8) area and, thus imply a Younger Dryas age for the ML there.

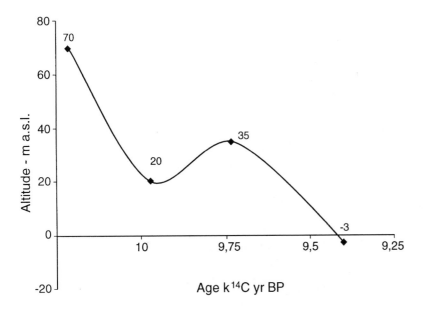

Fig. 3.12 The development of Younger Dryas – Preboreal relative sea-level changes in the Dalir district in western Iceland (Based on: Norðdahl and Ásbjörnsdóttir, 1995).

3.5.2 The Reykjavík Area, Southwestern Iceland

The area under consideration covers the coastal area between the town of Hafnarfjörður in the southwest and Mt. Esja in the northeast (Fig. 3.13). Fossiliferous sediments of Younger Dryas and Preboreal ages are widespread in the area below the ML shorelines, which are found at increasingly higher altitudes from 32 m asl in Hafnarfjörður to 60 m asl below Mosfell (Ingólfsson *et al.*, 1995). Out of 30 formally published dates on samples collected in this area 14 of them, yielding ages between 10,985 ± 250 and 9,435 ± 160 ^{14}C yr BP, are from the Alleröd Fossvogur Layers, while nine samples are related to raised marine shorelines in the area.

Reconstruction of isobases for the ML shorelines in the Reykjavík area defines a plane rising towards the northeast with a gradient of about 1.15 m km[-1] (Ingólfsson *et al.* 1995). Two samples of whole shells, plates, and fragments of *Balanus balanus* collected from a sandy diamicton at about 20 m asl at Helgafellsmelar (Figs. 3.13, 3.14) have been dated to 10,215 ± 90 and 10,415 ± 115 ^{14}C yr BP, respectively with a weighted mean age of 10,315 ± 105 ^{14}C yr BP (Table 3.4; 16-17). These samples date the formation of an ice-contact delta graded to about 60 m asl and a glacier margin in the mouth of the Mosfellsdalur valley (Fig. 3.13), and the ML shorelines in the Reykjavík area (Ingólfsson *et al.*, 1995). Prominent raised beaches at altitudes below the ML shorelines have been mapped at about 20 m asl in Hafnarfjörður, at 25-30 m asl in Garðabær and Kópavogur, at 20-25 m asl at Austurströnd on the Reykjavík peninsula, and at about 40 m asl at Helgafellsmelar and Álfsnes northeast of Reykjavík (Fig. 3.13). A single sample of marine shells collected from about 15 m asl at Helgafellsmelar and two samples from beach deposits reaching 20-25 m asl at the Austurströnd locality have yielded ages between 9,815 ± 150 and 9,925 ± 210 ^{14}C yr BP (Table 3.4; 18-20) and a weighted mean age of 9,865 ± 170 ^{14}C yr BP. Fossiliferous littoral and sublittoral sediments reaching some 10-15 m

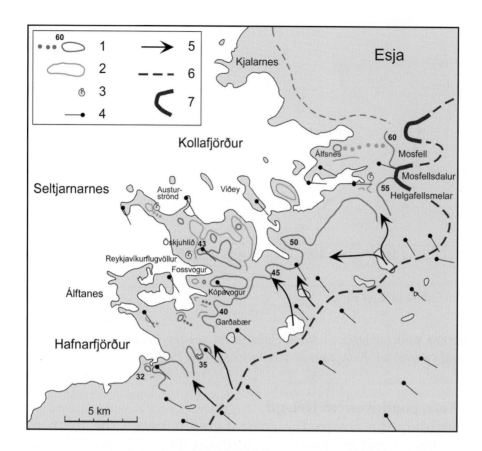

Fig. 3.13 Radiocarbon-dated marine samples in the Reykjavík area in southwestern Iceland with the approximate position of a Younger Dryas glacial margin and raised shorelines with place names mentioned in the text (based on: Ingólfsson et al., 1995). 1) Younger Dryas marine limit shoreline. 2) Preboreal raised shoreline. 3) Radiocarbon-dated marine sample. 4) Glacial striae (Kjartansson, 1960). 5) Apparent topographically controlled flow of ice. 6) Approximate position of Younger Dryas glacial margin. 7) Younger Dryas moraines.

asl at Reykjavíkurflugvöllur (Einarsson, 1964) are related to raised shorelines at about 20-25 m asl and dated to about 9,895 ± 125 ^{14}C yr BP, a weighted mean age of 4 samples (Table 3.4; 21-24). The 7 Preboreal dates, averaging to about 9,875 ± 90 ^{14}C yr BP, are regarded as an approximation to the age of the lower set of prominent raised shorelines found below the ML shorelines in the Reykjavík area. Since the formation of the lower set of shorelines at 9,875 ± 90 ^{14}C yr BP, RSL was apparently lowered to a position at least 2.5 m below present sea-level at about 9,400 ^{14}C yr BP. The lowermost position of RSL at about –30 m was apparently reached at about 8,500 ^{14}C yr BP (Ingólfsson et al., 1995).

Earlier Kjartansson (1952) tried to explain these two sets of raised shorelines in Hafnarfjörður as the result of a Lateglacial transgression of RSL reaching the ML at 32 m asl there (Fig. 3.13). During a subsequent regression freshwater sediments were accumulated at about 10 m asl until RSL transgressed again reaching between 15 m and 30 m asl. Þórarinsson (1956c) also suggested RSL changes for the Reykjavík area with two transgressions; the first reaching the ML at 43 m asl at Öskjuhlíð and the second

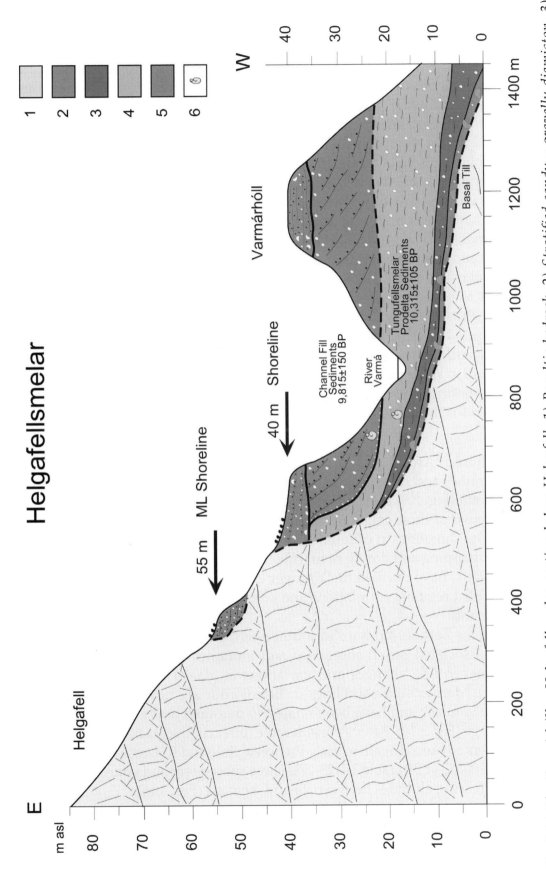

Fig. 3.14 The Varmárhóll – Helgafellsmelar section below Helgafell. 1) Basaltic bedrock. 2) Stratified sandy – gravelly diamicton. 3) Poorly stratified sand and gravel. 4) Fossiliferous sandy prodelta sediments. 5) Fossiliferous channel fill and foreset sediments. 6) Radiocarbon dated sample of marine shells.

one reaching some 20 m asl before it regressed to a position at least 8 m below present sea-level. Þórarinsson (1956c) proposed a pre-Alleröd age for the ML shorelines and an age of 9,000 or 7,500 BC for the second transgression.

Because of an extended excavation of a ML ice-contact delta in Mosfellsdalur, it has been possible to follow its prodelta sediments into the excavated Helgafellsmelar and Varmárhóll terraces below Helgafell (Figs. 3.13, 3.14). There, a channel was eroded into the fine grained Tungufellsmelar prodelta sediments when base-level, i.e. RSL, was lowered from the ML at about 55 m asl to a position close to, or below, the 20 m level (Fig. 3.15). During a subsequent transgression of RSL base-level was again raised and the erosional channel was filled in with a 17-20 m thick unit of cross-bedded delta foreset beds of sorted sand and gravelly sand sediments (Figs. 3.14, 3.15). The delta foreset beds are discordantly overlain by an up to 8 m thick unit of sand and well rounded pebble-cobble rich gravel, and delta topset beds reaching about 40 m asl in the Helgafellsmelar and Varmárhóll terraces (Figs. 3.14, 3.15). Finally RSL regressed from the 40 m level and below the –2.5m level at about 9,400 ^{14}C yr BP (Ingólfsson *et al.* 1995). The foreset beds, dipping 15°-20° in a northwesterly direction, indicate that the delta foreset beds and, thus, the Helgafellsmelar – Varmárhóll delta was formed by the former Varmá river (Figs. 3.14, 3.15). The sample collected from Helgafellsmelar and dated to 9,815 ± 150 ^{14}C yr BP (Table 3.4; 18) probably dates this younger marine phase and formation of the 40 m Helgafellsmelar – Varmárhóll delta. The stratigraphical sequence described above confirms and dates the RSL changes Kjartansson (1952) and Þórarinsson (1956c) suggested for the Hafnarfjörður and Reykjavík area, respectively.

3.5.3 The Berufjörður Area, Southeastern Iceland

Survey of moraines and raised shorelines in eastern Iceland, between Hornafjörður and Melrakkaslétta (Fig. 3.9) by Hjartarson *et al.* (1981) and Norðdahl and Einarsson (1988), revealed two or three consecutive phases of glacier advance and formation of concurrent marine shorelines at different altitudes in the area (Norðdahl, 1990). Since then additional mapping and dating of marine mollusc and barnacle shells (Norðdahl and Einarsson, 2001) have clarified the Younger Dryas – Preboreal situation in eastern Iceland and enabled regional correlations and comparison with other dated sediment sequences in Iceland.

In and around the Berufjörður fjord in southeastern Iceland (Fig. 3.16) dated marine sediments and associated raised shorelines and moraines have revealed a complex relationship between RSL changes and variations in glacier extent (Norðdahl and Einarsson, 2001). Marine limit shorelines in Berufjörður are now found at altitudes between 35 and 58 m asl, and have been related to moraines and a glacial advance of Younger Dryas age (Fig. 3.16). After a considerable retreat of the glaciers and regression of RSL, the glaciers advanced again in early Preboreal time and RSL rose and culminated at about 9,880 ± 100 ^{14}C yr BP (Table 3.4; 25-27). These subsequent and younger shorelines are found at increasing altitudes between 24 and 40 m asl (Fig. 3.16).

A reconstruction of the glacier extent in Younger Dryas and Preboreal times (Fig. 3.17), based on the data presented above, reveals a situation with contemporaneous alpine valley glaciers and outlet glaciers occupying the fjords and valleys of eastern Iceland (Norðdahl and Einarsson, 2001). In Younger Dryas time, during formation of

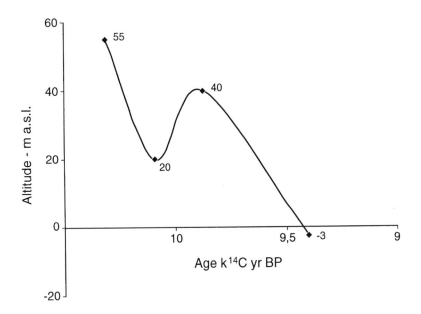

Fig. 3.15 The development of Younger Dryas – Preboreal relative sea-level changes in the Reykjavík area in southwestern Iceland.

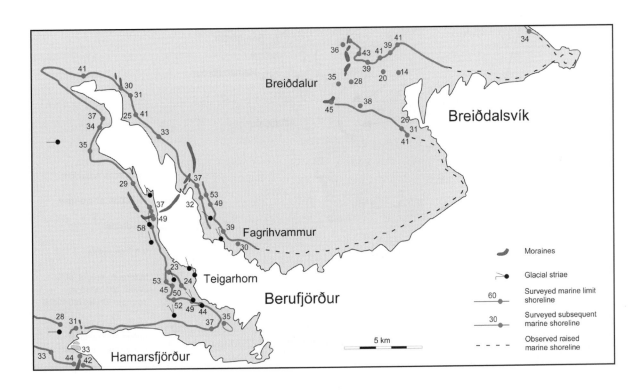

Fig. 3.16 The Berufjörður area and its vicinity in eastern Iceland with raised shorelines, moraines and striae (modified after: Norðdahl and Einarsson, 2001).

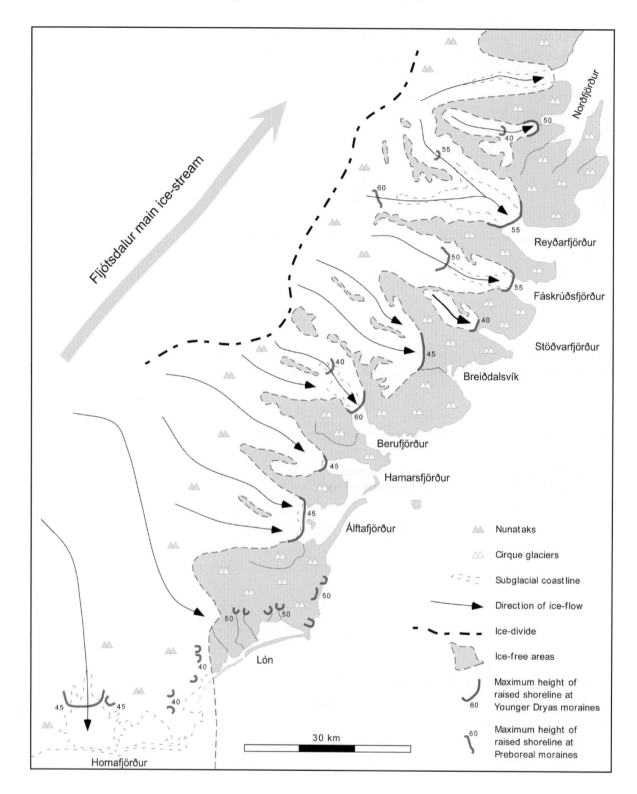

Fig. 3.17 The extent of glaciers in eastern and southeastern Iceland in Younger Dryas and Preboreal times (from: Norðdahl and Einarsson, 2001).

the ML shorelines, outlet glaciers reached the outer part of the fjords, and thus limited the extent of contemporaneous ML shorelines, which end at conspicuous end moraines. In early Preboreal times, younger and more extended shorelines were formed when the East Icelandic glaciers re-advanced and reached a temporary maximum position some 10-23 km inside the preceding Younger Dryas moraines (Fig. 3.17).

3.5.4 The South Icelandic Lowland

The Lateglacial research history of the South Icelandic Lowland has previously been outlined e.g. by Hjartarson and Ingólfsson (1988), Norðdahl (1990) and Geirsdóttir *et al.* (1997) in all cases with the main emphasis on the formation of the Búði moraine proper at the river Þjórsá (Fig. 3.18). According to Kjartansson (1943, 1958) the Búði moraine can be correlated with other moraine ridges to form 'a belt of terminal moraine ridges' that have been traced to Vatnsdalsfjall, some 25 km towards the southeast across the South Lowlands, and northwestwards for about 16 km to Grámelur at river Hvítá. Later, Kjartansson (1962b, 1970) extended the Búði moraine to Efstadalsfjall, about 28 km northwest of the Búði moraine at the river Þjórsá (Fig. 3.18). Jóhannesson (1985) emphasized that the Búði moraine is a complex of up to seven more or less parallel ridges and suggested a considerable age difference between the outermost and innermost ridges. A recent study shows that the Búði terminal zone is made of two sets of end moraines some 5 km apart (Fig. 3.18) (Axelsdóttir, 2002). The inner (younger) moraine crosses the river Þjórsá at the Búði waterfall and can be traced for about 15 km towards the southeast where it disappears underneath a postglacial lava flow. It is the inner moraine that has been followed from the Búði waterfall to Efstadalsfjall. The outer (older) moraine has been followed from Mt. Vatnsdalsfjall and for about 25 km towards the northwest where it disappears in the Holt area (Fig. 3.18).

In the late 1980s Hjartarson and Ingólfsson (1988) concluded, on the basis of structural and sedimentological data collected at the river Þjórsá (Búðaberg and Þrándarholt) and at the river Stóra-Laxá (Hrepphólar), that a glacier had advanced across glaciomarine sediments to form moraine-like ridges consisting of disturbed beds of glaciomarine diamicton. Four samples of marine shells collected from sediments in close relation to the Búði moraine at Þjórsá and Stóra-Laxá have yielded a weighted mean age of 9,855 ± 90 ^{14}C yr BP (Table 3.4; 28-31) for the formation of the moraine and, hence, the advance of the glacier. Geirsdóttir *et al.* (1997) interpreted the depositional history of the Búði moraine proper as reflecting a more or less continuous late Younger Dryas retreat of a partly marine-based glacier and a subsequent early Preboreal sedimentation in a high-energy glaciomarine environment characterized by repeated subglacial meltwater outbursts. Furthermore, a basal stratified diamicton in the Búðaberg section is interpreted as an ice-marginal formation, postdated by a sample of barnacles yielding an age of 9,855 ± 90 ^{14}C yr BP (Table 3.4; 31) (Geirsdóttir *et al.*, 2000). Fragments of barnacles, collected from a diamicton with ice-rafted clasts in the upper parts of the Búðaberg section, have yielded dates of 9,925 ± 140, 9,755 ± 150, and 9,765 ± 170 ^{14}C yr BP (Table 3.4; 32-34) (Geirsdóttir *et al.* 1997), confirming that the formation of the Búði moraine at the river Þjórsá was apparently completed in early Preboreal time (Hjartarson and Ingólfsson, 1988; Geirsdóttir *et al.*, 1997, 2000). The seven samples of marine shells collected from the Búði moraine, yielding a weighted mean age of 9,845 ± 75 ^{14}C yr BP

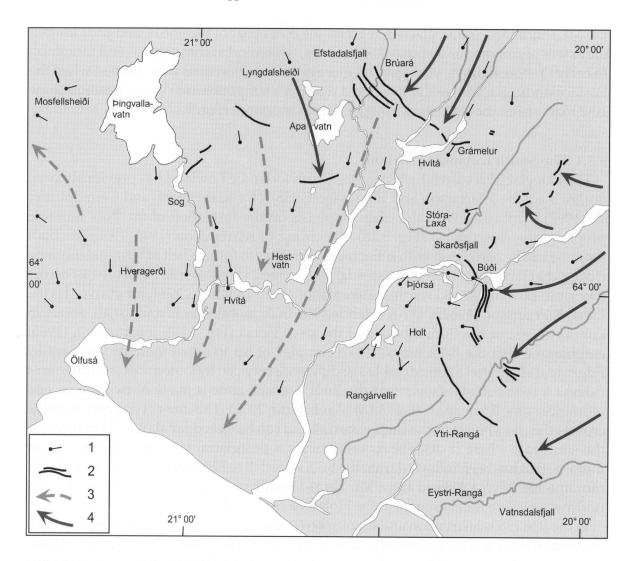

Fig. 3.18 The South Icelandic Lowland with 1) Striae (Kjartansson, 1962b), 2) End-Moraines (Kjartansson, 1943, 1958, 1962b, 1970; Jóhannesson, 1985; Axelsdóttir, 2002), 3) Older and 4) Younger pattern of ice-flow, with place names mentioned in the text.

(Table 3.4; 28-34), which also indicates an early Preboreal age to the formation of the inner moraine in the Rangárvellir area (Fig. 3.18). The outer set of moraines was, therefore, formed during a glacier event preceding the formation of the Búði moraine proper.

Kjartansson (1958) determined the altitude of ML shorelines in the Holt area and found it to be about 110 m asl, and about 100 m asl below Skarðsfjall, just north of the Búði moraine proper. On the geological map of South-Central Iceland (Jóhannesson *et al.*, 1982) the ML shorelines are shown at the 105 m level. Hjartarson and Ingólfsson (1988) concluded that RSL was well above the 65-70 m level both prior and subsequent to a glacier advance and the formation of the Búði moraine. Geirsdóttir *et al.* (1997) envisaged that the South Lowlands probably had been continuously submerged since late Younger Dryas times and into the Preboreal Chronozone indicating high RSL with ML shorelines at about 100 m asl.

Hjartarson and Ingólfsson (1988) drew attention to the fact that the oldest radiocarbon dated marine material from southern Iceland has yielded an age of 10,075 ± 90 BP (Table 3.4; 35). They explained this apparent early Preboreal – Preboreal maximum age of marine shells by a glacier extending beyond the present coastline in Lateglacial times, thus preventing accumulation of glaciomarine sediments there. On the other hand, Geirsdóttir *et al.* (1997, 2000) assumed a more restricted extent of a late Younger Dryas glacier with an ice margin situated only about 1000 m west of the Búði moraine proper at the river Þjórsá. This assumption is further supported by studies of sediments in the Lake Hestvatn basin (Fig. 3.18) (Geirsdóttir *et al.*, 2000; Harðardóttir *et al.*, 2001) indicating that the basin has not been invaded by glaciers since late Younger Dryas times.

3.5.5 The Eyjafjörður – Skjálfandi Area, Northern Iceland

At the LGM major outlet glaciers occupied the Eyjafjörður fjord and the Skjálfandi bay in northern Iceland (Fig. 3.9), reaching far out and onto the shelf off northern Iceland (Norðdahl, 1983, 1991; Norðdahl and Hafliðason, 1992; Andrews *et al.*, 2000). These outlet glaciers had two principal sources of ice, from the main Icelandic ice sheet covering the central highland south of the Eyjafjörður area, and from corrie and valley glaciers within the Tröllaskagi massif (Fig. 3.9). Consequently, the Eyjafjörður outlet glacier was a composite glacier fed both by the central ice sheet and local glaciers. The occurrence of the Lateglacial Skógar-Vedde Tephra in clastic sediments in the area has greatly aided stratigraphical correlations within the Eyjafjörður – Skjálfandi area (Norddahl, 1983; Norðdahl and Hafliðason, 1992). Absolute age determinations of the Skógar-Vedde Tephra in western Norway indicate an age of about 10,300 ^{14}C yr BP (Bard *et al.*, 1994) for the tephra fall-out throughout the North Atlantic Region.

Subsequent to a considerable retreat, the outlet glaciers in Eyjafjörður and Skjálfandi again reached an advanced position. The Skjálfandi glacier terminated just south of Húsavík where raised shoreline features containing shards of the Skógar-Vedde Tephra are situated at about 50 m asl (Fig. 3.19). In Eyjafjörður, the outlet glacier terminated just south of the island of Hrísey in the outer parts of the fjord leaving distinct lateral moraines on the eastern side of the fjord (Fig. 3.19). An apparent marine limit beach ridge at about 30 m asl, close to a lateral moraine, marks the highest known position of RSL in the outer parts of Eyjafjörður. At that time, the Eyjafjörður outlet glacier sealed off the Dalsmynni valley (Fig. 3.19) and caused the formation of the extensive Austari-Krókar ice-lake in the Fnjóskadalur valley (Norðdahl, 1983). Considerable quantities of the Skógar-Vedde Tephra were trapped in the ice-lake, dating it and a concurrent formation of the Hrísey terminal zone in Eyjafjörður to about 10,300 ^{14}C yr BP (Norðdahl and Hafliðason, 1992).

A marginal position of the Eyjafjörður outlet glacier close to Espihóll, some 50 km south of the Younger Dryas terminal zone (Fig. 3.19), represents a substantial retreat of the glacier and inundation of the sea (Pétursson and Norðdahl, 1999; Norðdahl and Pétursson, 2000). Raised shorelines correlated with this position of the glacier are found at decreasing altitudes from about 40 m asl at Espihóll to about 10 m asl at Svarfaðardalur. At this time glaciers extended out into the mouth of Svarfaðardalur and Hörgárdalur, forming marginal deltas there at about 10 m asl and 20-25 m asl, respectively,

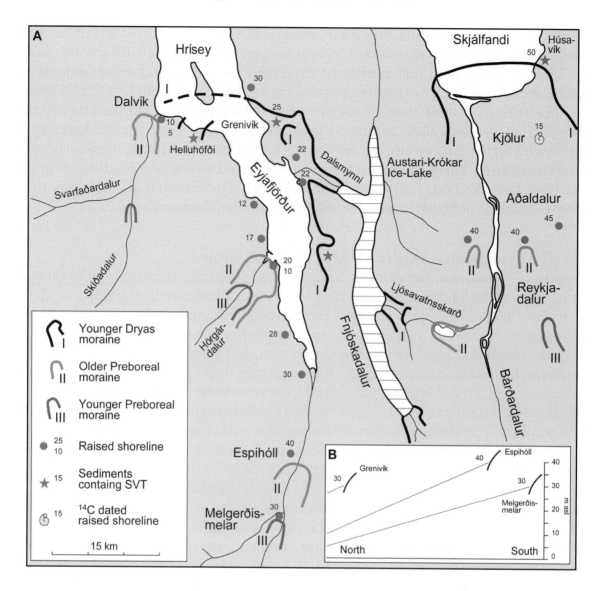

Fig. 3.19 A: An outline of the deglaciation of the Eyjafjörður – Skjálfandi area in northern Iceland with altitude of raised marine features and place names mentioned in the text. B: Approximate extent and gradient of three successive raised shorelines and respective ice margins in the Eyjafjörður area.

contradicting the conclusion of Häberle (1991a) and Stötter (1991a) of limited glacier extent in these valleys at that time. A subsequent and lower set of raised shorelines in Eyjafjörður is found at altitudes between 5 m asl at the mouth of Svarfaðardalur to about 30 m asl at Melgerðismelar in southern Eyjafjörður (Fig. 3.19).

Raised shorelines and marginal features south of the proposed Younger Dryas terminal zone in Eyjafjörður are younger than the 10,300 ^{14}C yr BP old Skógar-Vedde Tephra and were most likely formed during a Preboreal glacier advance and transgression of RSL. Using arguments about decreasing shoreline gradients with time in Norðdahl (1981, 1983) but taking new data dates into consideration i.e. the Austari-Krókar ice-lake (Fig. 3.19) containing the 10,300 ^{14}C yr BP old Skógar-Vedde Tephra (Norðdahl and Hafliðason,

1992) and that glacio-isostatic uplift was depleted at about 8,500 ^{14}C yr BP (Ingólfsson *et al.*, 1995; Norðdahl and Einarsson, 2001) we are able to estimate the age of the Espihóll and Melgerðismelar marginal position of the Eyjafjörður outlet glacier to be about 9,800 and 9,700 ^{14}C yr BP, respectively (Fig. 3.20). Comparing this development with other regions in Iceland shows that during a period of about 500 ^{14}C years the outlet glacier in Eyjafjörður retreated some 50 km. This great retreat was most likely caused by a considerable negative mass-balance of the ice sheet and the local glaciers leading to a rapid thinning of the Eyjafjörður outlet glaciers. The withdrawal of a rapidly thinning glacier was probably enhanced by transgression of RSL which broke up and floated the glacier ice away from the Eyjafjörður fjord (Norðdahl and Pétursson, 2000).

A whalebone recovered from a littoral formation some 10-15 m asl at Kjölur in Aðaldalur in the Skjálfandi area (Fig. 3.19) has been dated to be 10,165 ± 80 ^{14}C yr BP (Table 3.4; 36) and it predates ice-contact deltas situated at 40-45 m asl in the Aðaldalur and Bárðardalur valley. Taking into account knowledge from other places in Iceland, it is our conclusion that the raised deltas and related marine shorelines in the southernmost parts of the Skjálfandi area should be correlated with Preboreal glacier advances recognized elsewhere in Iceland (Norðdahl *et al.*, 2000).

3.5.6 Northeastern Iceland

The region under consideration is the coastal area between the Vopnafjörður fjord in the east and the Öxarfjörður fjord in the north (Fig. 3.21). There, mapping raised shorelines and other marine features has revealed shorelines mainly at altitudes between

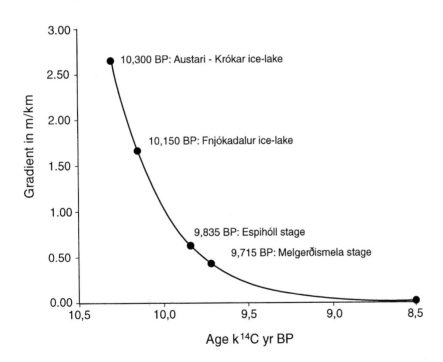

Fig. 3.20 Decreasing shoreline gradient with time in the Eyjafjörður area (based on: Norðdahl, 1981, 1983).

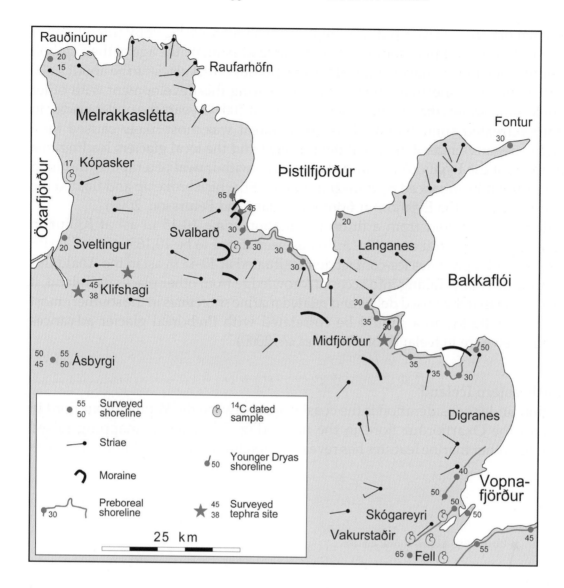

Fig. 3.21 An outline of the deglaciation of northeastern Iceland with Younger Dryas and Preboreal raised marine terraces, and moraines, and place names mentioned in the text (compiled from: Norðdahl and Hjort, 1995; Pétursson, 1986, 1991; Pétursson and Norðdahl, 1994; Sæmundsson, 1977; Sæmundsson, Þ., 1995).

10 and 50 m asl. Seven samples of marine mollusc shells of Younger Dryas and Preboreal age have been collected from raised marine sediments between 5 and 60 m asl; five samples in Vopnafjörður (Sæmundsson, Þ., 1995), one from the Þistilfjörður area (Richardson, 1997), and one at Kópasker on Melrakkaslétta (Pétursson, 1991). Dating these samples has yielded ages between $9,615 \pm 95$ and $10,205 \pm 80$ [14]C yr BP (Table 3.4; 37-43). The occurrence of the Skógar-Vedde Tephra (Norðdahl and Pétursson, 1998) and the Miðfjörður Tephra (Norðdahl and Hjort, 1995) in clastic sediments has aided stratigraphical correlations within the area.

3.5.6.1 The Vopnafjörður Area

Raised shorelines at increasing altitude, from about 45 m asl at the head of Vopnafjörður (Fig. 3.21), have been traced inland for about 25 km and up to about 75 m asl in the Hofsárdalur valley. Comparable raised shoreline features in the Vesturárdalur valley rise from about 50 m asl to about 65 m asl at Vakurstaðir over a distance of about 16 km. In the Selárdalur valley raised shoreline features have been traced inland for some 11 km, rising from about 40 m asl to about 50 m asl (Sæmundsson, Þ., 1995). These distinctly tilted ML shorelines in the Vopnafjörður area have all been related to marginal features in Hofsárdalur, Vesturárdalur and Selárdalur, respectively (Fig. 3.21). In Hofsárdalur and Vesturárdalur fossiliferous fine grained marine sediments have been correlated with the ML shorelines. Radiocarbon-dated samples of marine shells have yielded a weighted mean age of 9,655 ± 70 ^{14}C yr BP (Table 3.4; 37-41) for the formation of the ML shorelines and, thus, also for the concurrent position of glacier termini in the Vopnafjörður area (Sæmundsson, Þ., 1995).

According to Sæmundsson, Þ. (1994, 1995) littorally reworked deposits, from about 55 m asl and downwards at the head of Vopnafjörður (Fig. 3.21), were originally deposited at, or close to, a glacier terminus indicating a temporary standstill of the Hofsárdalur glacier in late Younger Dryas times, a standstill most likely caused by topographical narrowing of the Vopnafjörður trough. Furthermore, the deglaciation of Hofsárdalur is suggested to have been rapid and mainly by calving as topographical threshold conditions were released (Sæmundsson, Þ., 1995). The rapid withdrawal of ice from Hofsárdalur may also have been aided by transgression of RSL concurrently with a termination of a glacier advance in the Hof-Teigur terminal zone.

3.5.6.2 The Langanes Area

More or less continuous raised shoreline features at 30-35 m asl in the Langanes area (Fig. 3.21) represent the regional marine limit there (Norðdahl and Hjort, 1995). Stratigraphical succession of sediments, e.g. in the Miðfjörður terrace in the Bakkaflói area, reveals a glacier margin situated well inside the present coastline, that was coupled with lowering of RSL. Furthermore, this sedimentary sequence also illustrates a subsequent change from an ongoing regression to a transgression of RSL, which eventually reached the regional marine limit at 30-35 m asl. At this moment the glacier margin was situated some 5 km inland from the regional marine limit shorelines (Norðdahl and Hjort, 1995). The discovery of the Miðfjörður Tephra on top of the Miðfjörður terrace (Fig. 3. 21) offers possible tephrochronological correlations with other areas as the tephra is chemically equivalent to the S-Tephra in the Lake Mývatn area (Norðdahl and Hjort, 1995). Sæmundsson (1991) has suggested an early Holocene age for the S-Tephra, which postdates the formation of the 30-35 m shorelines in the Bakkaflói area. A sample of marine mollusc shells, collected at Svalbarð at about 7 m asl from the fine grained lower part of a gravel terrace reaching some 30 m asl in the Þistilfjörður area, has yielded a ^{14}C age of 10,035 ± 75 ^{14}C yr BP (Table 3.4; 42) (Richardson, 1997).

Raised shorelines at 40-50 m asl with very limited lateral extent in the western part of Þistilfjörður and in the eastern part of Bakkaflói (Fig. 3.21), were most likely formed when glaciers reached beyond the present coastline and into the fjords. These shorelines are older than the regional marine limit shorelines and are tentatively correlated with the Younger Dryas Chronozone (Norðdahl and Hjort, 1995). At that time the Langanes peninsula was covered by glaciers only with the Fontur promontory protruding beyond the glaciers (Fig. 3.21) (Pétursson and Norðdahl, 1994; Norðdahl *et al.*, 2000).

3.5.6.3 The Öxarfjörður Area

The western part of Melrakkaslétta is a part of the northern volcanic zone of Iceland and both the structure and topography of the area reflect associated active volcanism, visualized in postglacial lava flows, subglacially formed hyaloclastite ridges, and prominent north-south orientated normal faults (Pétursson, 1991). It has, in spite of the postglacial neo-tectonism, been possible to map and correlate raised shoreline features and, thus, outline the history of RSL changes in the Öxarfjörður area (Pétursson, 1986; Norðdahl and Pétursson, 1998).

Surveyed shoreline features are located along the entire western coast of Melrakkaslétta from the Rauðinúpur cape in the north to the Ásbyrgi canyon some 17 km south of the present coast (Figs. 3.21, 3.22). An equidistant diagram, constructed along NW-SE orientated plane of projection, reveals raised shoreline features at an increased altitude towards the southeast – towards increased glacial loading. The limitation of the southward extent of raised shoreline features in the area around Ásbyrgi

Fig. 3.22 The raised marine limit shoreline at Sveltingur (Fig. 3.21) about 15 km south of Kópasker in northeastern Iceland.

is due to the increased altitude of the land and to extensive Holocene lava flows reaching below the altitude of the marine limit.

A sample of marine mollusc shells collected from littoral-sublittoral gravel deposits at about 10 m asl near Kópasker (Fig. 3.21), yielding a ^{14}C age of 10,205 ± 80 ^{14}C yr BP (Table 3.4; 43), is related to the marine limit situated at about 17 m asl (Pétursson, 1986, 1991). Further south at Klifshagi, the Skógar-Vedde Tephra has been detected in glacio-marine sediments tentatively correlated with the marine limit there at about 45 m asl (Fig. 3.21). On the basis of the ^{14}C dated sample and the occurrence of the Skógar-Vedde Tephra we conclude that the marine limit in the Öxarfjörður area – rising from about 15-20m at Rauðinúpur to about 50-55m asl near Ásbyrgi in the south (Fig. 3.21) – was formed in late Younger Dryas time at about 10,300 ^{14}C yr BP. Lower and subsequently younger shorelines in this area were most likely formed in Preboreal times.

Pétursson (1986) concluded from his studies that RSL transgressed from an altitude below the 10 m level up to the marine limit altitude at about 17 m asl near Kópasker. The sedimentary sequence at Klifshagi in the southern part of the area also supports Pétursson's (1986) transgression of RSL prior to the formation of the marine limit there, situated at about 45 m asl. Furthermore, deposition of sediments related to an intermittent RSL at about 38 m asl at the Klifshagi section was preceded by distinct lowering of base-level and erosion of the marine limit sediments and, hence, RSL must have transgressed up to the 38 m shoreline there. Comparison with other investigated sites in Iceland support a Preboreal age to be attributed to the 38 m raised shoreline – rising from about 10-15 m at Rauðinúpur to about 45-50 m asl near Ásbyrgi (Fig. 3.21).

3.6 DISCUSSION AND CONCLUSIONS

Stratigraphical data from the Quaternary lava-pile of Iceland have revealed repeated stadial–interstadial conditions in the early part of the Weichselian Stage (Sæmundsson, 1991, 1992). In Middle Weichselian times coastal parts of Iceland were ice-free and at least partly inundated by the sea during an interstadial that has been dated between 34.7 and 20.3 k ^{14}C yr BP, a period that is probably comparable with the Ålesund Interstadial in Norway (Norðdahl and Sæmundsson, 1999; Sæmundsson and Norðdahl, 2002). Therefore, the LGM extent of the Icelandic ice sheet most likely occurred at or shortly after, 20.3 k ^{14}C yr BP when ice streams from a central ice sheet reached out onto the shelves around Iceland. The actual extent of the ice sheet has been arrived at on the basis of basal core dates which either post- or predate the LGM extent, leaving it only indirectly and inexactly known.

Considering the present oceanographic circulation around Iceland (Valdimarsson and Malmberg, 1999) with warm water-masses arriving at southeastern Iceland (Fig. 3.1), the sea off western Iceland must have warmed up before the warm sea-current rounded northwestern Iceland and reached the area off northern Iceland (Ingólfsson and Norðdahl, 2001). Basal dates of about 12.7 k, 15.4 k and 13.6 k ^{14}C yr BP from the shelf off western, northwestern, and northern Iceland (Jennings *et al.*, 2000; Eiríksson *et al.*, 2000a) are, therefore, considered minimum dates for the deglaciation of the Iceland shelf. By comparing these dates to the age of the oldest dated marine shells in Iceland, about 12.6 and 12.7 k ^{14}C yr BP in western and northeastern Iceland, the Iceland shelf seems to have been completely deglaciated during a period of 1000 to 3000 ^{14}C years. The innermost parts of the shelf such as the Faxaflói bay may have been deglaciated in about 400 ^{14}C years, showing an extremely high rate of glacier retreat (Ingólfsson and Norðdahl, 2001).

The dating of raised shorelines in western Iceland enables us to determine the amount of glacio-isostatic depression of the coastal areas at some 12.6 k ^{14}C yr BP. Adding the height of the ML shoreline (150 m asl) and the position of eustatic sea-level (ESL) at that time (Fairbanks, 1989; Tushingham and Peltier, 1991) gives an isostatic depression of about 225 to 250 m. Ingólfsson and Norðdahl (2001) have demonstrated that the high elevation of the ML shoreline and great mobility of the Icelandic crust must indicate an extremely rapid deglaciation of the shelf off western Iceland. A slow ice retreat from the shelf area along with gradual melting and thinning of the glacier would quickly have been compensated for by continuous glacio-isostatic rebound, resulting in a ML shoreline at a much lower altitude than 150 m asl. Consequently, greatly elevated ML shorelines of early Bölling age in western Iceland could, therefore, not be formed unless the deglaciation of the shelf and coastal areas in western Iceland progressed very rapidly (Ingólfsson and Norðdahl, 2001).

3.6.1 Lateglacial Relative Sea-Level Changes

A general conclusion concerning the initial deglaciation of Iceland, based on dates from the Iceland shelf, western and northeastern Iceland, is that ice covering the shelf around Iceland during the LGM was very rapidly withdrawn from the shelf and onto present-day dry land. The deglaciated areas were consequently submerged by the sea and greatly elevated early Bölling ML shorelines were formed in coastal parts of Iceland (Fig. 3.5). Subsequent to the formation of the ML shorelines the rate of glacio-isostatic

Table 3.5. Age and altitude (Alt) of Younger Dryas ML shorelines and Preboreal intermittent shorelines in Iceland, with the extent of known transgressions (T) and regressions (R) of relative sea-level. Association of end moraines (M) indicated.

| | Lateglacial RSL Changes | | | | | | | | | |
| | Younger Dryas ML | | | | | Preboreal SL | | | | |
	^{14}C age BP	T	Alt	M	R	^{14}C age BP	T	Alt	M	R
Reykjavík	10,315 ± 105	—	60	•	-40	9,875 ± 90	+25	40	—	•
Dalir	10,215 ± 100	—	70	•	-45	9,735 ± 50	+15	35	—	•
Berufjörður	~10,300	—	60	•	•	9,880 ± 100	•	35	•	•
South Iceland	—	—	—	•	—	9,850 ± 75	—	100	•	•
Eyjafjörður	10,300 (SVT)	—	30	•	•	9,775 (Grad.)	•	40	•	•
Vopnafjörður	—	—	—	—	—	9,655 ± 70	—	75	•	•
Langanes	>10,035 ± 75	—	50	•	-35	~10,035 ± 75	+15	35	•	•
Axarfjörður	10,205 ± 80 (SVT)	+10	55	•		Pb (<SVT)	•	50	—	•
Skagi	~10,100	+6	45		-25	~9,750	+4	25	—	•

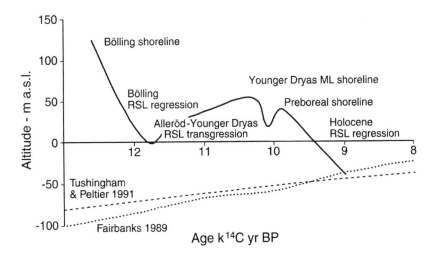

Fig. 3.23 The development of Lateglacial and early Holocene relative sea-level changes in western and southwestern Iceland with the eustatic curves of Fairbanks (1989) and Tushingham and Peltier (1991).

recovery exceeded the rate of ESL rise and RSL regressed as evidenced by late Bölling shorelines at lower altitudes (Fig. 3.23) (Magnúsdóttir and Norðdahl, 2000).

Data revealing the development of RSL changes and extent of the Icelandic inland ice sheet in Younger Dryas and Preboreal times have been collected in regions covering the majority of coastal Iceland. Data from the western part of northern Iceland, northwestern Iceland, and southeastern Iceland are not available at the moment. The most observable fact in these regions is the occurrence of two sets of prominent raised shorelines, the Younger Dryas marine limit shorelines and the subsequent lower set of Preboreal shorelines (Fig. 3.23), except in southern Iceland and in Vopnafjörður, where the marine limit shorelines are of Preboreal age (Table 3.5).

3.6.1.1 Bölling – Alleröd RSL Changes

Examining the available data on the extent and development of the Bölling – Alleröd marine environment in Iceland we see that sediments of that age (Fig. 3.6) are mostly found in lowlying areas in southwestern, western, and northeastern Iceland. This apparent limitation of distribution of Bölling – Alleröd sediments might be due to a later expansion and advance of the ice sheet.

The greatly elevated apparent ML shorelines in Iceland were formed in early Bölling times (Fig. 3.23) subsequent to an extremely rapid deglaciation of the inner shelf and coastal areas (Ingólfsson and Norðdahl, 2001). The relatively large number of dated samples (26) yielding Bölling ages between 11.9 and 12.7 k ^{14}C yr BP for marine sediments, allows us to partially reconstruct the history of glacier retreat and RSL changes during the Bölling Chronozone. In western Iceland the ML shoreline was formed at about 12.6 k ^{14}C yr BP at 'northerly' increasing elevation between 105 m and 150 m asl (Fig. 3.23). General lowering of RSL at the end of the Bölling Chronozone shows that glacio-isostatic recovery of the coastal area had positively exceeded the rise of ESL. RSL was consequently lowered to a position below the ML shorelines in the area and at about 12.2 k ^{14}C yr BP RSL was most likely situated close to the 50 m level in western Iceland (Fig. 3.23). Much the same development of RSL changes are seen in the data from Melrakkaslétta with an early Bölling (12.7 k ^{14}C yr BP) ML shoreline at 50 m to 60 m asl and subsequent late Bölling (12.1 k ^{14}C yr BP) shorelines situated between 20 m and 30 m asl.

In the Melabakkar-Ásbakkar Cliffs in Western Iceland the deposition of the late Bölling (12.1 k ^{14}C yr BP) Ásbakkar diamicton was terminated by an extensive erosion, deformation and deposition of stratified gravels, sands and diamictons of the Ás beds when, according to Ingólfsson (1988) and Hart (1994), glaciers approached and eventually overran the area. This explanation is contradicted by the fact that the 12.6 k ^{14}C yr BP ML shorelines in the area have not been overridden by glaciers (Magnúsdóttir and Norðdahl, 2000). An alternative explanation would, therefore, account for a considerable lowering of RSL and subaerial glaciofluvial erosion and deposition of the Ás beds between 12.2 k and 11.5 k ^{14}C yr BP, when RSL was lowered down to or slightly below present sea-level (Fig. 3.23). On Melrakkaslétta RSL may at that time have been lowered from 20-30 m asl to a position below the 10 m level (Pétursson, 1986).

The earliest Alleröd sediments in western Iceland are the glaciomarine Látrar beds in the Melabakkar – Ásbakkar Cliffs with a mean age of about 11.5 k ^{14}C yr BP. All dated samples yielding Alleröd ages in this area were collected at altitudes between 2 and 7 m asl, while the marine sediments reach as high as 25 m asl in the cliffs (Ingólfsson, 1988), thus providing a 25 m minimum altitude for RSL at that time (Fig. 3.23). At about 11.3 k ^{14}C yr BP a mollusc fauna containing the high-arctic species *Portlandia arctica* appeared in the sediments at the Saurbær and Melar terraces in the eastern parts of Breiðafjörður, indicating the arrival of cold arctic sea water. At that time RSL was situated close to the 25-30 m level and apparently rising (Bárðarson, 1921) towards a temporary maximum together with apparently increasing water temperatures. Later, water depth seems to have decreased with apparent lowering of water temperatures and increased freshwater influence on the marine environment, most likely from melting glaciers in the eastern parts of Breiðafjörður (Ásbjörnsdóttir and Norðdahl, 1995).

Close to the end of the Alleröd Chronozone, at about 11.1 k ^{14}C yr BP the high-arctic species *Portlandia arctica* and *Buccinum groenlandicum* appeared in a transitional faunal assemblage between boreal-arctic and high-arctic assemblages in the Súluá interstratified sand and silt facies of the Látrar beds in western Iceland (Ingólfsson, 1988). This faunal change shows that at this time arctic sea water had reached south of the Snæfellsnes peninsula and to the coastal parts of the Faxaflói area. At that time RSL was situated above the 35 m level rising towards a temporary Younger Dryas maximum in the Melabakkar-Ásbakkar area (Fig. 3.23) (Ingólfsson, 1988). Radiocarbon-dated marine shells from the Fossvogur Layers in Reykjavík, show that these sediments were deposited between 11.4 k and 10.6 k ^{14}C yr BP (Sveinbjörnsdóttir *et al.*, 1993). During the formation of the Fossvogur Layers RSL was situated at least some 20 m above present sea-level and most likely rising. Diamictites on top of those layers indicate an increased glacio-marine character of the environment, probably due to an approaching glacier at the transition between the Alleröd and Younger Dryas Chronozones (Geirsdóttir and Eiríksson, 1994).

The Bölling – Alleröd development of RSL changes in Iceland was characterized by a very rapid lowering of RSL in Bölling times (Fig. 3.23) when global ESL was rising (Fairbanks, 1989; Tushingham and Peltier, 1991). This occurred because rapid glacier unloading and consequent glacio-isostatic rebound was occurring at a higher rate then the rise of ESL (Ingólfsson and Norðdahl, 2001). The apparent increased water depth towards the end of the Alleröd Chronozone (Fig. 3.23) was most likely due to expansion and readvance of the Icelandic inland ice sheet.

3.6.1.2 *Younger Dryas and Preboreal RSL Changes*

Tephrochronologically and ^{14}C dated Younger Dryas marine limit shorelines have an average age of about 10,240 ^{14}C yr BP (Table 3.5). The marine limit altitude is the altitude of the highest known shoreline feature in each of the different regions, i.e. the altitude of a proximal shoreline (Table 3.5). This altitude varies from some 30 m in Eyjafjörður to about 70 m in the Dalir area in western Iceland and this difference reflects a regional difference in glacio-isostatic depression of the crust, where configuration and thickness of the ice sheet, topography and geothermal heat flux play an important role. The greatly elevated ML shorelines in the Dalir area are situated in an area that was affected by the load of two separated glaciers, the main Icelandic ice sheet and the ice cap on top of the northwestern peninsula (Fig. 3.24). The low altitude ML shorelines in the Eyjafjörður area (Table 3.5) was probably due to the considerable Bölling – Alleröd extent of corrie and valley glaciers in the mountainous Tröllaskagi peninsula preventing formation of greatly elevated shorelines but not hampering the glacio-isostatic rebound of the crust there. Other low altitude Younger Dryas ML shorelines were formed in a more proximal position than the more distal and low altitude shorelines (Table 3.5). The formation of the Younger Dryas marine limit shorelines at about 60 m asl in the Reykjavík area is a manifestation of a temporary equilibrium between the rate of eustatic rise of sea-level and glacio-isostatic uplift, a situation that was most likely preceded by transgression of RSL (Fig. 3.23). Studies of isolation basins on the Skagi peninsula in northern Iceland (Rundgren *et al.*, 1997) and lithostratigraphical studies on Melrakkaslétta (Pétursson, 1986) have revealed about 6 m and 10 m transgression of RSL, respectively culminating at the Younger Dryas marine limit. Subsequently glacio-isostatic uplift exceeded eustatic rise of sea-level and RSL regressed from the marine limit.

Similarly, the formation of an intermittent raised shoreline – [14]C and tephrochronologically dated to about 9,800 ^{14}C yr BP (Table 3.5) – clearly shows that again there was a period with an approximately equal rate of glacio-isostatic uplift and eustatic sea-level rise in the coastal regions of Iceland. In the Reykjavík, Dalir, and Langanes areas sedimentological studies have revealed that subsequently to the formation of the Younger Dryas marine limit shorelines, RSL was lowered by some 25-45 m and then again raised by some 15-25 m up to the regional Preboreal shorelines (Fig. 3.23, Table 3.5). The extremely high altitude of the Preboreal shorelines in the South Icelandic Lowlands and in the Vopnafjörður area is probably best explained by their proximal position compared to the altitude of the much more distal shorelines in other parts of Iceland (Table 3.5). The Preboreal temporary maximum height of RSL at about 40 m asl in the Reykjavík area was followed by a regression towards present sea-level and e.g. in Reykjavík RSL was situated 2.5 m below present-day sea-level at about 9,400 ^{14}C yr BP (Figs. 3.15, 3.23) (Ingólfsson *et al.*, 1995)

The areal extent of the two sets of raised shoreline features, the Younger Dryas marine limit and the Preboreal intermittent shorelines, was in some places limited by glaciers reaching into the sea at these times. Dating the relevant shorelines makes it possible to assign absolute ages to the glacier margins and related shorelines (Norðdahl, 1990). Such an association of raised shorelines and marginal features enhances the value of morphological correlations within and between different regions in Iceland (Norðdahl and Einarsson, 2001). Transgression of RSL by some 15-25 m up to the intermittent

Preboreal shorelines shows (Fig. 3.23, Table 3.5) that isostatic loading must have increased again. At about 9,800 ^{14}C yr BP the rate of ESL rise was about 0.8 cm yr^{-1} (cf. Tushingham and Peltier, 1991) or about 1.9 cm yr^{-1} (cf. Fairbanks, 1989) compared to an approximate rate of RSL rise of about 5-10 cm yr^{-1}. Consequently, to achieve such a rate of RSL rise – about six times the ongoing rise of ESL – increased isostatic loading of the crust must have been occurring at the rate of about –4.8 cm yr^{-1}. The only plausible increase in loading must have been due to increased glacier load, i.e. a positive mass-balance change of the Icelandic inland ice sheet and contemporaneous advance of its glaciers, culminating at about 9,800 ^{14}C yr BP. We have every reason to think that a comparable course of events preceded the formation of the Younger Dryas marine limit shorelines in Iceland.

Above we have argued that a transgression of RSL – amounting to some 20- 25 m – culminated in early Preboreal times at about 9,800 ^{14}C yr BP (Fig. 3.23). The available data show that subsequent to the culmination of the transgression, RSL regressed towards, and eventually below, the altitude of present sea-level. In Reykjavík we have the only reliable data on when this occurred preserved in a submerged peat formation where the beginning of organic sedimentation started about 9,400 ^{14}C yr BP, when RSL passed below the –2.5 m level (Fig. 3.15) (Ingólfsson *et al.*, 1995). Thors and Helgadóttir (1991) proposed a maximum lowering of RSL in the Faxaflói Bay, some 20-25 km west off Reykjavík, to 30-35 m below present sea-level on the basis of erosional unconformity and inferred coastal deposits. Two dated samples of dredged peat from a depth of about 17-30 m have yielded an age of 9,460 ± 100 and 9,120 ± 180 ^{14}C yr BP with a weighted mean age of 9,375 ± 90 ^{14}C yr BP. As the sampling of the peat was by dredging and there is no control of the deposits, it cannot be excluded that the peat originally drifted from an already emerged land. These dates are, therefore, regarded as an absolute maximum age for the lowermost position of RSL (Ingólfsson *et al.*, 1995).

Extrapolation of the RSL-curve for the Reykjavík area intersects the ESL-curve of Tushingham and Peltier (1991) at a depth of about 44 m and about 8,940 ^{14}C yr BP (Fig. 3.23). Furthermore, ESL rose above the –30 m level ca 7,000 ^{14}C yr BP leaving ample time for formation of submerged coastal features at 30-35 m below present sea-level. Using the ESL-curve from Fairbanks (1989) the intersect occurred at about – 36 m and about 9,020 ^{14}C yr BP (Fig. 3.23) and ESL rose above the –30 m level at about 8,600 ^{14}C yr BP leaving some 400 ^{14}C years for the formation of the submerged coastal features in Faxaflói. On average, the minimum position of RSL in the Reykjavík area was, therefore, most likely reached at about 8,980 ^{14}C yr BP some 40 m below the present sea-level (Fig. 3.23, Table 3.6).

Applying the same line of arguments, with RSL regressing below present sea-level at about 9,400 ^{14}C yr BP and intersecting the ESL-curves of Tushingham and Peltier (1991) and Fairbanks (1989) respectively, the minimum position of RSL in other parts of Iceland was probably reached between 8,925 and 9,220 ^{14}C yr BP. At that time RSL was situated some 40-43 m below present sea-level (Table 3.6). In Eyjafjörður in northern Iceland Thors and Boulton (1991) have presented a hypothetical curve for RSL changes during the deglaciation of the area, with a prominent submerged delta formation some 40 m below sea-level. They estimated the time of formation to a period between 10,800 and 10,000 ^{14}C yr BP or between 12,000 and 10,000 ^{14}C yr BP. Extrapolation of our RSL data from Eyjafjörður with a Preboreal shoreline at about 40 m asl shows a RSL-curve intersecting the ESL-curves at 42 m below sea-level – the depth of the Lower-Hörgá delta (Thors and Boulton, 1991) – at ca 9,000 ^{14}C yr BP (Table 3.6).

In southern Iceland the great Þjórsárhraun lava flow, originating some 140 km inland from the present coastline, entered the sea at about $7,800 \pm 60$ [14]C yr BP (Hjartarson, 1988), when RSL was situated at about –15 m (Einarsson, 1978). According to Fairbanks (1989) ESL was at that time situated at about –18 m. Again we have a geological formation related to RSL at a time when glacio-isostatic uplift was apparently at an end. Therefore a combination of the Bölling – Alleröd sea-level data and the Reykjavík Younger Dryas – Preboreal sea-level data results in a tentative Relative Sea-Level Curve for western and southwestern Iceland (Fig. 3.23).

The further development of sea-level changes in Iceland must, therefore, have been a reflection of ESL changes alone. In northeastern Iceland a shoreline, the Main Raised Shoreline at 3 to 5 m asl, has been dated to a Mid-Holocene transgression at about 6,500 ^{14}C yr BP (Richardson, 1997). Two sets of beach ridges, at 3 and 5 m asl, in the northern-most part of Hvammsfjörður in western Iceland represent repeated Holocene transgressions of sea-level. A black tephra layer deposited some 4000 ^{14}C yr BP (Indriðason, 1997) covers the lower and younger set of beach ridges. Consequently, the upper set is older than 4,000 ^{14}C yr BP and may possibly be compared with the Mid-Holocene Main Raised Shoreline in northeastern Iceland. Finally, there are indications that sea-level in southern Iceland has reached as high as 2- 6 m asl on three occasions; 3,200, 2,700, and 2,200 ^{14}C yr BP (Símonarson and Leifsdóttir, 2002).

Table 3.6. Age and altitude (Alt) of Preboreal intermittent shorelines in Iceland with the depth and age of the minimum position of relative sea-level around Iceland.

Region	Preboreal RSL Changes					
	Preboreal Max SL				Preboreal Min SL	
	[14]C age BP	Alt.	Rgrs		Depth	[14]C age BP
Reykjavík	$9,875 \pm 90$	40	•		-40	8,980
Dalir	$9,735 \pm 50$	35	•		-42	9,055
Berufjörður	$9,880 \pm 100$	35	•		-40	8,935
South Iceland	$9,850 \pm 75$	100	•		-43	9,220
Eyjafjörður	9,775 (Grad.)	40	•		-42	9,055
Vopnafjörður	$9,655 \pm 70$	75	•		-44	9,220
Langanes	$>10,035 \pm 75$	35	•		-42	9,090
Axarfjörður	Pb (<SVT)	50	•		—	—
Skagi	~9,750	25	•		-40	8,925

3.6.2 Younger Dryas and Preboreal Glacier Extent

Deterioration of the Bölling – Alleröd marine environment at the end of the Alleröd Chronozone, witnessed e.g. by the appearance of high-arctic mollusc species as *Portlandia arctica* and *Buccinum groenlandicum* in western Iceland, also initiated a positive mass-balance change of the Icelandic glaciers. Rising RSL at the end of the Alleröd and beginning of the Younger Dryas Chronozones (Fig. 3.23), was both caused by ongoing global rise of ESL and increased glacio-isostatic load in the coastal regions of Iceland following an expansion of the inland ice sheet and advance of the glaciers (Norðdahl, 1990; Norðdahl & Ásbjörnsdóttir, 1995; Rundgren *et al.*, 1997; Norðdahl and Einarsson, 2001). Thus the Bölling – Alleröd marine environment was rapidly changed from a moderate low-arctic/high-boreal to an arctic/high-arctic environment characterized by deposition of glaciomarine sediments frequently barren of mollusc shells and foraminifera tests. Furthermore, a significant decline of inferred mean summer temperature in the North Atlantic seaboard regions (NASP members, 1994), reduced inflow of Atlantic water to the Norwegian sea (Lehman and Keigwin, 1992) and to the waters around Iceland (Eríksson *et al.*, 2000a; Jennings *et al.*, 2000), and a southward displacement of the North Atlantic Polar Front (Ruddiman and Macintyre, 1981), all occurred around the Alleröd – Younger Dryas transition, leading to the onset of the Younger Dryas cold spell.

3.6.2.1 Younger Dryas Ice Sheet

The available data on absolutely dated Younger Dryas glacier marginal features and correlations with other undated marginal features in Iceland allows us to reconstruct the Younger Dryas extent of the Icelandic ice sheet (Fig. 3.24). In the Reykjavík area the glaciers terminated at, or very close to a concurrent sea-level in the eastern part of the area, while raised shorelines indicate that glaciers in western Iceland terminated in valleys in the Borgarfjörður region. The extent of the assumed Younger Dryas marine limit shorelines around Hvammsfjörður in the Dalir area is an approximation of the extent of the inland ice sheet and its separation from the ice cap in northwestern Iceland. Through detailed mapping of an ice-lake in Fnjóskadalur containing the Skógar-Vedde Tephra, and from occurrences of the tephra outside the ice-lake, the Younger Dryas glacier extent has been determined in Eyjafjörður and Skjálfandi areas in northern Iceland (Fig. 3.24). For the western part of northern Iceland we have based our conclusions on the fact that the Skagi peninsula only carried local cirque and valley glaciers from an early Alleröd deglaciation of lake basins in the northern parts of the peninsula (Rundgren *et al.*, 1997), and on our distinguishing between different sets of raised shorelines in the area. Spatial extent of the Younger Dryas marine limit shorelines in the Öxarfjörður area shows that the glacier margin was at that time situated inside the coastline. Accumulation of the Skógar-Vedde Tephra in an ice-lake basin in the highland above Öxarfjörður (Fig. 3.21) also points to a limited glacier extent in the central parts of Melrakkaslétta. Laterally truncated shorelines in the Langanes area of an inferred Younger Dryas age represent glaciers which at that time reached beyond the present coastline. Further south a reconstruction of a glacier margin during the formation of the marine limit shorelines shows outlet glaciers reaching out into the fjords of eastern Iceland (Fig. 3.24) (Norðdahl and Einarsson, 2001). Haraldson (1981) correlated a moraine

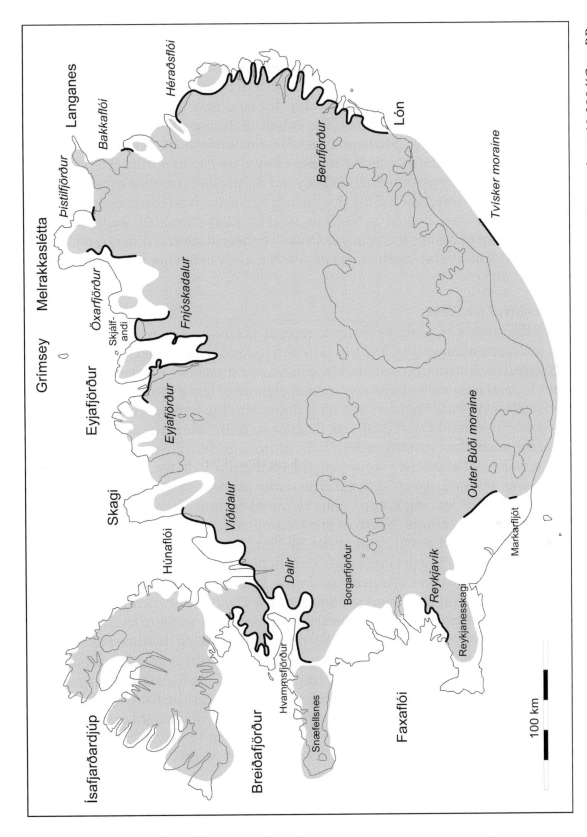

Fig. 3.24 Reconstruction of the extent of the Icelandic inland ice sheet in Younger Dryas time at about 10,300 ^{14}C yr BP.

below the surface of the Markarfljót Sandur with the "outer" Búði moraine north of Vatnsdalsfjall. The outer Búði moraine in southern Iceland (Fig. 3.24) has tentatively been related to a Younger Dryas glacier event there (Axelsdóttir, 2002).

Based on these data we attempt to outline the extent of a Younger Dryas ice sheet in Iceland when only small pieces of present-day dry land may have protruded beyond the margin of the ice sheet (Fig. 3.24). The extent of Younger Dryas glaciers on the Reykjanes and Snæfellsnes peninsulas is still not known in any detail. A first approximation towards the extent of a Younger Dryas ice cap in northwestern Iceland indicates that the glacier margin extended beyond the present coastline in central parts of the Breiðafjörður north coast and into the main fjords on the west coast of northwestern Iceland when RSL was situated at 60-70 m asl (Fig. 3.24) (Norðdahl and Pétursson, 2002). This attempt to outline the Younger Dryas ice sheet in Iceland differs from earlier attempts (Norðdahl, 1991a; Ingólfsson *et al.*, 1997) mainly in having the margin drawn in greater detail.

3.6.2.2 Preboreal Ice Sheet

In much the same way as above we try to reconstruct the areal extent of the Icelandic ice sheet in Preboreal times (Fig. 3.25). In southern Iceland the ice sheet reached into the sea and formed the Inner Búði moraine. Mapping raised shorelines and marginal features in eastern Iceland has enabled a reconstruction of more or less a continuous ice margin in the area showing outlet glaciers reaching down into the inner parts of valleys and fjords of eastern Iceland (Norðdahl and Einarsson, 2001). Hummocky moraines just above the Héraðsflói bay in northeastern Iceland are considered to reflect the Preboreal extent of a major outlet glaciers occupying the Fljótsdalur valley and entering the sea when RSL was situated about 30 m asl. The outlet glaciers in the Vopnafjörður area reached into the sea leaving conspicuous terminal features showing the Preboreal position of the glacier termini. North of there, marginal features situated some 3-5 km within a regional raised coastline at about 30-35 m asl, have been traced across the Langanes area representing the Preboreal extent of the glaciers there. In the Öxarfjörður area the glacier margin was situated within the Preboreal coastline at an unknown position in the highland south and southeast of the Öxarfjörður fjord (Fig. 3.25). The rather detailed reconstruction of a Preboreal glacier margin in northern Iceland is mainly based on tephrochronological and morphological correlations. At this time Fnjóskadalur was without an ice-dammed lake suggesting a considerable glacier retreat in Eyjafjörður while glaciers in tributary valleys such as Hörgárdalur and Svarfaðardalur, respectively reached the sea at the mouth of these valleys (Fig. 3.25). In the Skagafjörður area a major outlet glacier, reaching as far to the north as the head of the fjord, receded in late Preboreal times. A raised glaciofluvial delta formed at that time was graded to a base level close to the 40-50 m level. Further west, a reconstruction of an assumed post-Younger Dryas glacier margin is partly based on a survey of raised shorelines in the Skagi area (Moriwaki, 1991; Rundgren *et al.*, 1997) and partly on our yet unpublished data on raised shorelines and marginal features in western parts of northern Iceland. In western Iceland we still lack reliable data on the Preboreal extent of the inland ice sheet except that its margin did reach into the sea in the southern Dalir area, and that an outlet glacier terminated in the innermost part of Hvalfjörður with RSL situated at about 65 m asl.

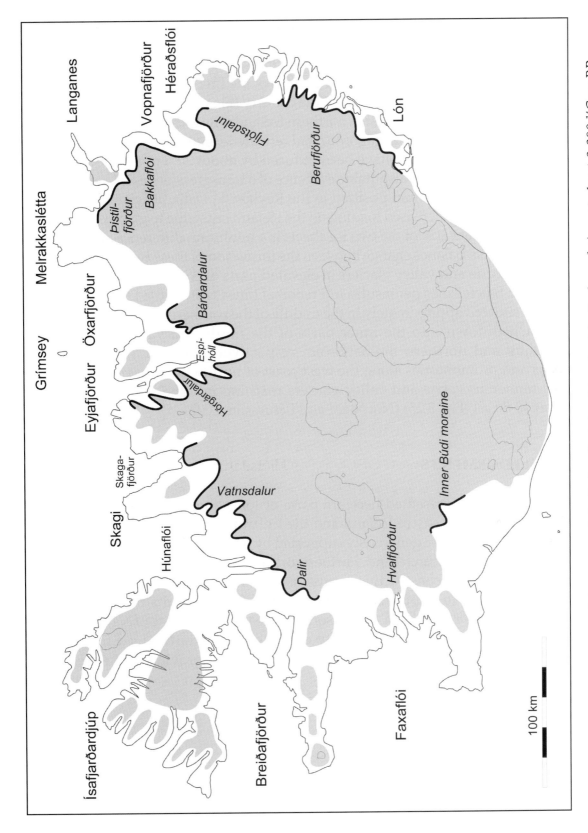

Fig. 3.25   Reconstruction of the extent of the Icelandic inland ice sheet in Preboreal time at about 9,800 ^{14}C yr BP.

Based on data presented above, we have produced the first reconstruction of the spatial extent of an Icelandic Preboreal (ca 9,800 ^{14}C yr BP) ice sheet. Differences in the configuration and extent between the Younger Dryas and the Preboreal ice sheets are not very marked with the exception of the margin of the Preboreal ice sheet which had retreated a bit further inland in all parts of the country (Figs. 3.24, 3.25). The difference, though, seems to have been greatest in northern Iceland where the major outlet glaciers in Skjálfandi, Eyjafjörður and Skagafjörður had retreated some 30-50 km. On the whole, the Icelandic ice sheet seems to have been reduced by about 20% in a period of about 500 ^{14}C years. Furthermore, indisputable evidence of a transgression of RSL temporarily reaching a Preboreal maximum position in the Reykjavík, Dalir, Skagi, and Langanes area (Table 3.5), reflects increased loading on the Iceland crust due to glacier expansion. Evidently, the different size of the two ice sheets is a minimum difference and it is only reflecting a net mass-balance change between the formation of these two ice sheets. The extent of local corrie and valley glaciers in elevated parts of coastal Iceland and on the Reykjanes and Snæfellsnes peninsulas in Preboreal times is still little known.

An ice cap situated approximately in the middle of the southern part of northwestern Iceland reached down into the inner parts of the fjords and valleys in southern Ísafjarðardjúp and northern Breiðafjörður, respectively (Fig. 3.25). Mountainous peninsulas and promontories along the west coast of northwestern Iceland carried at that time numerous corrie and valley glaciers reaching into the sea when RSL was situated at 40-50 asl (Fig. 3.25) (Norðdahl and Pétursson, 2002).

ACKNOWLEDGEMENTS

Some of the results presented here are parts of long time research projects at the Science Institute, University of Iceland and the Icelandic Institute of Natural History. Parts of these research were generously supported by Rannis, the Icelandic science fund and by the University research fund. Furthermore, we would like to thank many of our colleagues in Iceland for constructive discussions about the deglaciation of Iceland and for permitting access to unpublished material like ^{14}C dates and stratigraphical and morphological data. Dr. Kristinn J. Albertsson at the Icelandic Institute of Natural History and Dr. Chris Caseldine at the University of Exeter are acknowledged for critically reading the manuscript, making valuable comments and improving the English.

4. Recent developments in oceanographic research in Icelandic waters

Steingrímur Jónsson[1,2] and Héðinn Valdimarsson[1]

[1]Marine Research Institute, Skúlagata 4, IS-101 Reykjavík, Iceland

[2]University of Akureyri, Glerárgata 36, IS-600 Akureyri, Iceland

The ocean around Iceland has much to offer in terms of interesting physical oceanographic phenomena. Among these phenomena are great contrasts in water mass properties, strong atmospheric forcing in the area, both dynamic and thermodynamic, and the fact that a large part of the thermohaline circulation of the world ocean has its origin there or passes over the ridges on either side of the country. Icelandic waters also show large interannual and decadal variability. In recent years, research has been more oriented towards quantitative studies. In this paper we will illustrate some of those properties using examples from recent research projects carried out in the area. Oceanographic research in the area around Iceland has a long history. In the 19th and the beginning of the 20th century, the research was mostly conducted by other nations. Now the oceanographic research is done mostly by Icelandic institutions, often in close co-operation with foreign colleagues.

4.1 INTRODUCTION

Various authors have described the physical oceanography of Icelandic waters in some detail, (e.g. Stefánsson, 1962; Swift, 1980; Malmberg and Kristmannsson, 1992; Vilhjálmsson, 1997). Therefore only a brief description of the most important aspects will be given here.

Iceland is situated just south of the Arctic Circle in the North Atlantic (Fig. 4.1a). It is the largest part of the Mid-Atlantic Ridge that rises above sea level with an area of 103,000 km² and a coastline of about 6,000 km. Except for the south coast, the coastline is indented by a variety of bays and fjords of various shapes and sizes. The shelf area around Iceland, when defined by the 200 m depth contour, covers 115,000 km². The shelf is narrowest off the south coast, whereas it gradually broadens to the west. It is broadest off the west and north coast, where it extends over 100 km offshore.

Iceland lies at the junction between two submarine ridges, the Mid-Atlantic Ridge and the Greenland-Scotland Ridge. Especially important is the exchange of water across the Greenland-Iceland Ridge, through the Denmark Strait, and over the Iceland-Faroe Ridge since they both prevent direct connection of waters below their threshold depth

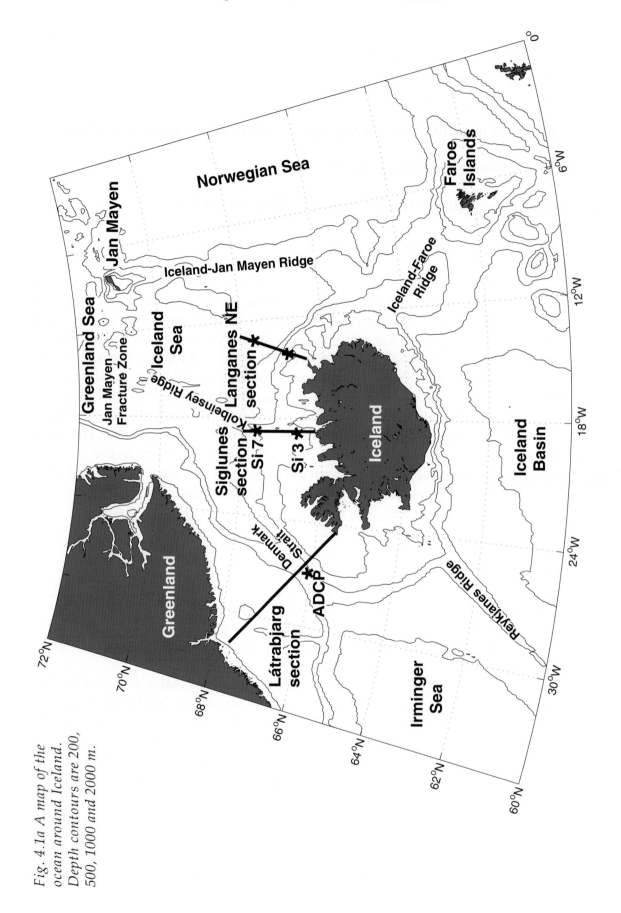

Fig. 4.1a A map of the ocean around Iceland. Depth contours are 200, 500, 1000 and 2000 m.

which is 620 m and 550 m respectively. To the southwest of Iceland, the Mid-Atlantic Ridge is called the Reykjanes Ridge and it gradually increases in depth as it stretches southwards. To the north of Iceland the Kolbeinsey Ridge is a continuation of the Mid-Atlantic Ridge which stretches to the Jan Mayen Fracture Zone that lies between Jan Mayen and Greenland. South from Jan Mayen, the Iceland-Jan Mayen Ridge extends to the Iceland-Faroe Ridge and separates the Iceland Sea from the Norwegian Sea to the east. The Jan Mayen Fracture Zone marks the northern limit of the Iceland Sea and separates it from the Greenland Sea. To the south, the Greenland-Iceland Ridge provides a natural boundary of the Iceland Sea in the Denmark Strait area. The Iceland Sea is split in two by the Kolbeinsey Ridge. To the south of Iceland and east of the Reykjanes Ridge is the Iceland Basin and west of the ridge lies the Irminger Sea.

The ocean around Iceland is dominated by two primary water masses of very different origins and properties (Fig. 4.1b). Warm and saline Atlantic Water originating far south in the Atlantic is brought to the area south of Iceland where it splits into two branches. One branch flows east towards the Iceland-Faroe Ridge while the other flows as the Irminger Current westwards and north along the west coast. Off the west coast most of this water turns west towards Greenland and subsequently flows along the slope off Greenland to the southwest. A smaller branch continues northwards into the north Icelandic shelf area as the North Icelandic Irminger Current.

The other primary water mass is Polar Water originating in the Arctic Ocean. It is rather fresh and very cold and flows out of the Arctic Ocean through the Fram Strait between Spitsbergen and Greenland as the East Greenland Current. It carries with it sea ice that can extend into Icelandic waters especially during late winter and spring and sometimes conditions are favourable for formation of sea ice (Malmberg, 1969). Mixing and cooling of different proportions of these two primary water masses forms all other water masses in the area. The only exception to this rule is the dilution of water by freshwater run-off from land, especially during summer, that creates a coastal water mass that generally circulates clockwise around the country.

The bathymetry of the ocean bottom influences the oceanic conditions around Iceland to a large extent. It limits the exchange of water across ridges, but it also directs the ocean currents as they have a tendency to follow isobaths. This topographic steering is also decisive in determining the position of the fronts separating the water masses in the area. In Fig. 4.2, that shows the average of the satellite-derived sea surface temperature over a ten day period, it can be seen that the maximum gradient in temperature to a large extent follows the Greenland-Scotland Ridge. Since it is the average temperature over a ten day period, the fronts seem to be broader than an instantaneous picture would show. The fronts not only mark a separation between water masses, but also in the biology because the living conditions on either side of the fronts can be very different with different species occupying the different water masses. Productivity is also different and in general the Atlantic water has a higher productivity than the colder and more vertically stable water masses on the other side (Þórðardóttir, 1984).

The main weather systems affecting Iceland and its immediate surrounding oceans are the Icelandic Low and the Greenland High. A parameter or an index often used to describe the conditions over the North Atlantic is the so-called North Atlantic Oscillation

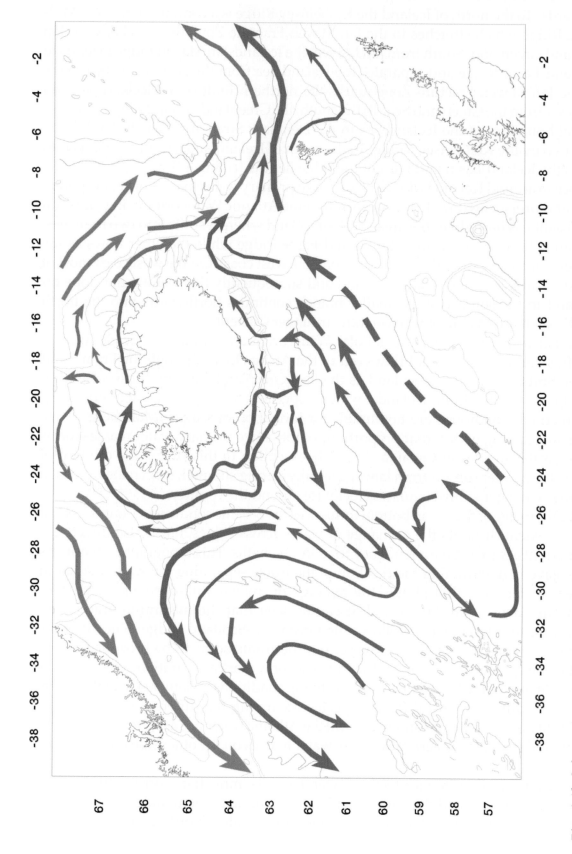

Fig. 4.1b Schematic presentation of the surface circulation in Icelandic waters. Red arrows: Atlantic water; Blue arrows: Polar water; Green arrows: mixed waters. Depth contours are 200, 500, 1000, 2000 and 3000 m (from Valdimarsson and Malmberg (1999). Values around the margin are in degrees of latitude and longitude (negative values are degrees west, positive are degrees north)

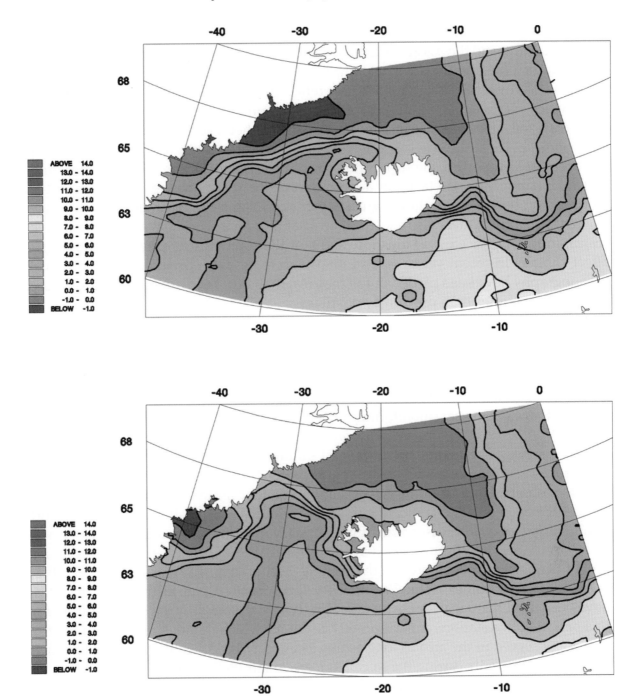

Fig. 4.2 Sea surface temperature in the ocean around Iceland in February 1994 (above) and February 1995 (below). Values around the margin are in degrees of latitude and longitude (negative values are degrees west, positive are degrees north)

index (NAO) (Hurrell, 1995). This describes various forms of the pressure difference between Iceland and the Azores and thus represents the gradient in air pressure between the Icelandic Low and the Azores High. This gradient is a measure of the strength of the westerlies over the Atlantic; it does not reflect directly the strength of the Greenland High which exerts a strong influence in the area around Iceland. The track of the low as it traverses the Atlantic is also of importance e.g. whether it goes west or east of Iceland.

What makes Icelandic waters special for physical oceanography are the great contrasts in water mass properties, the strong atmospheric forcing in the area, both dynamic and thermodynamic, and the fact that a large part of the thermohaline circulation in the world ocean has its origin in the area or passes over the ridges on either side of the country. Icelandic waters also show large variability on interannual and decadal time scales. In this paper we will illustrate some of these properties especially those that affect climate. In recent years, several international research projects on the physical oceanography have been carried out in the area, that have been oriented towards more quantitative studies than previously possible.

4.2 CLIMATIC VARIABILITY AND SURFACE LAYERS

Climatic variability of Icelandic waters has been observed since 1970 on seasonal (three to four times a year) monitoring cruises of the Icelandic Marine Research Institute (MRI). Prior to 1970 the observations were mostly carried out in spring (May/June) and time series of annual temperature and salinity observations on standard sections exist from 1949. These are the longest regular records of ocean climate in the area and therefore observations in spring e.g. on station 3 on the Siglunes section (Fig. 4.1a) have frequently been used to indicate climatic variability on the shelf north of Iceland (Malmberg et al., 1996). Seasonal and inter-annual variability of the northward extension of Atlantic Water (or southward extension of Polar Water) can be seen on a time series plot of temperature at station 7 on the Siglunes section (Fig. 4.1a) for 1990 to 1999 (Fig. 4.3). This spans the time that the Conductivity-Temperature-Depth instruments, or CTDs, have been used regularly for observations in Icelandic waters. Surface temperature for this period at this station ranges from -0.7 to 7.9 °C, which demonstrates the gradients that are observed in the area as long as this location is not covered with sea ice, which happened repeatedly during the so-called ice years in 1965-1970.

Frequent clouds have hampered the use of satellites or remote sensing to monitor changes in temperature. However a technique developed during the European Research Satellite 1 (ERS-1) mission which started in 1991, has offered some improvement over earlier sensors in the form of the Along Track Scanning Radiometer (ATSR) (Anon., 1992). Winter measurements of sea surface temperatures (SST) are valuable, as this is the season when the mixed layer is deepest and the temperature at the surface is more representative of the distribution of colder and warmer water masses. Climatic variability of Icelandic waters is reflected in Fig. 4.2. February 1994 was observed as having close to medium temperature distribution, while February 1995 was showing one of the lowest surface layer temperatures in the northern areas for decades (Malmberg and Jónsson, 1997). The ATSR data compare well with ship observations at the time (Anon., 1995).

Knowledge of surface currents around Iceland until the last decade was mainly built

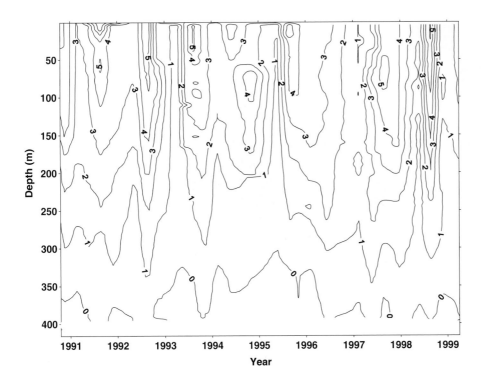

Fig. 4.3 Time series of temperature on a profile on station 7, Siglunes section north of Iceland.

on observations of the temperature and salinity fields and geostrophic calculations. In addition information from drift bottle experiments early in the 20th century was used to strengthen the picture of surface currents. Experiments with satellite-tracked surface drifters started in 1991 in Icelandic waters. In 1995 increased use of these experiments started as a co-operation between scientists from Iceland and the USA in connection with the Global Drifter Project (GDP), a part of the World Ocean Circulation Experiment (WOCE). For three years 10 drifters were deployed four times a year, in all 120 drifters. This gave a rather good coverage, in space and time, of drifting buoys around Iceland sampling its waters for more than four years. In addition a coastal drifter project was started by MRI where the sampling of surface currents on the shelf was continued. Over the ten year period used by Poulain *et al.* (2001) drifters from other experiments in connection with the GDP drifted into Icelandic waters. This resulted in almost a decade of current observations which are still being analyzed. Description of these GDP projects can be found in Poulain *et al.* (1996) and Valdimarsson and Malmberg (1999).

Mean surface currents are frequently computed by averaging over selected rectangles of latitude longitude (bin size) (Poulain *et al.*, 1996). Fig. 4.4 shows the average current speed and direction for each bin. The minimum number of observations needed for averaging is set to 40, but in general the number of observations is at least 10 times that number and considered to result in stable means. The variability of the currents' speeds and directions is shown in Fig. 4.5. As expected the highest velocities are along the fronts between the cold waters from the north and the warm waters from the south. The same areas stand out with the highest variability in surface currents. This reflects the meandering nature of the fronts and also the climatic variability of the location of the frontal areas.

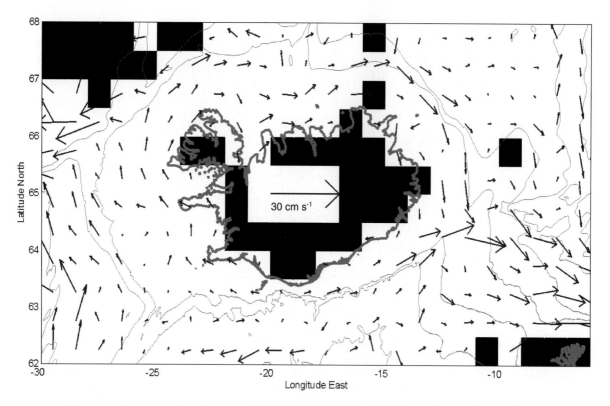

Fig. 4.4 Mean surface circulation binned from all drogued data 1990–1999 (from Poulain et al., 2001).

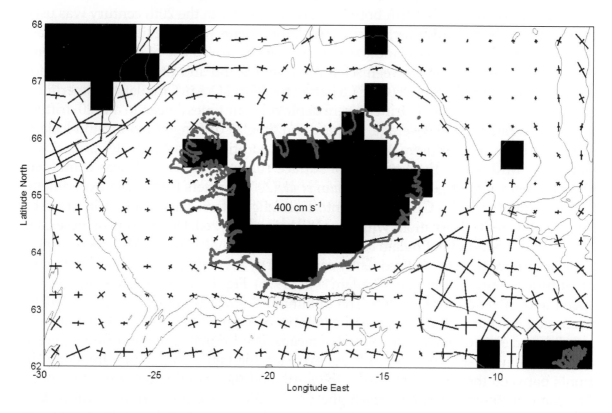

Fig. 4.5 Principal axes of velocity variance for the same data as in previous figure (from Poulain et al., 2001).

The average and variability maps reveal the stronger and more variable currents over the shelf breaks and in the Iceland Basin south of the Icelandic shelf. This variability in the Iceland Basin is linked to the eddy activity in the area and possibly seasonal variability in the surface currents as well. There seems to be a tendency to a stronger cyclonic vorticity in the western Iceland Basin in winter (Valdimarsson and Malmberg, 1999). In general a decade of experiments with satellite-tracked surface drifters has strengthened the ties between bottom topography and the surface currents. An example of this is the tendency of surface currents to follow depth contours to the southwest along the Reykjanes Ridge on the eastern side and northeastwards on the western side. A certain tendency to follow topography is also seen north of Iceland at the western side of the Kolbeinsey Ridge although that pattern is less consistent, perhaps because of fewer observations.

4.3 FRESHWATER CIRCULATION AND CONVECTION

One of the processes in the ocean around Iceland that influences the climate is the flow of freshwater within the East Greenland Current and the subsequent leakage of parts of it into the convective gyres in the Iceland and Greenland Seas. The amount of this leakage is very variable and it affects the climate in two ways. The freshwater affects the properties of the surface layer by making it fresher and more stable and since this water has a polar origin it is rather cold. A second effect of this can be a diminishing or prevention of deep convection within the gyres thus affecting the thermohaline circulation.

The amount of freshwater advected through the Fram Strait was estimated by Aagaard and Carmack (1989) to be 0.125 Sv (1 Sv = 10^6 m^3 s^{-1}). This was calculated relative to a salinity of 34.93, which they took as an average salinity for the Greenland, Iceland and Norwegian Seas. A part of this freshwater enters the Greenland and Iceland Sea gyres within the Jan Mayen Current and the East Icelandic Current respectively. However most of it passes through the Denmark Strait into the North Atlantic uninterrupted. It was shown by Jónsson (1992) that most of the freshwater in the Iceland Sea has its origin in the East Greenland Current. The EU project "Variability of Exchanges in the Northern Seas" (VEINS) had as one of its goals to estimate the flux of freshwater within the East Icelandic Current. For this purpose two current meter moorings were deployed on a section from Langanes towards Jan Mayen (Fig. 4.1a), one over the slope, and one further out over the Iceland Plateau. Salinity and temperature were measured on the section 4 times per year. Using geostrophic calculations referenced to the current measurements, the freshwater flux above 170 m has been estimated for 5 different coverings of the section from May 1997 – June 1998 (Fig. 4.6). The average of the flux was 5500 m^3 s^{-1} or 0.0055 Sv - about 4.4% of the freshwater flux through the Fram Strait. Most of the flux occurs over the slope where the current is strongest and the geostrophic shear is largest. At 80 m depth over the slope the average current for the year was 10 cm s^{-1}. Over the plateau the shear is very weak and there seems to be a barotropic current of 1-2 cm s^{-1}. No estimates of the similar freshwater flux within the Jan Mayen Polar Current in the Greenland Sea exist, but it is probably of the same order of magnitude. Aagaard and Carmack (1989) estimated that about 3% of the freshwater flux through the Fram

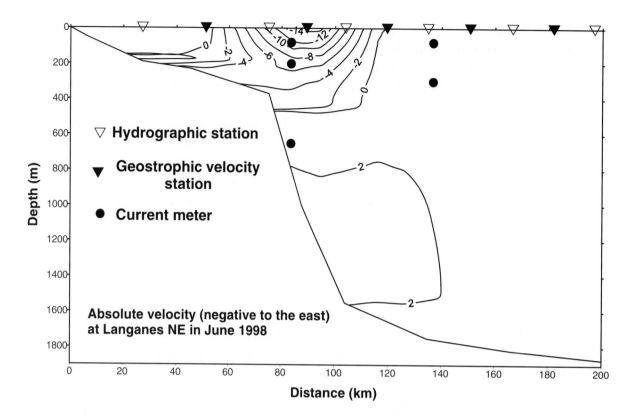

Fig. 4.6 The geostrophic velocity field on the Langanes-Jan Mayen section referenced to the measured velocity at 80 m depth. Negative values indicate flow to the east.

Strait entered the convective gyres of the Greenland and Iceland Seas so our estimate is substantially greater than that if one also takes into account the Jan Mayen Current. However they only considered the central parts of the gyres and it is still true, as they stated, that the convective gyres in the Greenland and the Iceland Seas are very sensitive to an increased freshwater transport into them. A corollary to this is that if it was possible to measure in a similar way the freshwater flux within the Jan Mayen Current, one could obtain an estimate for the freshwater flux through the Denmark Strait, which has proved difficult to measure directly.

The freshwater content in the Iceland Sea has been estimated in May/June each year as the freshwater thickness above 150 m, relative to a salinity of 34.93 at the fourth hydrographic station from land on the Langanes-Jan Mayen section and it is shown in Fig. 4.7. It is quite variable and was rather high in both 1997 and 1998; thus the flux of freshwater calculated here is probably higher than the long term average.

During the years 1965-1970 large amounts of freshwater probably exited through the Fram Strait and during this period large quantities of freshwater also accumulated in the Iceland Sea (Fig. 4.7). This increased freshwater subsequently flowed out through the Denmark Strait and created one of the largest climatic signals ever directly observed in the North Atlantic Ocean. It was advected through the subpolar gyre of the North Atlantic from 1968 until it arrived again in north Icelandic waters in 1982 after the signal had traveled through the whole subpolar gyre. This is now known as the Great Salinity Anomaly (GSA) and has been described in detail by Dickson *et al.* (1988).

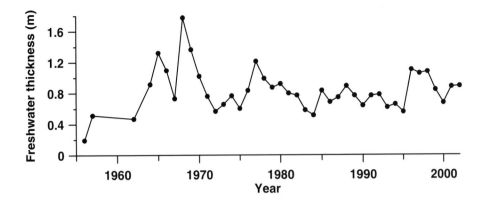

Fig. 4.7 The freshwater thickness at Langanes 4 above 150 m, relative to a salinity of 34.93.

4.4 THERMOHALINE CIRCULATION

One way in which the ocean redistributes heat and salt is through the thermohaline circulation, which is of primary importance in the global climate system. The thermohaline circulation is mainly driven by heat loss from the ocean to the atmosphere with the result that water at high latitudes sinks and flows towards lower latitudes at deep levels and is replaced at the surface by warmer water from lower latitudes. In the northern hemisphere the main inputs to the thermohaline circulation are from the Arctic Ocean, the Greenland Sea, the Iceland Sea and the Labrador Sea. In the southern hemisphere various sources are located on the continental shelf of Antarctica. One of the most important of the northern sources is the Denmark Strait Overflow Water (DSOW) that comprises 30% of the total overflow-generated tongue flowing southwards around the Grand Banks, including waters from all the northern sources (Swift, 1984). The ocean around Iceland is very important in this respect since deep and intermediate waters formed in the seas north of Iceland subsequently flow into the North Atlantic over the ridges between Greenland and Scotland. To be able to assess the possible impacts of different climatic conditions on the strength of the thermohaline circulation one must at least know where and how the different water masses contributing to the circulation are formed and by which way they enter the deep ocean. Since Cooper (1955) suggested that the overflow through the Denmark Strait was a significant contributor to the deep water of the North Atlantic, there have been different views on the origin of this water and how it reaches the Denmark Strait and this debate still continues.

As a part of the Nordic WOCE project a 150 kHz Acoustic Doppler Current Profiler (ADCP) was deployed in November 1996 close to the bottom just south of the sill at 660 m depth (Fig. 4.1a). It was recovered in September 1997 and a 75 kHz ADCP was deployed at the same position which was recovered in September 1998. The temperature on a section over the Denmark Strait from Látrabjarg over the ADCP mooring and towards Greenland is shown in Fig. 4.8. Warm Atlantic water flowing northwards lies over the Icelandic shelf and partly over the slope. A sharp sloping front separates this from colder waters of various types that are flowing southwards into the North Atlantic Ocean. The cold water in the surface layer is Polar Water from the north and below that is warm Atlantic Water from the south, especially pronounced over the banks. The

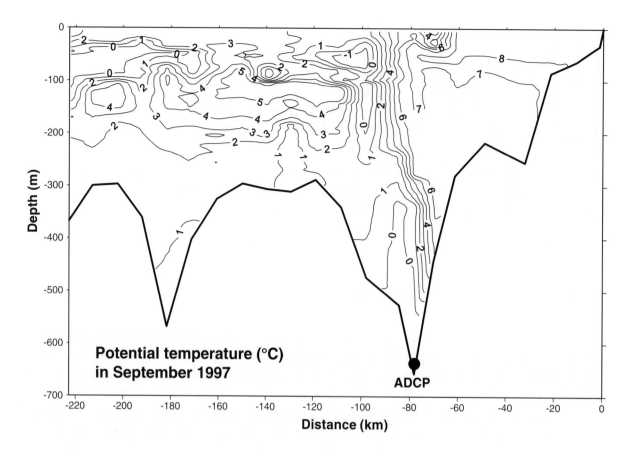

Fig. 4.8 Potential temperature from Látrabjarg to Greenland over the Denmark Strait. Also shown is the position of the ADCP on the section.

DSOW is the cold water in the deepest channel on the section where the ADCP is positioned and it exits through this channel (see also Fig. 4.1a).

Fig. 4.9 shows the monthly average of the along channel velocity at 60 m height above the bottom in 1996-1998, negative towards the North Atlantic. The mean flow is about 50 cm s^{-1} and is somewhat smaller for the latter period. The plume in 1996-1997 appeared to be mainly barotropic and its height usually exceeded 300 m which was the upper limit of the 150 kHz ADCP. With the 75 kHz ADCP the measurements reached up to the surface and preliminary analysis reveals that the plume usually does not exceed 400 m in thickness. Although the monthly means show little variability, without any clear sign of seasonal variability, the measurements reveal very energetic short period variability, with maximum velocity of over 2.5 m s^{-1} towards the North Atlantic. Together with the variability in the water masses present at the mooring site this makes it a challenge to estimate the transport of DSOW, and a better knowledge of the horizontal structure of the plume is necessary. These measurements will help in identifying the need for future measurements of the DSOW plume. The temperature of the DSOW as measured at the instrument 10 m above bottom is shown in Fig. 4.9 where it is seen that it is about 0°C most of the time.

As an example of the short term variability of the flow Fig. 4.10 shows the N-S component of the velocity at 60 and 210 m above the bottom and the temperature as recorded at the ADCP for a five day period in March 1997. Positive values of the velocity

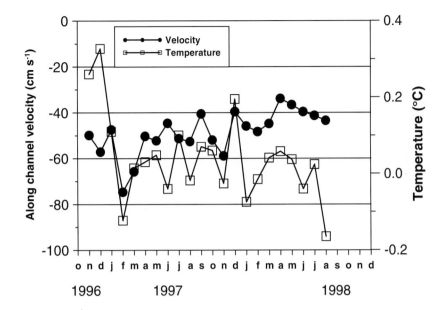

Fig. 4.9 Monthly averages of the along channel velocity and temperature from an ADCP situated just south of the sill. Negative values indicate flow to the southwest.

indicate currents to the northeast while negative velocities are to the southwest. During the first day the current changes from southwestwards to northeastwards and when it has been flowing towards the northeast for some time the temperature suddenly rises from 0°C to almost 3°C. A few hours later the temperature falls again to 0°C and the current turns southwestwards. The rise in temperature indicates that the front between the Atlantic Water to the east and the cold DSOW to the west has been displaced over the mooring. There is thus a lateral shift in the front and this is quite frequent in the records although most of the time the flow is to the southwest.

The lack of seasonal variations in the velocity of DSOW observed by the ADCP seems to be characteristic for this flow as it has also been reported further downstream by Dickson and Brown (1994) and Aagaard and Malmberg (1978). If the thermohaline circulation is the dominant driving force for the flow over the Greenland-Scotland Ridge, a similar lack of seasonal variability should also be observed in the Atlantic Water flowing into the Nordic Seas. If the flow was mainly driven by wind one would expect to see a clear seasonal signal in the flow since the seasonal signal is dominant in the wind field. This has been investigated by Hansen *et al.* (2000) who concluded that there were few seasonal variations in the total inflow. This indicates that probably most of the inflow is governed by the thermohaline circulation and is not induced by wind. Thus the formation of the deep and intermediate water in the Arctic Ocean and in particular the Greenland and Iceland Seas is probably crucial for the climate of the Nordic Seas.

4.5 DISCUSSION

We have outlined briefly some aspects of recent developments in climate-related physical oceanographic research in the past years. Although major steps have been taken by projects such as the Greenland Sea Project, the Nordic WOCE and VEINS projects, further studies are needed. More effort should be put into making quantitative studies of the various complicated oceanographic features in the ocean around Iceland. This is true for example for the fluxes of various water masses and their heat content

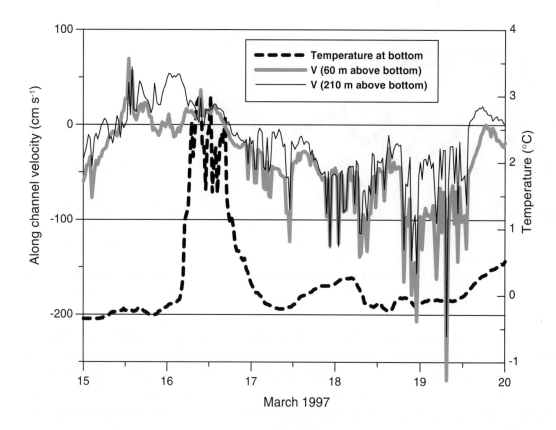

Fig. 4.10 The along channel velocity at 60 and 200 m above the bottom and the temperature as measured at the ADCP for a 5 day period in March 1997. Negative values indicate flow to the southwest.

and salinity. It is also necessary to obtain longer time series of these variables from crucial points of the system in order to be able to study the variability and the responses of the system to the forces affecting it, such as the atmospheric forces. This would further the understanding of the processes that affect the climate in the area. The seasonal CTD cruises, although they have good spatial coverage, are not revealing the short-term variability of the hydrography and it would be preferable to increase temporal resolution with continuous measurements of hydrographic variables. This could be combined with measurements of biological parameters and with modern technology the data can be transmitted in real time. This would increase our knowledge of short-term changes and add valuable information for the management of the ecological system in the ocean around Iceland.

ACKNOWLEDGEMENTS

We would like to thank our colleagues at the Marine Research Institute for their collaboration with collecting and processing the data used in this work and discussions over the years. Thanks are also due to the crews of the research vessels of the institute for help at sea with data collection. We also thank P.-M. Poulain for his co-operation with drifter data analysis.

5. The glacier-marginal landsystems of Iceland

David J.A. Evans

Department of Geography and Topographic Science, University of Glasgow, Glasgow, G12 8QQ, Scotland, UK (now Department of Geography, Science Site, University of Durham, DH1 3LE, UK)

5.1 INTRODUCTION

Landsystems differentiate terrain through the identification of recurrent patterns in both surface form and underlying sedimentary structures. Landsystems and process-form models have been applied to a wide range of glaciated basins where they have aided in assessments of the genesis of landform-sediment assemblages, particularly in the ancient terrestrial glacial record (e.g. Clayton and Moran, 1974; Boulton and Paul, 1976; Boulton and Eyles, 1979; Eyles, 1983a,b; Benn and Evans, 1998; Evans and Rea, 1999; Evans et al., 1999a). The accuracy of the reconstructions of palaeo-ice dynamics that stem from such research are essentially calibrated by contemporary analogues where ongoing process-form relationships can be demonstrated unequivocally (e.g. Boulton and Eyles, 1979; Krüger, 1987,1994; Owen and Derbyshire, 1989,1993; Evans et al., 1999a; Evans and Rea, 1999; Evans and Twigg, 2002). This paper reviews landsystems models for the range of glacial depositional settings that presently exist in Iceland, representing the main glacier morphologies typical of the historical period of glaciation.

The glaciers of Iceland (Fig. 5.1) are classified as temperate or warm-based in thermal terms (Björnsson, 1979). Temperate glacier snouts are mainly wet-based for at least part of the year, and a narrow frozen zone often develops at the margin during the winter due to the penetration of the seasonal cold wave from the atmosphere. Debris-rich basal ice sequences are typically thin or absent beneath temperate glaciers (Hubbard and Sharp, 1989), although considerable concentrations of debris are observed in the basal ice facies of glaciers whose subglacial meltwaters are influenced by supercooling in overdeepenings (e.g. Alley et al., 1998, 1999b; Lawson et al., 1998; Evenson et al., 1999). Any debris-rich ice that does exist can be elevated to the glacier surface by compressive flow in the snout. In Iceland a large volume of fine grained tephra emerges at the glacier snout after its transport from the accumulation zone. Additionally, the elevation of small quantities of subglacial debris to englacial and supraglacial positions may be achieved where marginal folding and thrusting takes place. The morphology of a glacier and its physiographic setting dictate the characteristics of the landforms and sediments associated with its margins. Four major glacier morphologies dominate the historical glacial legacy in Iceland: the plateau piedmont, lowland surging, debris-charged and plateau icefield types. Each morphology is associated with a distinctive set of landforms and sediments (landsystem), which are now reviewed in turn.

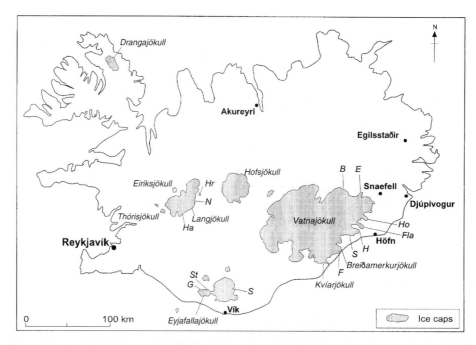

Fig. 5.1 Map of the glacier cover of Iceland, showing individual glacier snouts used as examples in this paper (Hr - Hrútfell; N - Nyðriskriðjökull; Ha - Hagafellsjökull; St - Steinholtsjökull; G - Gígjökull; S on Mýrdalsjökull - Sandfellsjökull; F - Fjallsjökull; S on Vatnajökull - Skálafellsjökull; H - Heinabergsjökull; Fla - Fláajökull; Ho - Hoffellsjökull; E - Eyjabakkajökull; B - Brúarjökull).

5.2 THE PLATEAU PIEDMONT GLACIAL LANDSYSTEM

Previous models of landform evolution at temperate glacier margins (e.g. Boulton and Eyles, 1979; Gustavson and Boothroyd, 1987; Krüger, 1987; Benn and Evans, 1998) have identified the major process-form relationships based upon research on specific landform-sediment associations (e.g. Harris and Bothamley, 1984; Kruger and Thomsen, 1984; Humlum, 1985; Krüger, 1985,1993,1994,1997; Boulton, 1986; Boulton and Hindmarsh, 1987). The lowland plateau outlet glaciers of Iceland generally transport relatively small quantities of supraglacial and ice-marginal debris compared to the upland glaciers in other temperate regions. Consequently latero-frontal moraines and supraglacial landform associations, such as hummocky moraine and widespread ice stagnation topography, are uncommon around their margins. Exceptions to this general rule can be explained by localized factors; the piedmont lobe Kötlujökull has produced considerably more supraglacial debris in its outer snout zone compared to nearby outlet glaciers from the same ice cap (Krüger, 1994), probably as a result of complex marginal oscillations that have led to proglacial thrusting and the re-incorporation of stagnating ice in a fashion similar to the surging glacier landsystem (see below). Previous studies have highlighted the dominance of three depositional domains on the forelands of the Icelandic glaciers, specifically the marginal moraine, glacifluvial/glacilacustrine and subglacial domains (Fig. 5.2; Evans, 2003).

The maritime, cold-temperate climate of Iceland is probably responsible for the late winter readvances of the receding glacier margins particularly in the southeast of the island (Boulton, 1986). This active recession results in the construction of numerous

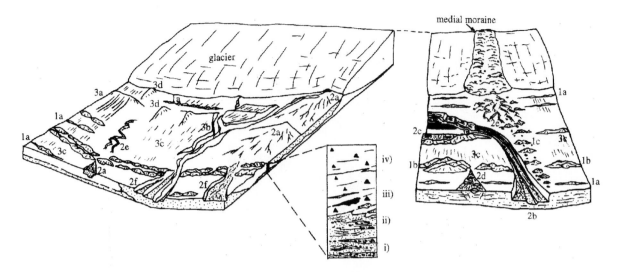

Fig. 5.2: The active temperate glacial landsystem (after Evans & Twigg 2002 & Kruger 1994). Landforms are numbered according to their domain (1 = morainic domain; 2 = glacifluvial domain; 3 = subglacial domain): 1a) small, often annual, push moraines; 1b) superimposed push moraines; 1c) hummocky moraine; 2a) ice-contact sandur fans; 2b) spillway-fed sandur fan; 2c) ice margin-parallel outwash tract/kame terrace; 2d) pitted sandur; 2e) eskers; 2f) entrenched ice-contact outwash fans; 3a) overridden (fluted) push moraines; 3b) overridden, pre-advance ice-contact outwash fan; 3c) flutes; 3d) drumlins. The idealized stratigraphic section log shows a typical depositional sequence recording glacier advance over glacifluvial sediments, comprising: i) undeformed outwash; ii) glacitectonized outwash/glacitectonite; iii) massive, sheared till with basal inclusions of pre-advance peat and glacifluvial sediment; iv) massive sheared till with basal erosional contact.

recessional push moraines in the marginal moraine domain (Fig. 5.3a,b). Local characteristics indicate that the push moraines are produced either by ice-marginal squeezing of water-soaked subglacial sediment and/or pushing of proglacial materials, especially where small scale glacitectonic disturbance is evident (Howarth, 1968; Price, 1970; Krüger, 1987). The push moraines are constructed from material derived largely from the glacier foreland and they often record annual recession of active ice (Sharp, 1984; Boulton, 1986; Krüger, 1995). Although few exposures through the moraines are available they comprise a wide range of sediments from matrix-rich diamictons to bouldery clast-rich diamictons with large numbers of striated and facetted clasts, reflecting the nature of the local materials from which they were derived. The plan-form of these push moraines reveals crenulated, lobate or saw-tooth crests whose re-rentrants coincide with longitudinal crevasses at the glacier margin. Coalescence or partial superimposition of moraines indicates a winter readvance by the glacier that was more extensive than that of the previous year. Stationary ice fronts can construct larger push moraines (Krüger, 1993) either by the stacking of sediment slabs frozen on to the bed during the winter (Howarth, 1968; Humlum, 1985; Krüger, 1985,1987,1993,1994,1996; Matthews *et al.*, 1995), or as a result of the range of dump, squeeze and push mechanisms operating at the same location over several years (Evans and Twigg, 2002), or by incremental thickening of an ice-marginal wedge of deformation till (eg. Johnson and Hansel, 1999; Fig. 5.3c). The continuation of flutings over the proximal slopes and onto the crests of many push moraines indicates that subglacial

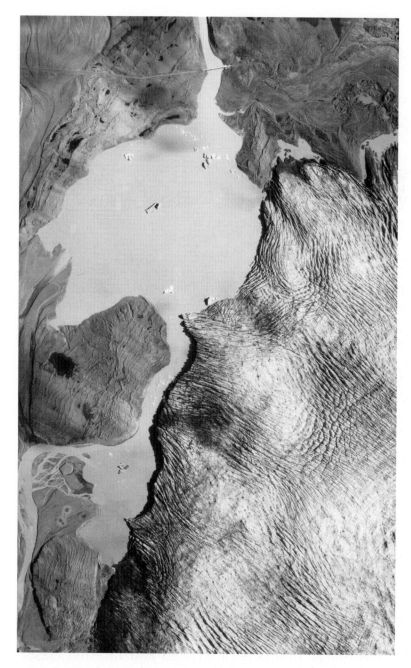

Fig. 5.3a

Fig. 5.3 *Typical landforms and sediments of the marginal moraine domain of plateau piedmont glacier margins: a) aerial photograph of part of the eastern foreland of Fjallsjökull, showing recessional push moraines with saw-tooth crests and associated flutings. Note the partial superimposition of some moraines recording near stationary glacier margins during moraine construction (University of Glasgow and Landmælingar Íslands 1965); b) modern push moraines at the margin of Fjallsjökull; c) large push moraines produced by thickening wedge of deformation till at the margin of Fláajökull, a glacier that has remained stationary for much of the last decade; d) aerial photograph of the foreland of Sandfellsjökull, showing overridden push moraines (op) superimposed by recessional push moraines (rp) (Landmælingar Íslands 1992); e) veil of boulders draped on the push moraines on the central foreland of Breiðamerkurjökull. The boulders represent the former location of the medial moraine Mávabyggðarönd which was lowered onto the substrate during glacier recession.*

features are the product of melt out of shallow glacier margins previously buried by glacifluvial sediment due to either lateral dumping of ice-dammed lakes or high sediment rates in marginal meltwater streams (e.g. Welch, 1967; Howarth, 1968; Price, 1969,1971,1973). Melting of large parts of the glacier snout beneath glacifluvial sediments explains the unusually large size of some individual kettle holes. With the exception of lateral margins where meltwater streams are forced to flow over parts of the glacier snout, pitted outwash is not extensively developed in front of actively receding temperate glaciers due to the absence of large tracts of stagnating ice. Pitted sandur fans at the margins of Icelandic glacier snouts are often explained as the products of jökulhlaups (e.g. Churski, 1973; Galon, 1973a,b; Klimek, 1973a,b; Olszewski and Weckwerth, 1997; Fay, 1999, 2002a,b). The development of numerous small kettle holes at the fan apex documents the melt out of individual ice blocks originally deposited by floodwaters on the fan (Maizels, 1977,1992). A jökulhlaup origin was proposed by Howarth (1968) and Bodéré (1977) for a pitted sandur fan on the distal side of the LIA terminal moraine at east Breiðamerkurjökull.

The most intensive study of eskers in Icelandic lowland glacier snouts has been associated with Breiðamerkurjökull, aided by surveying and mapping of the snout and proglacial area since 1945 (Welch, 1967; Howarth, 1968,1971; Price, 1969,1973,1982). The eskers are typically sharp-crested with steep sides and comprise predominantly coarse gravels. They are arranged into complex forms comprising single ridges and multiple, anabranched sections often resembling fans. The eskers on the west side of Jökulsárlón are located in the former positions of the large medial moraines of the glacier. The down-glacier ends of some eskers terminate at the apices of sandur fans; the ice-contact sandur fans on Breiðamerkursandur were constructed from the sediment transported by such esker systems between the 1930s and 1965. Historical surveys of the eskers reveal significant surface lowering, indicating that large parts of them have been deposited in englacial or supraglacial positions. This interpretation is supported by observations of esker ice cores and eskers emanating from the wasting glacier surface

Fig. 5.4d.

D.J.A. Evans

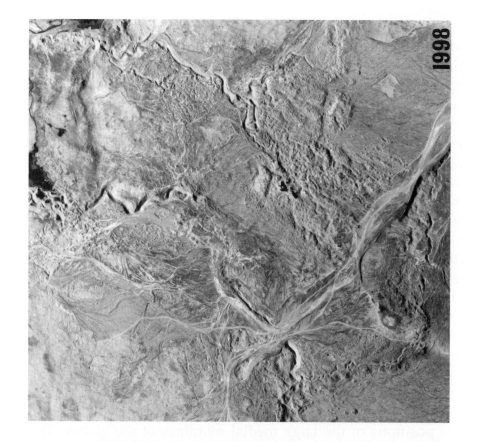

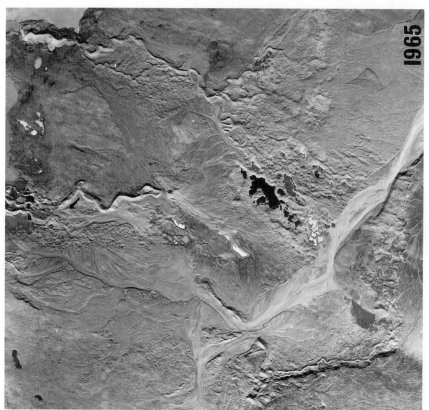

Fig. 5.4e.

Fig. 5.4f.

in the 1960s (Welch, 1967; Howarth, 1968,1971; Price, 1969). Aerial photographs record the evolution of large tracts of 'kame and kettle topography' (Welch, 1967; Howarth, 1968; Price, 1969) associated with the Breiðamerkurjökull eskers in the 1960s. For example, eskers and a pitted outwash surface on 1945 aerial photographs evolved into eskers and 'kame and kettle topography' by 1965. They then developed into a complex anabranched system of esker ridges arranged in a fan shape by 1998 (Fig. 5.4e), indicating that a considerable expanse of glacier ice lay below a thin cover of glacifluvial outwash and that the larger volumes of esker sediment within the ice ensured esker preservation after melt out. Such complex esker networks clearly develop in parts of the glacier snout that occupy topographic depressions, where a thin cover of glacifluvial sediment can accumulate over the top of glacier ice containing eskers.

Ice-dammed and proglacial lakes also are common features around the receding lowland glacier snouts of Iceland (Þórarinsson, 1939; Björnsson, 1976), particularly where the glacier has uncovered an overdeepening as for example at Breiðamerkurjökull, Fjallsjökull, Heinabergsjökull, Hoffellsjökull, Hagafellsjökull, Nyðriskriðjökull and southwest Langjökull (Howarth and Price, 1969; Björnsson, 1996; Bennett *et al.*, 2000). Their development gives rise to the accumulation of thick sequences of glacilacustrine sediments and shorelines and deltas, the surfaces of which can be heavily pitted or extensively deformed by ice melt out in glacier-contact settings. Lake sediments also constitute ideal material for push and thrust moraine development (Howarth, 1968; Evans *et al.*, 1999b; Bennett *et al.*, 2000) and the production of glacitectonite and deforming bed tills (Benn and Evans, 1996; Evans *et al.*, 1998,1999c; Evans and Twigg, 2002). Valleys dammed by plateau outlet glaciers can be the cause of substantial jökulhlaups (Björnsson, 1992; Tweed and Russell, 1999). The most notable of these lake systems is Grænalón at the margin of Skeiðarárjökull, which empties over the space of nine days and attains peak discharges of 6,000m³s⁻¹ (Þórarinsson, 1939,1956b). Another striking example is Vatnsdalur at the margin of Heinabergsjökull. The cyclical drainage of this lake has been responsible for jökulhlaups at the margin of Heinabergsjökull,

Fig. 5.4g.

resulting in the raising of the proglacial lake level and increases in discharge in the spillway on its southern shore. At the end of each lake dumping event icebergs are stranded on the floor of the shoreline-covered lake bottom of Vatnsdalur (Fig. 5.4f). A similar lake existed to the south in Dalvatn when Heinabergsjökull stood at its Little Ice Age maximum (Evans *et al.*, 1999b). Thick sequences of ice-proximal glacilacustrine sediments and flights of extensive gravelly shorelines characterize Dalvatn, documenting the existence of the short-lived Little Ice Age ice-dammed lake (Fig. 5.4g). Pitting of glacilacustrine valley fills in such settings documents the melt-out of stranded icebergs.

The subglacial domain occurs on the land surfaces between ice-marginal morainic and glacifluvial depo-centres and is dominated by subglacial landform assemblages of flutings, drumlins and overridden push moraines (Fig. 5.5) in addition to areas of discontinuous, striated and polished bedrock often characterized by roches moutonnées and indicative of widespread abrasion and quarrying at the ice/bedrock interface (Kruger and Thomsen, 1984; Kruger, 1987; Benn, 1995; Hart, 1995; Evans *et al.*, 1999a,b). The flutings, drumlins and tills of the Icelandic lowland snouts are central to contemporary subglacial deformation theory. The ubiquitous lodged and striated boulders on the forelands of lowland glaciers often occur at the heads of flutings (Fig. 5.5b), prompting the most popular explanation of fluting genesis involving the squeezing of till into the cavities that develop on the down-glacier sides of lodged boulders (Boulton, 1976; Benn, 1994). Small isolated patches of crevasse-squeeze ridges, identifiable by their cross-cutting ridge networks (e.g. Sharp, 1985a; Evans and Rea, 1999), occur in narrow, ice-front parallel bands, probably as a result of the squeezing of subglacial sediment into sub-marginal crevasses when marginal push/squeeze moraines were being constructed (e.g. Price, 1970).

Experiments conducted beneath the west lobe of Breiðamerkurjökull demonstrated that the subglacial till was emplaced by deformation and that it possessed a two-tiered structure in response to the ductile flow of an upper dilatant layer (A-horizon) and the

Fig. 5.5a, for caption see over.

brittle or brittle-ductile shearing of a lower stiff layer (B-horizon; Boulton and Dent, 1974; Boulton, 1979; Boulton and Hindmarsh, 1987). The two-tiered structure was observed in deposited tills by Dowdeswell and Sharp (1986) at Skálafellsjökull and at the subglacial experimental site at west Breiðamerkurjökull by Benn (1995) after it had been exposed by glacier recession. Boulton (1987) proposed a model of erosion and deposition at the margins of temperate lowland glaciers based upon Breiðamerkurjökull. He suggests that net deposition occurs over a sub-marginal zone that stretches for a distance of 400 m in from the terminus. Erosion dominates inside that zone and therefore recently deglaciated sites that have been located up-ice of the depositional zone should display stratigraphic evidence of erosion followed by deposition.

Fig. 5.5b.

Such evidence exists in the forelands of the Icelandic lowland glaciers as deformed proglacial outwash and localized proglacial lake sediment and peat beds capped by glacitectonites and deformation tills (Fig. 5.5c); smeared inclusions of the stratified sediments and the peat occur in the overlying till (Howarth, 1968; Boulton, 1987; Benn and Evans, 1996,1998; Evans, 2000; Evans and Twigg, 2002). The glacitectonites comprise folded and faulted stratified sediments, equivalent to Boulton's (1987) 'B$_1$ horizon', and the deformation tills are represented by the A horizon. Although glacifluvial features are cannibalized to produce the deforming layer and the drumlins of temperate lowland glaciers (Hicock and Dreimanis, 1989,1992; Benn and Evans, 1996,1998; Evans *et al.*, 1998,1999c; Evans, 2000), the forms of overridden ice-contact fans are preserved relatively intact beneath streamlined till surfaces.

Assessments of sedimentology and stratigraphy on the deposits of Icelandic lowland glaciers suggest that tills in such settings were emplaced by sub-marginal processes of deformation and lodgement, and comprise materials derived from pre-existing stratified sediments in addition to localized abrasion of rock surfaces. Tills are generally thin (less than 2 m in thickness), but thicker sequences have been constructed by the sequential plastering of several till layers onto stratified sediments and bedrock. This is similar to the rheologic superposition proposed by Hicock (1992), Hicock and Dreimanis (1992) and Hicock and Fuller (1995), and the till/stratified interbed successions of Eyles *et al.* (1982), Evans *et al.* (1995) and Benn and Evans (1996). Vertical dykes of sediment in the tills on the foreland of Sléttjökull have been interpreted by van der Meer *et al.* (1999) as the products of subglacial meltwater discharges. Larger push moraines and complex till sequences are constructed where glacier margins become stationary for substantial periods. Due to their location near the limit of the LIA maximum, some pockets of Holocene sediment and peat have survived as glacitectonized and partially cannibalized sub-till stratigraphic units.

Fig. 5.5c.

Fig. 5.5 Typical landforms and sediments of the subglacial domain of plateau piedmont glacier margins: a) aerial photograph of the east foreland of Breiðamerkurjökull, showing push moraines, flutings and drumlins superimposed on overridden push moraines (University of Glasgow and Landmælingar Íslands 1965); b) a lodged and striated boulder at the head of a fluting on the foreland of Skálafellsjökull; c) section through glacitectonized outwash and overlying deformation till on the surface of an overridden moraine at east Breiðamerkurjökull (after Evans and Twigg, 2002).

5.3 THE SURGING GLACIER LANDSYSTEM

Unlike the climatically-driven oscillations of the lowland plateau outlet glaciers of Iceland, surging by some of the islands glaciers represents a cyclic flow instability which is triggered from within the glacier system (Raymond, 1987), probably by geothermal activity and concomitant reorganizations in the subglacial drainage system (Þórarinsson, 1969; Björnsson, 1975,1992; Clarke *et al.*, 1984; Kamb *et al.*, 1985; Clarke, 1987; Fowler, 1987; Kamb, 1987). The active phase of a surge involves a rapid advance of the glacier front followed by a period of slow flow or quiescence during which the snout stagnates. Evans and Rea (1999, 2003) have demonstrated that, regardless of the trigger mechanism, the landform-sediment assemblages produced by surging glaciers appear to be consistent and predictable. Those landform-sediment assemblages are reviewed in turn.

Rapid ice advance into unfrozen sands and gravels will produce high compressive proglacial stresses. Such sediments will develop only low pore water pressures, making brittle failure and shearing and stacking likely. At the margins of the Icelandic surging glaciers Brúarjökull and Eyjabakkajökull, pre-surge peat layers and thick sequences of stratified sediments have been vertically stacked in a series of thrust overfolds to produce thrust block moraines (Sharp, 1985b,1988; Croot, 1988; Fig. 5.6a). Although thrust block moraines are the most spectacular constructional features produced by surging glacier margins, they can be produced only in areas where sufficient deformable sediment is available for glacitectonic thrusting, folding and stacking. For example, the wide surging margin of Brúarjökull has produced thrust blocks only in areas that lie down flow from topographic depressions large enough to accumulate sediments during the quiescent phase. Elsewhere, the glacier constructs low amplitude push moraines by bulldozing and/or sub-marginally squeezing the veneer of till or peat that drapes the proglacial land surface.

Conspicuous ice-moulded hills on the proglacial forelands of the Icelandic surging glaciers lie down-ice of topographic depressions from which the hills were originally displaced by thrusting. The surfaces of these features appear extensively fluted or drumlinized and their internal structures comprise glacitectonized outwash or lake sediments capped by glacitectonite (Benn and Evans, 1996,1998) and truncated by subglacial till. These ice-moulded hills are interpreted as overridden thrust block moraines or cupola hills. The sediments displaced by thrust block construction are

Fig. 5.6 Opposite and following pages. Typical landforms and sediments of the surging glacier landsystem: a) thrust block moraine composed of pre-surge stratified outwash and peat layers at the AD 1890 surge margin of Eyjabakkajökull; b) aerial photograph of the west foreland of Brúarjökull showing concertina esker in association with other surge geomorphology (flutings, crevasse squeeze ridges) produced during a surge in 1963 (Landmælingar Íslands 1993); c) crevasse-squeeze ridges on the east side of Eyjabakkajökull; d) long flutings on the west foreland of Brúarjökull produced during the 1963 surge (back pack for scale); e) hummocky moraine, with localized evidence of ongoing melt out of buried ice, inside the AD 1890 surge limit on the west foreland of Brúarjökull; f) pitted and channelled outwash fan located on the shallow west central snout of Brúarjökull.

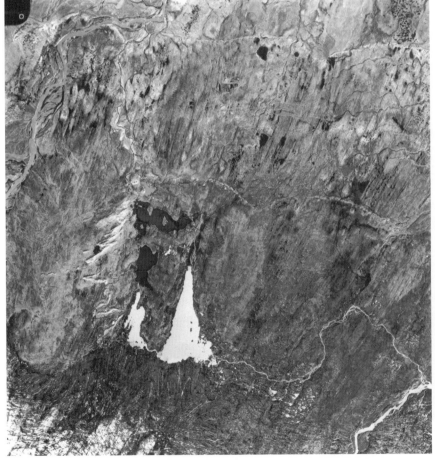

*Above Fig. 5.6a,
left Fig. 5.6b.*

Above Fig. 5.6c, below Fig. 5.6d.

deposited during the surge quiescent phase when proglacial lakes and meltwater streams extensively modify the foreland. Each thrust block demarcates the former glacier margin during a surge, overriding taking place either during the same surge or during a later, more extensive surge.

In addition to single, sinuous forms, eskers also display a 'concertina' plan-form on aerial photographs of the margins of the Icelandic surging glaciers Brúarjökull and Eyjabakkajökull (Knudsen, 1995; Fig. 5.6b). An englacial or supraglacial origin for the concertina eskers is indicated by extensive exposures of glacier ice and widespread evidence of melt out of buried ice (Evans and Rea, 1999; Evans *et al.*, 1999a). It has been argued by Evans and Rea (2003) that the concertina eskers must be formed in the immediate post-surge period when substantial quantities of pressurized water stored behind the surge front are discharged periodically (e.g. Humphrey and Raymond, 1994) through high levels of the glacier, establishing a new englacial and/or supraglacial drainage system by exploiting the extensive network of crevasses created during the surge.

Crevasse-squeeze ridges (Fig. 5.6c) are best known from the contemporary surge-type glaciers of Brúarjökull and Eyjabakkajökull, Iceland and have been used as diagnostic criteria of surging activity (Sharp, 1985a, 1985b; Evans and Rea, 1999, 2003; Evans *et al.*, 1999a). They comprise cross-cutting diamicton ridges that can be traced from the foreland and into crevasse systems in the snouts. Crevasse squeeze-ridges will form when there is soft, readily deformable sediment at the ice-bed interface and the base of the glacier is highly fractured or crevassed. The extreme stresses experienced during a surge leave the glacier surface highly crevassed (e.g. Kamb *et al.*, 1985; Raymond *et al.*, 1987, Herzfeld and Mayer, 1997) and therefore, if the glacier overlies soft, readily deformable sediment, this will promote the development of crevasse-squeeze ridges.

Flutings occur on the forelands of many glaciers and are certainly not diagnostic of glacier surging. However, fluting length may provide important evidence for rapid advances over substantial distances and excellent examples of this exist on the foreland of Brúarjökull where regularly spaced flutings are continuous for more than 1km inside the 1964 surge moraine (Fig. 5.6d), constituting individual flow sets analogous to the isochronous mega-lineations observed at the former beds of ice sheets (Clark, 1999). Additionally, the association of flutings and crevasse-squeeze ridges is an important aspect of the subglacial geomorphology of surging glaciers. Sharp (1985a,b) notes that fluting crests at Eyjabakkajökull rise to intersect the crevasse-squeeze ridge crests, indicating that subglacially deformed till was squeezed into basal crevasses as the glacier settled onto its bed at the end of the surge. The contemporaneous production of the flutings and crevasse-squeeze ridges at Eyjabakkajökull renders them diagnostic of glacier surging in the landform record.

Conspicuous mounds of hummocky topography on the down-ice sides of topographic depressions on the forelands of the Icelandic surging glaciers are characterized by extensive evidence of on-going melt-out of buried ice which has disturbed the faint fluting patterns that occur on the hummock surfaces (Fig. 5.6e). The hummocks comprise intensely glacitectonized, fine-grained stratified sediments and diamictons or poorly-sorted gravels. Small pockets of stratified sediments interbedded with diamictons occur in small depressions on the hummocky topography. These sediments have been contorted into low amplitude folds by the melt-out of underlying ice. All of these

Above Fig. 5.6e, below Fig. 5.6f.

characteristics indicate that the hummocky topography evolves during surging when proglacial lake sediments and outwash lying over pre-existing stagnant ice are thrust, squeezed and bulldozed. The faint flutings on the hummock surfaces indicate that the sediment and stagnant ice were overridden by the surging snout. Supraglacially reworked sediments are locally deposited over the bulldozed sediments as they emerge from beneath the melting ice, leading to the deposition of the small pockets of stratified sediments. The post-surge melt-out of the older buried ice results in the production of chaotic hummocky terrain upon which flutings may still be observed as discontinuous linear ridges, at least during the early stages of melt-out. Mapping of the hummocky moraine at Brúarjökull demonstrates that it occurs in discrete pockets on the foreland. These are located immediately down-glacier from extensive depressions that have been partially filled with proglacial outwash and glacilacustrine sediments since glacier recession.

Proglacial outwash tracts and glacilacustrine depo-centres at the margins of Icelandic surging glaciers have accumulated over the shallow stagnant margins of the snouts during surge quiescence (Evans and Rea, 1999; Fig. 5.6f). The outwash occurs as ice-contact fans fed by subglacial and englacial meltwater portals during periodic discharges of pressurised meltwater (e.g. Humphrey and Raymond, 1994). It is clear from the concertina eskers that such discharges can carry large debris loads and therefore may dump considerable volumes of glacifluvial sediment over wide areas beyond their exit portals. This implies that those parts of the glacier snout which occupy topographic hollows are prone to burial by large outwash fans and, in more distal locations, glacilacustrine sediments. Gradual melting of the stagnant snout results in the formation of kettles, followed by the collapse of tunnels in the stagnant ice to form ice-walled channels. The final stage in the evolution of the ice-cored outwash involves the production of chaotic sand and gravel hummocks within which sinuous eskers may be recognizable.

None of the above landform-sediment assemblages can be used as a sole diagnostic criterion for surging activity, but a landsystem for surging glaciers has been proposed by Evans *et al.*, (1999a) and Evans and Rea (1999, 2003) based upon the Icelandic glacier snouts. The geomorphology is arranged in three overlapping zones (Fig. 5.7). The outer zone (zone A) represents the limit of the surge and is composed of weakly consolidated pre-surge sediments, proglacially thrust or pushed by rapid ice advance. The intermediate zone (zone B) consists of patchy hummocky moraine located on the down-glacier ends of topographic depressions and often draped on the ice-proximal slopes of the thrust block and push moraines. The inner zone (zone C) consists of subglacial deformation tills and long, low amplitude flutings, produced by subsole deformation during the surge, and crevasse-squeeze ridges, documenting the filling of basal crevasses at the beginning of the surge quiescent phase. Concertina eskers can also occur in this zone where they are draped over the flutings and crevasse-squeeze ridges. Some diagnostic forms of surging are intrazonal, because either they are palimpsests of older surges (e.g. overridden moraines), or they relate to the courses of proglacial outwash fans and streams (ice-cored, collapsed outwash), or they occur in ponded topographic depressions on the foreland (collapsed lake plains). Specifically, collapsed glacilacustrine sediment bodies and ice-contact fans may occur within topographic depressions where the stagnating glacier snout became buried during the early part of the quiescent phase.

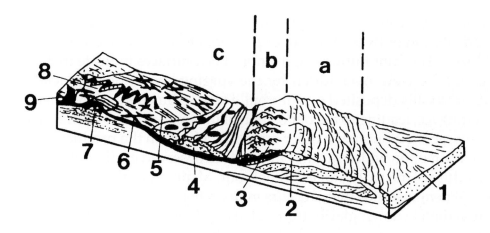

Fig. 5.7   Landsystems model for surging glacier margins (from Evans and Rea, 2003, after Evans et al., 1999b and Evans and Rea, 1999): a) outer zone of proglacially thrust pre-surge sediment which may grade into small push moraines in areas of thin sediment cover; b) zone of weakly developed chaotic hummocky moraine located on the down-ice sides of topographic depressions; c) zone of flutings, crevasse-squeeze ridges and concertina eskers; 1 = proglacial outwash fan; 2 = thrust block moraine; 3 = hummocky moraine; 4 = stagnating surge snout covered by pitted and channeled outwash; 5 = flutings; 6 = crevasse-squeeze ridge; 7 = overridden and fluted thrust block moraine; 8 = concertina esker; 9 = glacier with crevasse-squeeze ridges emerging at surface.

5.4 THE DEBRIS-CHARGED GLACIAL LANDSYSTEM

Glaciers nourished in precipitous upland terrains characterized by extensive bedrock cliffs and scree slopes transport large volumes of debris to their ablation zones. The majority of this debris is sourced by extraglacial mass wasting. If this occurs in the accumulation zone the debris re-emerges as debris-charged folia at the glacier surface in the ablation zone due to compressive flow in the snout, constituting controlled supraglacial moraine. Alternatively, if rockfall debris falls directly onto the ablation zone, which it predominantly does in Iceland at present, it may blanket the snout as an uncontrolled moraine cover and fall into crevasse networks to be reworked by supraglacial streams. Sediment may accumulate at the margins of glaciers in high relief terrain as inwash from surrounding valleys and slopes (e.g. Evenson and Clinch, 1987), leading to the construction of substantial kame terraces. Such debris-charged glaciers construct large latero-frontal moraine loops and extensive supraglacial hummocky moraine (e.g. Kvíárjökull; Fig. 5.8). They also develop into rock glaciers due to high rates of debris supply and reduced ablation rates in the debris-covered portions of their snouts (e.g. Snæfell; Fig. 5.9). In addition to restricting the flow of glacier ice as a piedmont lobe, the latero-frontal moraines effectively restrict or dam proglacial meltwater streams, giving rise to the deposition of heavily reworked ice-contact glacifluvial sediment accumulations (terraced and kettled sandar and kame terraces) in association with hummocky moraine and patchy ice-proximal glacilacustrine deposits. In many respects

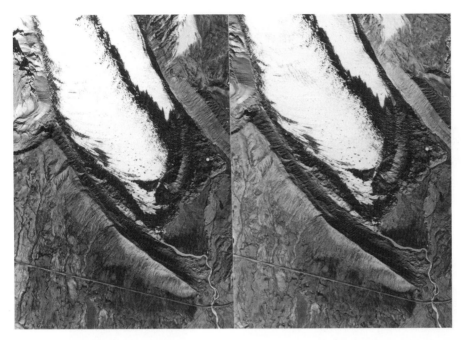

Fig. 5.8a.

Fig. 5.8b.

Fig. 5.8 Kvíárjökull and its proglacial geomorphology: above, a) aerial photograph stereopair (Landmælingar Íslands/University of Glasgow 1998) of Kvíárjökull, showing the large latero-terminal moraine loop, hummocky moraine, incised and kettled outwash and supraglacial debris patterns. Note the controlled supraglacial moraine on the west side of the glacier produced by the emergence of rockfall debris in rising folia; below, b) view over the hummocky moraine and outwash inside the latero-terminal moraine loop at Kvíárjökull. Note the linearity in the hummocky moraine and the tendency for abandoned outwash tracts to be located between hummocky moraine ridges. The moraine linearity is thought to be a product of a combination of supraglacial controlled moraine development and ice-marginal pushing (Spedding and Evans, 2002).

Fig. 5.9 The Snæfell rock glaciers, west Iceland. The distinct lithological composition of each rock glacier can be related to source rock outcrops in the cirque basins on the side of Snæfell.

these glacier systems resemble the glaciated valley landsystem of alpine terrains (Boulton and Eyles, 1979; Eyles, 1983a), although they are relatively uncommon in Iceland, being restricted to the dissected slopes of volcanoes such as Snæfell, Öræfi and Eyjafjallajökull. Numerous rock glaciers with glacier ice cores exist in the shallow, north-facing cirques that have been cut into the flood basalt plateaux in northern Iceland (Martin and Whalley, 1987; Martin *et al.*, 1991,1994; Whalley and Martin, 1994; Hamilton and Whalley, 1995; Whalley *et al.*, 1994,1995a,b). The high density of rock glaciers in areas such as the Tröllaskagi Peninsula is probably a function of the high rates of debris provision from the steep rockwalls that encircle the cirque glaciers located on the backwasting plateaux edges (Fig. 5.10). The balanced to slightly negative mass budget of the glacier components ensures that debris accumulates on their surfaces, protecting the ice from further melting and thereby initiating a negative feedback. Snow blow ensures that the plateaux remain ice-free at present, suggesting that small debris-covered/rock glaciers located on the margins of basalt plateaux are the precursors to plateau icefield outlet lobes. The corollary is that at least some of the bouldery till cover deposited by plateau icefield outlet glaciers is sourced from debris-covered cirque/rock glaciers that were incorporated during plateau icefield expansion (see below). Moreover, the present day rock glaciers and debris-charged cirque glaciers may be relict outlets that became detached from their parent icefield as it receded from the plateau edge. An excellent example of this particular origin for debris-charged snouts/rock glaciers in Iceland is provided by the west margin of Þórisjökull, a plateau icefield that has receded from the plateau edge leaving a series of small glaciers with debris-charged snouts on the lower cliffs (Fig. 5.11).

There are some aspects of the glaciology of the debris-charged glaciers of Iceland that distinguish them from the normal glaciated valley landsystem. This has been demonstrated by Kirkbride and Spedding (1996) and Spedding and Evans (2002) based upon Gígjökull, Steinholtsjökull and Kvíárjökull (cf. Eyles, 1979,1983b). These glaciers contain debris that can be sub-divided into four categories: angular rockfall debris, volcanic ash, englacial debris bands containing more rounded clasts, and crudely-stratified basal ice. The characteristics of the englacial debris bands and the stratified basal ice suggest that debris is not transported in the Icelandic debris-charged glaciers merely by passive supraglacial modes as proposed by the glaciated valley landsystem model ('supraglacial morainic till' of Boulton and Eyles 1979 and Eyles 1979). Rather, large volumes of sediment appear to have been transported to the debris-charged glacier snouts by englacial meltwater conduits, thereby explaining the prominence of more rounded clasts in the glacier and in the large latero-frontal moraines. Additionally, marginal zones of upwelling turbid meltwater and high water levels in moulins in the debris-charged glacier indicate high subglacial or englacial water pressures. Such high water pressures in the snout match the theoretical expectations of water in overdeepenings and suggest that the stratified basal ice may be the product of supercooling similar to that proposed for the Matanuska Glacier in Alaska by Alley *et al.* (1998,1999b), Lawson *et al.* (1998) and Evenson *et al.* (1999). This process would be an additional effective method of transporting debris to the glacier surface, independent of the passive supraglacial transport pathways directly from extraglacial sources. Whether or not overdeepenings are associated only with debris-charged glaciers needs further research but such an association might be explained by the restrictions placed

Fig. 5.10 Aerial photographs of the Skjöldalur II rock glacier, Tröllaskagi (Landmælingar Íslands). Left - 1960. Right - 1946.

upon the lateral spreading of glacier snouts by their large latero-terminal moraine loops.

The size of the latero-frontal moraines at the margins of the debris-charged glaciers of Iceland appear to be related to lengthy periods of deposition. For example, Kvíárjökull's moraines (Fig. 5.8a) are thought to have originated around 3,200 ^{14}C yr BP and since have been constructed incrementally during repeated reoccupation of the area by the glacier. The evolution of the hummocky moraine (Fig. 5.8b), and kettled and incised outwash inside the latero-terminal moraine loop at Kvíárjökull is the product of reworking by topographic inversion, incremental ice-marginal stagnation and melt-out of ice buried by marginal glacifluvial sediments (e.g. Eyles, 1979). Hence the volume of debris turned over by such glaciers, by a combination of passive transport of rockfall debris, reworking by englacial streams, supercooling and supraglacial reworking (eg. Spedding and Evans, 2002), results in the construction of a landsystem characterized by extensive hummocky moraine and latero-frontal moraine ridges. Some degree of ice-marginal pushing is apparent at the margin of Kvíárjökull where small scale glacitectonic disturbance occurs in proglacial sand and gravel ridges; this process contributes locally to the production of linearity in hummocky moraine assemblages.

5.5 THE PLATEAU ICEFIELD LANDSYSTEM

The plateau origins of most glacier snouts in Iceland dictate the volume of debris transfer to the ice margins. The stapis or tuyas of Iceland constitute ideal physiographic features for the accumulation of small plateau icefields in regions that are at the limit of glacierization (e.g. Ives *et al.*, 1975). Glaciers nourished on plateau surfaces are unlikely to accumulate large volumes of supraglacial debris due the lack of extraglacial sources such as rock avalanches and debris flows (Evans, 1990; Rea *et al.*, 1998). However, substantial accumulations of rockfall debris characterize the lateral margins of some outlet glaciers where they descend through precipitous cliffs at the plateau edge (see above and Fig. 5.11). Unlike debris-charged systems, these debris accumulations are restricted to the areas that lie directly beneath those parts of the plateau edge where bedrock cliffs are steep and unstable. Consequently, lateral moraines are discontinuous and asymmetrical on individual outlet lobes. The plateau icefield landsystem is distinct from that of the debris-charged and lowland plateau outlet landsystems in that it is characterized by a glacier ice body whose flow patterns are largely characterized by radial flow on the plateau summit. Small outlet glaciers descend from the plateau edge into surrounding valleys but the geomorphology of the whole glacier system is dictated by the plateau physiography. As Rea *et al.* (1998) point out in their comprehensive review of plateau icefields, the physiography controls the glacier morphology and glacial depositional style until plateau icefields coalesce and a larger ice sheet inundates the region.

Excellent examples of plateau icefields in Iceland include Þórisjökull, Eiríksjökull and Hrútfjell, which surround the larger Langjökull icecap, Drangajökull in the northwest and Torfajökull in the south. The distribution of ice on the plateau summits is typically asymmetrical, with ice-free areas occurring on the southwest corners and the more extended outlet lobes descending into neighbouring valleys on the north and

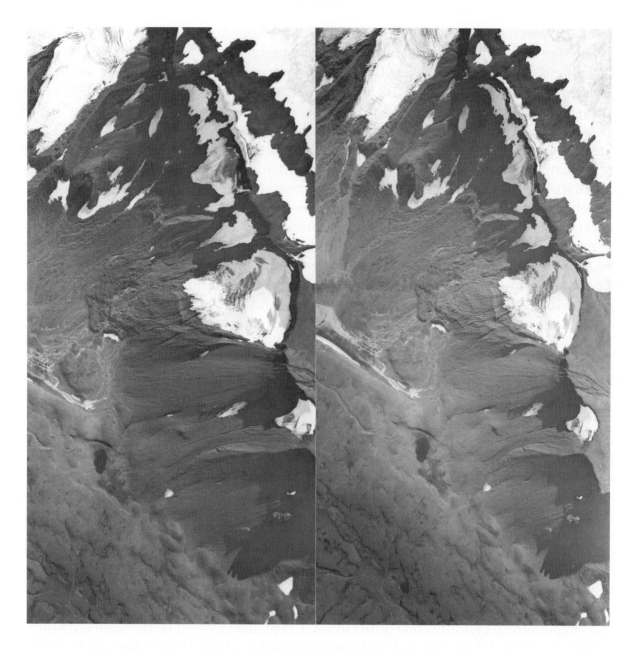

Fig. 5.11 Aerial photograph stereopair (Ísgraf/Loftmyndir and University of Glasgow 1999) of the west margin of the Þórisjökull plateau, showing a small remnant glacier at the cliff base and remnants of its former debris-charged/rock glacierized snout.

northeast corners. While on the plateau summit the debris supply to the glacier is restricted to the weathered regolith and quarried bedrock (e.g. Gellatly *et al.*, 1988; Rea and Whalley, 1996; Rea *et al.*, 1996). Once outlet lobes begin their descent into surrounding valleys the debris supply is augmented by rockfall debris from plateau edge cliffs and then by the unlithified sediments of the valley bottoms. Considerable thicknesses of rockfall and avalanche debris accumulate on the lower slopes surrounding the plateaux, augmenting the debris supply to the plateau ice as it descends from the plateau edge. Early stages of descent by the plateau glacier are characterized by ice

6. Subglacial volcanic activity in Iceland

Magnús T. Guðmundsson

Science Institute, University of Iceland, Hofsvallagötu 53, IS-107 Reykjavík, Iceland

Volcanic activity within glaciers is an important feature of the geological record in Iceland. In historical times subglacial and emergent eruptions have been most frequent in Vatnajökull, notably in Grímsvötn, and in Mýrdalsjökull (Katla). Eruptions causing ice melting and jökulhlaups have also occurred in historical times in the stratovolcanoes Öræfajökull, Eyjafjallajökull and Hekla. The majority of eruptions within glaciers in Iceland start as subglacial, but soon break through the ice cover and become emergent phreatomagmatic eruptions. In some eruptions most of the volcanic material piles up above the vents, forming hyaloclastite ridges and mounds. In other cases, a large fraction of the erupted material may be transported away with meltwater while a part is dispersed by an eruption plume. The very high rates of heat transfer from magma to ice causes rapid melting and large volumes of meltwater may rapidly form. Jökulhlaups commonly occur soon after eruptions start, since conditions for ponding of water at the glacier base usually do not exist at the eruption sites. Exceptions to this are eruptions into pre-existing subglacial lakes like Grímsvötn. The most conspicuous landforms of subglacial volcanism are the steep-sided tuyas (table mountains) and ridges mainly composed of pillow lava and hyaloclastite, formed in eruptions under ice during Pleistocene glaciations. Many of the tuyas appear to be monogenetic and some are very large (up to 50 km³). Thin and wide pillow lava formations that have flowed subglacially some kilometres from the vents have been described in Cental Iceland. Although most of the hyaloclastite formations are basaltic, rhyolitic ridges, mounds and tuyas also occur.

6.1 INTRODUCTION

Iceland was glaciated to a varying degree throughout the Pleistocene and at present 11% of the island is ice covered (e.g. Björnsson, 1979). During the maximum extent of Pleistocene glaciations Iceland may have been almost completely covered by ice sheets (e.g. Thoroddsen, 1905-06; Hoppe, 1982; Norðdahl, 1990). This implies that subglacial volcanism has for long periods been the main form of volcanic activity; in some parts of the volcanic zones the landscape is dominated by hyaloclastite ridges, mounds and tuyas (table mountains) formed in subglacial eruptions during the Pleistocene (e.g.

Kjartansson, 1959; Sæmundsson, 1979; Chapman *et al.*, 2000). Since the time of settlement of Iceland about 1100 years ago subglacial volcanic activity has been an important part of volcanism in Iceland with frequent eruptions in Vatnajökull and Mýrdalsjökull (Þórarinsson, 1974b, 1975; Larsen *et al.* 1998; Larsen, 2000). These eruptions have produced numerous tephra layers preserved in the soil outside the glaciers and in the glaciers themselves (Larsen *et al.*, 1998). The eruptions have also caused large jökulhlaups, both directly when meltwater was drained from an eruption site to the edge of a glacier and also indirectly from the subglacial caldera lake at Grímsvötn, when meltwater accumulated in the lake before its release in swift high-magnitude jökulhlaups (Þórarinsson, 1974b; Björnsson, 1988; Guðmundsson and Björnsson, 1991). Very high magnitude jökulhlaups have occurred repeatedly in historic times, accompanying eruptions in Katla, Mýrdalsjökull (Þórarinsson, 1975; Tómasson, 1996; Larsen, 2000). The large sandur plains in southern Iceland are considered to have formed mainly by sediment deposition in repeated high magnitude jökulhlaups caused mainly be volcanic activity in Mýrdalsjökull and Vatnajökull (e.g. Maizels, 1991). In addition, ice melting by volcanic activity has occurred at partly ice-covered stratovolcanoes, such as Öræfajökull, Eyjafjallajökull and Hekla (Þórarinsson and Sæmundsson, 1979; Kjartansson, 1951). Prehistoric Holocene jökulhlaups, considered to have been caused by volcanic activity in Vatnajökull, carved out the large canyon of Jökulsá á Fjöllum in north Iceland (e.g. Tómasson, 1973).

A standard approach in the study of volcanic eruptions is to compare observations during eruptions with observations of the deposits left by the eruptions. This also applies to subglacial volcanism but the connections between processes and deposits are less direct than for most subaerial eruptions. This is because of concealment of subglacial edifices by the ice cover in present-day eruptions. Research into the processes and products of eruptions within glaciers can be broadly divided into three categories. These are, firstly, direct monitoring of present-day eruptions under glaciers, secondly, geological mapping of exposed and partly dissected volcanoes formed under past glaciers, and thirdly, theoretical studies of various processes occurring during subglacial erptions.

The study of present-day eruptions under glaciers provides fundamental data on how eruptions and glaciers interact and rates of processes can be derived from direct observations. Parameters that can be studied include heat transfer, ice melting and deformation, drainage of water from eruption sites, and jökulhlaups resulting from eruptions (e.g. Guðmundsson and Björnsson, 1991; Guðmundsson *et al.*, 1997, 2003; Björnsson, 2001, Smellie, 2002). However, the subglacial volcanic products are usually poorly exposed and internal stratification can only be inferred from indirect methods such as geophysical prospecting and constraints provided by melting rates (e.g. Guðmundsson *et al.*, 2002; Behrendt *et al.*, 1995). Comparisons with observations of subaqueous and emergent eruptions such as the Surtsey eruption in 1963-1967 should also be included in this category since they provide important constraints on eruption mechanisms, internal structure and post-eruption processes such as palagonitization (e.g. Þórarinsson *et al.* 1964; Jakobsson, 1978; 1979; Jakobsson and Moore, 1982).

The exposed hyaolclastite ridges, mounds, tuyas and other features formed in volcano-ice interaction during earlier periods of more extensive glacial cover have been the main source of information on the morphology and form of volcanic activity under glaciers (Kjartansson, 1943, 1959, 1964; Mathews, 1947; Bemmelen and Rutten, 1955; Sigvaldason, 1968; Jones, 1969, 1970; Nelson, 1975; Moore and Calk, 1991; Smellie *et al.*, 1993; Skilling, 1994, 2002; Werner and Schmincke, 1999, Smellie, 2000). Stratigraphic mapping and lithofacies analysis of individual units that make up the subglacially-formed volcanoes provide the basis of our present understanding of subglacial volcanism. However, such studies do not provide good constraints on the rates of processes, and limited information can be obtained on the glacier response and meltwater drainage.

Analysis of processes during subglacial eruptions based on physical principles, material properties of magma, tephra, ice and water yield important insight. Most such studies have been on heat transfer in subglacial eruptions (Einarsson, 1966; Allen, 1980; Höskuldsson and Sparks, 1997; Wilson and Head, 2002; Guðmundsson, 2003; Guðmundsson *et al.*, 2003) and studies of magma-water interaction, including laboratory experiments, provide understanding of magma fragmentation processes (e.g. Wohletz, 1983; Zimanowski, 1998). The influence of ice rheology on eruption behaviour has received less attention although it has been touched upon by some workers (Einarsson, 1966; Tuffen *et al.* 2002; Guðmundsson *et al.*, 2003). Much work remains to be done in this field.

The aim of this paper is to give a brief overview of subglacial volcanic activity in Iceland. The largest section describes recent eruptions; the space allocated to this topic reflects the fact that data from the Gjálp eruption in 1996 and other recent volcanic activity has thrown new light on ice-volcano interaction and considerable research on various aspects of subglacial eruptions has taken place (e.g. Guðmundsson *et al.* 1997, 2002, 2003; Höskuldsson and Sparks, 1997; Björnsson *et al.*, 2001; Wilson and Head, 2002; Smellie and Chapman, 2002). Historic and pre-historic volcanic activity within glaciers in Iceland is briefly mentioned and a short overview of Pleistocene subglacial formations is given.

6.2 PRESENT-DAY AND RECENT SUBGLACIAL VOLCANIC ACTIVITY

6.2.1 Subglacial and emergent/subaerial eruptions

The term subglacial eruption has been used rather loosely to describe eruptions within glaciers. However, very few exclusively subglacial eruptions, i.e. where the vents remained under ice cover during the entire eruption, have been identified. Most historically known eruptions within glaciers break through the ice cover with an emergent, subaerial, usually surtseyan phase, producing airborne phreatomagmatic tephra (e.g. Larsen, 1998). In some cases the subaerial explosive eruption is a major part of the overall activity with a sizeable part of the volcanic material being transported out of the vent area through an eruption plume and dispersed as tephra over surrounding areas. This applies to several Grímsvötn eruptions in recent times (Þórarinsson, 1974b; Guðmundsson and Björnsson, 1991). In contrast, the Gjálp eruption in 1996 was mostly subglacial; only a few percent of the material erupted was carried by the plume (Guðmundsson *et al.*, 1997).

In an exclusively subglacial eruption the erupted material is completely concealed by the ice cover. This implies that it may not be easy to define whether a particular event is caused by a small subglacial eruption or local increase in hydrothermal activity without magma erupted onto the glacier floor. An event of this type may include a sudden jökulhlaup coupled with subsidence of an ice cauldron that did not exist before. Several eruptions are known to have occurred within the Grímsvötn caldera in Vatnajökull (Þórarinsson, 1974b; Grönvold and Jóhannesson, 1984; Guðmundsson and Björnsson, 1991), a large depression in western central Vatnajökull with a subglacial lake covered by a 150-300 m thick ice shelf (e.g. Björnsson, 1988; Guðmundsson *et al.*, 1995). The problem of which events should be regarded as volcanic eruptions within the Grímsvötn area has been addressed by defining the effects of confirmed eruptions (Guðmundsson and Björnsson, 1991). These effects were: (1) layers of tephra were spread over the ice surface, (2) openings were melted in the glacier directly above the vent/vents, (3) exposed open water or a depression in the ice surface persisted for years after the eruption as a result of increased heat flux, and (4) a pile of volcanic material is built on the pre-eruption bedrock at the eruption site.

Although these criteria have proved useful to understand observations in Grímsvötn, it is clear that an exclusively subglacial eruption cannot satisfy (1) and (2). Moreover, other indications such as seismic tremors may suggest the occurrence of a small subglacial eruption. However, since bursts of volcanic tremor are known to occur in volcanic areas without being associated with eruptions (e.g. Leet, 1988) such events, occurring without further supporting evidence, cannot be regarded as reliable signs of eruptions under glaciers.

6.2.2 Ice-volcano interaction in the 20th century

Fig. 6.1 shows the part of Iceland where historical subglacial eruptions have taken place while Table 6.1 lists known eruptions within glaciers and eruptions that have caused jökulhlaups in the 20th century. The total number of volcanic eruptions in Iceland in the 20th century is not exactly known, not least because of uncertainty in classifying events within Vatnajökull prior to 1950 (e.g. Þórarinsson, 1974b; Björnsson and Einarsson, 1990). However, a likely number for 20th century eruptions is 40-45, including 8 minor eruptions in Askja in 1921-1929 (Sigvaldason *et al.*, 1992) and the 9 small individual eruptions during the Krafla fires in 1975-1984 (Einarsson, 1991a). Out of the remaining 25 or so, 16 are listed in Table 6.1. This demonstrates how large a role subglacial eruptions and ice-volcano interaction plays in volcanic activity in Iceland at present. The 20th century was in no way unusual; on the basis of tephrochronology and written contemporary sources, Larsen (2002) concluded that eruptions within glaciers account for more than half of known eruptions in historical times.

The largest events within glaciers are the Katla eruption in 1918, and the eruptions at Gjálp in 1996 and 1938. Out of the 16 listed eruptions, only 3 are exclusively subglacial: the Gjálp eruption in 1938 and the two minor eruptions in Mýrdalsjökull (Katla) in 1955 and 1999. One stratovolcano eruption is included in Table 6.1, that of Hekla in 1947. Melting of glaciers and snow fields led to a jökulhlaup in the river Rangá (Kjartansson, 1951), probably caused by melting by pyroclastic surges or flows during the short-lived plinian phase at the beginning of the eruption (Höskuldsson, 2000).

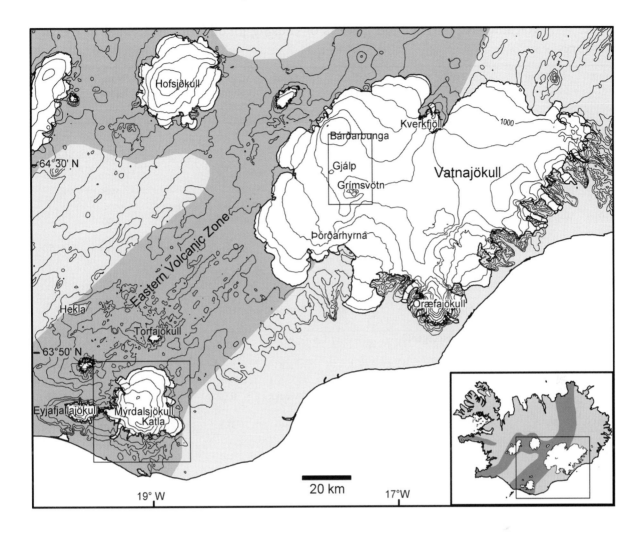

Fig. 6.1 Southeast and central Iceland. Most eruptions within glaciers in Iceland in historic times have occurred within the ice caps of Vatnajökull and Mýrdalsjökull. Ice-volcano interaction has also occurred at Öræfajökull, Eyjafjallajökull and Hekla. Pleistocene subglacial volcanic activity occurred throughout the volcanic zones. The areas covered in the maps on Figs. 6.2 and 6.6 are indicated.

6.2.2.1 Gjálp 1996

The eruption was preceded by an earthquake swarm in Bárðarbunga, beginning 33 hours before the onset of the continuous tremor that signalled the start of the eruption (Einarsson *et al.*,1997). The eruption site was midway between the central volcanoes of Bárðarbunga and Grímsvötn, in an area where the ice thickness varied from 550 to 750 m, close to the divide between the water drainage basins of Grímsvötn, the eastern Skaftá Cauldron and Dyngjujökull (Fig. 6.2). During the first 36 hours, two cauldrons in the ice surface were formed. These cauldrons were about 2 km wide and 100 m deep after 16 hours of eruption. The eruption broke through the 600 m thick ice in 31 hours and an emergent surtseyan type eruption commenced (Fig. 6.3). The erupted tephra was basaltic icelandite (Steinþórsson *et al.*, 2000). During the second day of the eruption, a new cauldron to the north of those formed earlier began to develop, implying that the eruptive fissure had by then reached a length of 6 km. The force of the eruption

Table 6.1. Volcanic eruptions within glaciers or causing jökulhlaups in Iceland in the 20th century

Year	Location	Type	Initial ice thickness (m)	Duration	Magma volume (km³)	Volume meltwater (km³)	Jökul-hlaup	References
1902-1903	Dyngjuháls *	—	—	—	—	—	yes	1, 2
1903	Þórðarhyrna*	subgl.-emergent	—	—	—	—	yes	1
1910	W-Vatnajökull*	subgl.-emergent	—	several weeks	—	—	no	1
1918	Katla	subgl.-emergent	~400	3 weeks	~1	1-8	yes	3, 4, 5
1919	Grímsvötn	subgl.-emergent	—	—	small	—	no	6
1922	Grímsvötn	subgl.-emergent	100-200	3 weeks	~0.1	—	yes	1, 7
1927	(Esjufjöll?)	subgl.-emergent?	—	—	—	—	yes	2
1933	(Grímsvötn?*)	subgl.-emergent?	—	—	—	—	no	8
1934	Grímsvötn	subgl.-emergent	100	2 weeks	0.03	~0.05	no	1, 9, 10
1938	Gjálp	subglacial	500-600	—	0.3	2.5	yes	1, 10
1947	Hekla	surface melting of icefields	0	—	—	0.003	yes	11
1955	Katla**	subglacial?	400-450	a few hours	~0.003	0.03	yes	12, 13, 14
1983	Grímsvötn	subgl.-emergent	100-150	6 days	0.01	0.01	no	15
1984	(Grímsvötn?*)	subglacial?	—	1 hour?	—	—	no	2, 10
1996	Gjálp	subgl.-emergent	600-750	13 days	0.45	4	yes	16, 17
1998	Grímsvötn	subgl.-emergent	50-150	10 days	0.05	0.15	no	18
1999	Katla**	subglacial?	450	a few hours	~0.002	0.02	yes	—

* eruption site unknown (?) Uncertain eruption ** Draining of a geothermally formed cauldron a possible alternative

References:
1: Þórarinsson (1974b)
2: Björnsson and Einarsson (1990)
3: Sveinsson (1919)
4: Þórarinsson (1975)
5: Tómasson (1996)
6: Larsen and Guðmundsson (1997)
7: Þorkelsson (1923)
8: Jóhannesson (1984)
9: Áskelsson (1936)
10: Guðmundsson and Björnsson (1991)
11: Kjartansson (1951)
12: Þórarinsson and Rist (1955)
13: Tryggvason (1960)
14: Björnsson et al. (2000)
15: Grönvold and Jóhannesson (1984)
16: Guðmundsson et al. (1997)
17: Einarsson et al. (1997)
18: Guðmundsson et al. (2000)

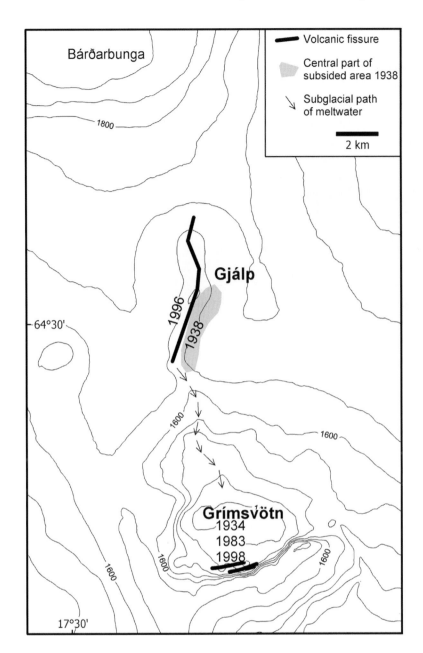

Fig. 6.2 Ice surface map of the Grímsvötn-Gjálp area in Vatnajökull. The eruptions in Grímsvötn in 1934, 1983 and 1998 all occurred under the southern wall of the Grímsvötn caldera. Grímsvötn acted as a reservoir for meltwater from the 1996 Gjálp eruption. The exact location of the fissure in 1938 cannot be determined with accuracy but the shaded area coincides with the location of depressions observed in 1938 and a subglacial ridge found in radio-echo soundings (Björnsson, 1988; Guðmundsson and Björnsson, 1991). Location of sections in Figs. 6.3 and 6.4 are shown.

diminished after the first four days and the eruption ended on October 13. Water drained from the eruption site towards the south, into the ice covered caldera lake at Grímsvötn (Fig. 6.2). The average melting rate over the first four days was 5000 $m^3 s^{-1}$ and data on water level in Grímsvötn and ice melting suggest continuous drainage throughout the eruption into Grímsvötn (Guðmundsson *et al.*, 1997, 2003). On October 13, about 3 km^3 of ice had melted and all the meltwater accumulated in Grímsvötn. The meltwater was released in a sudden jökulhlaup on November 5, three weeks after the end of the eruption.

The Gjálp eruption demonstrated an extremely fast rate of heat transfer from magma to ice. The heat flow rate was of order 5-6x10^5 W m^{-2} (Guðmundsson *et al.*, 2003), too high to be explained by cooling models of pillow lava (Höskuldsson and Sparks, 1997; Wilson and Head, 2002) but consistent with fragmentation of the magma to pyroclastic

Fig. 6.3 The Gjálp eruption on its third day. The surface was heavily crevassed and covered with tephra. The depression was 2-3 km wide at this stage of the eruption.

glass (Guðmundsson *et al.* 1997, 2003). The exposed top of the edifice, observed in June 1997, was primarily made of volcanic glass with a particle size ranging from 0.002 mm to several tens of mm (fine ash to lapilli) with the mean being 2 mm (Guðmundsson, 2003).

The volume of ice melted along the subglacial path of the meltwater, indicated that the temperature of the meltwater, as it left the eruption site, was 15-20°C (Guðmundsson *et al.*, 2003). Another observed phenomenon was surface drainage along a canyon, 150 m deep and 3.5 km long, formed in the ice surface above the central and southern parts of the volcanic fissure.

Flow of water along the glacier bed during the Gjálp eruption was down the gradient of the static fluid potential, defined on the basis of pre-eruption ice surface and bedrock topography (Björnsson, 1988; Björnsson *et al.*, 1992; Guðmundsson *et al.*, 2003). Thus, the pathways of meltwater along the bed were mainly controlled by the initial ice overburden pressure, in agreement with the theory (summarised by Paterson, 1994, pp. 110-114) which predicts that slopes in bedrock topography need to be about an order of magnitude greater than ice surface slopes if they are to control the flow of water along the bed. The applicability of the static fluid potential to large temperate ice caps (Björnsson, 1988) is of considerable importance, since it can be used to predict pathways of volcanic jökulhlaups in areas where ice surface and bedrock topography is known.

The ridge formed in the eruption was mapped in 1997-2000 with radio-echo soundings, gravimetry, and direct observations of the top part while it was exposed in June 1997 (Guðmundsson *et al.*, 2002). The ridge is 6 km long and up to 450 m high (Figs. 6.3, 6.4) and has a volume of about 0.7 km³. A further 0.07 km³ were transported with the meltwater into Grímsvötn. The total volume of erupted material was 0.45 km³ (dense

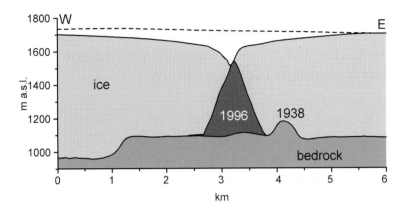

Fig. 6.4 *Cross-section over Gjálp after the eruption in 1996 (based on Guðmundsson et al., 2002). The ridge formed rises about 450 m above the pre-eruption bedrock. The top of the ridge was exposed for about a year after the eruption when it was buried by inflowing ice. The northern end of the ridge considered to have formed in the Gjálp 1938 eruption is indicated.*

rock equivalent) (Guðmundsson *et al.*, 2002). Gravity modelling suggests that the ridge is mainly composed of hyaloclastite, not pillow lava. The edifice is highest at the site of the subaerial crater near the centre of the fissure. The southern and central part is a steep, narrow ridge. In the northern part, where ice thickness was up to 750 m before the eruption, the ridge is relatively flat and wide. It has been suggested that these contrasting morphologies reflect different conditions during the formation of the two parts of the edifice (Guðmundsson *et al.*, 2002, 2003). High water pressure may have caused partial floating of the ice above the wide, low northern part, inhibiting confinement of the evolving edifice by the surrounding ice. In contrast, lower water pressure and greater ice confinement may have inhibited spreading of the volcanic material in the central and southern parts. The formation of the ice canyon above the central and southern parts of the eruption site, and supraglacial drainage along it supports this interpretation. Subglacial flow of meltwater from the northernmost part of the eruption site towards the south may have been impeded since the newly formed 400-450 m high edifice at the centre of the volcanic fissure may have acted as a dam (Guðmundsson *et al.*, 2002, 2003). Thus, water pressure in the northernmost part rose sufficiently for the water to drain along the canyon until it disappeared into the glacier, presumably reaching the original subglacial pathway of the meltwater at the glacier bed under the southern end of the canyon.

The ice-volcano interaction observed at Gjálp demonstrated a dynamic interplay between ice rheology and deformation on the one hand and the erupting volcano on the other. Firstly, a 2-3 km long crevasse observed at the southern end of the eruption site, aligned with the subsidence cauldrons and unrelated to any subsidence, can be explained by rupturing of the ice from below during intrusion of the feeder dyke. The dyke may have intruded the lower part of the glacier and formed the crevasse (Guðmundsson *et al.*, 2003). Secondly, analysis of deformation rates of the ice during cauldron formation reveals that pressure in a water vault underneath subsiding cauldrons must be considerably less than the load of the overlying ice. A part of the load appears to be carried by shear forces in the ice forming the sides of the cauldron; rough estimates suggest that this pressure reduction during early stages of subsidence may be 1-2MPa (equivalent to the load of 100-200 m of water) (Guðmundsson *et al.*, 2003). This pressure reduction may lead to early explosive activity.

6.2.2.2 Katla 1955

The name Katla traditionally refers to eruption sites in Mýrdalsjökull. Geographically Katla corresponds roughly to the eastern part of the large Mýrdalsjökull caldera (Björnsson *et al.*, 2000). Katla, together with Grímsvötn, has been the most active subglacial volcano in Iceland in historical times (Fig. 6.6), and Katla eruptions have been among the most destructive eruptions observed in Iceland, causing enormous jökulhlaups, and fallout of tephra has repeatedly caused damage to farmland (Þórarinsson, 1975; Larsen, 2000). The events that occurred at Mýrdalsjökull on June 25 in 1955 were relatively minor in comparison. Eight earthquakes of magnitude 2.4 to 3.6 were recorded from 14:32 hrs. to 19:00 hrs. GMT (Tryggvason, 1960) and a team, doing seismic depth soundings on Mýrdalsjökull in bad visibility, detected new crevasses forming in the afternoon. A swift jökulhlaup, issuing from the edge of Kötlujökull (Fig. 6.5), started at about 20:00 hrs. in the evening and reached a peak within an hour (Þórarinsson and Rist, 1955). The volume of water drained was 28×10^6 m³. Two cauldrons formed in the ice surface, the larger southern cauldron was about 1 km in diameter and 80 m deep (Rist, 1967). The volume of the cauldrons was similar to the volume of water drained in the jökulhlaup. Radio-echo soundings revealed a mound under the site of

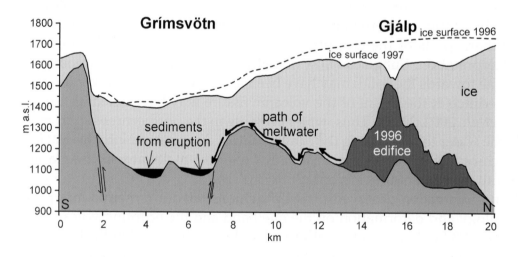

Fig. 6.5 A section through Grímsvötn and Gjálp based on radio-echo soundings and other geophysical data (after Guðmundsson et al., 2002). The cross-section on Fig. 6.3 is located over the highest central part of the ridge. The path of meltwater draining into Grímsvötn is shown. The existence of sediments accumulated in Grímsvötn is based on geophysical surveying before (Guðmundsson, 1989; Björnsson et al. 1992) and after the eruption. The volume of the sediments amount to about 0.07 km³, about 10% of total volume erupted.

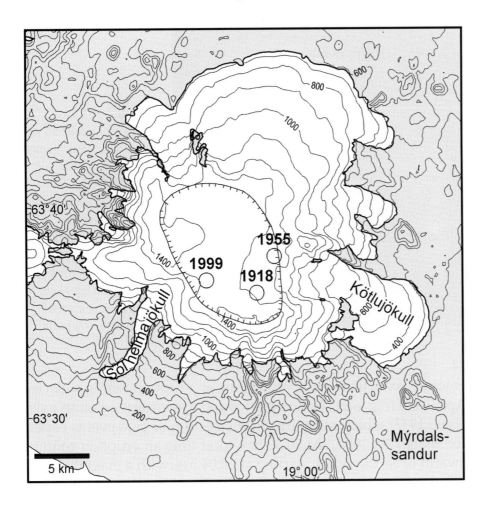

Fig. 6.6 The Mýrdalsjökull ice cap with the location of the eruption site of 1918, and the subsidence cauldrons formed in 1955 and 1999. The rim of the Katla caldera (Björnsson et al. 2000) is shown. The meltwater from the large 1918 eruption and the small, subglacial 1955 eruption drained towards east, to Mýrdalsandur while the cauldron formed in 1999 drained towards southwest, with the jökulhlaup emerging from the outlet Sólheimajökull.

the southern cauldron. The mound was 60 m high and with a diameter of about 300 m under 400 m thick ice (Björnsson *et al.*, 2000). In order to melt the 28×10^6 m³ of water, $2.5\text{-}3 \times 10^6$ m³ of basaltic magma is required; this is similar to the volume of the mound. Thus, the data are consistent with a short eruption on 25 July, 1955. Another possible explanation for the 1955 event would be more gradual melting by hydrothermal activity and storage of the meltwater at the site of the cauldrons. This explanation is unlikely, however, for the following reasons. Firstly, the earthquakes suggest considerable crustal unrest similar to that observed before the onset of several eruptions in Icelandic volcanoes (Einarsson, 1991b). Secondly, this was a single event without prior or later water accumulation of any significance. Thirdly, conditions for storage of water at the glacier bed at this location do not exist (Björnsson *et al.* 2000). A similar event occurred in Mýrdalsjökull on July 18, 1999, when a cauldron formed (Fig. 6.7) in association with a sudden jökulhlaup from the southern outlet glacier Sólheimajökull.

Fig. 6.7 A 1.5 km wide and about 50 m deep cauldron formed in Mýrdalsjökull on July 18 1999. Its formation coincided with a sudden jökulhlaup in Jökulsá á Sólheimasandi. A likely explanation for the formation of the cauldron is a small subglacial eruption.

6.2.2.3 Grímsvötn 1934

The first time that volcanic vents were observed during an eruption within a glacier was in Grímsvötn in April 1934 (Fig. 6.2). Hence, photos and eyewitness accounts exist (Áskelsson, 1936; Nielsen, 1937) which make the reconstruction of past events easier. The eruption started on March 30, 1934. At 20.30 hrs. local time an eruption column was observed and a swarm of earthquakes, starting at 20.04 hrs. with a magnitude 4.5 earthquake, was recorded on the seismometer in Reykjavík, with epicentral distance consistent with Grímsvötn (Tryggvason, 1960). It has been assumed that this earthquake swarm marked the start of the eruption (Tryggvason, 1960). The similar timing of the earthquake swarm and the sighting of a plume suggest a very short subglacial phase to this eruption. Indeed, it may well be the case that the 100-150 m of ice were ruptured by the dyke intrusion and an explosive phreatomagmatic eruption commenced without any subglacial phase. The 1934 eruption lasted 2 weeks. It melted 0.3-0.6 km wide openings in the ice shelf covering the Grímsvötn lake. Tephra was spread over most of Vatnajökull and Þórarinsson (1974b) estimated the volume of fallout away from the craters as 0.01-0.02 km³. The eruption was surtseyan in character and narrow crater rims rose 5-10 m above the water level in the openings in the ice shelf. The eruption was of moderate size, the total volume of erupted material is estimated as 0.03-0.04 km³ (Guðmundsson and Björnsson, 1991).

The 1934 eruption at Grímsvötn is considered to have been triggered by pressure release caused by lowering of the water level within the caldera. The eruption started when a jökulhlaup from Grímsvötn had been rising for 8 days (Áskelsson, 1936; Þórarinsson, 1974b). Þórarinsson´s hydrograph of the 1934 jökulhlaup, based on detailed descriptions by farmers living near the floodpath, together with data on water level and water volume in the Grímsvötn subglacial lake (Guðmundsson *et al.*, 1995), allows the fall in water level with time to be reconstructed (Fig. 6.8). The eruption commenced when the water level had fallen by 70-80 m, equivalent to a pressure release of 0.7-0.8

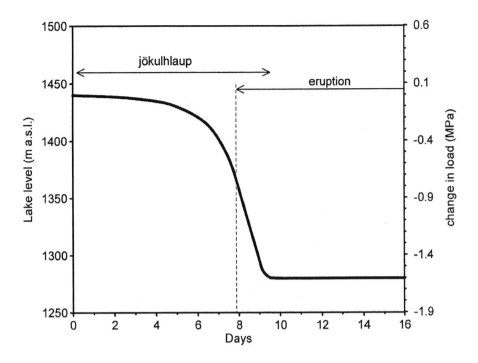

Fig. 6.8 Drop in the level of the Grímsvötn subglacial lake during the jökulhlaup in 1934, and the resulting pressure drop at the lake bottom. The subsidence curve is based on the hydrograph by Þórarinsson (1974b) and ice and water volume estimates by Guðmundsson et al. (1995). The drop in overburden pressure on the caldera floor is considered to have triggered the eruption.

MPa (7-8 bars). This triggering mechanism at Grímsvötn seems to have operated several times during the 19th and the early part of the 20th century (Þórarinsson, 1974b; Guðmundsson *et al.*, 1996; Larsen *et al.*, 1998).

6.2.2.4 Katla 1918

The eruption of Katla in 1918 is regarded as the largest subglacial eruption in Iceland in the 20th century; the total volume of erupted material has been estimated as 1.0 km^3 DRE (Tómasson, 1996). The first signs of an imminent eruption in 1918 were earthquakes at 1 p.m. on 12 October. An eruption plume was observed at about 3 p.m. At about the same time a jökulhlaup started on Mýrdalssandur, reaching a peak of ca. 3x10^5 m^{3}s^{-1} within 3 hours (Tómasson, 1996). Magma discharge seems to have been greatest in the first few hours and the eruption plume reached 14 km on the first day (Eggertsson, 1919; Larsen, 2000). The eruption lasted for 3 weeks, spreading tephra widely over surrounding areas. Sheep and horses were swept away by the jökulhlaup and many people narrowly escaped. A large part of the erupted material appears to have been carried away from the eruption site with the meltwater (Tómasson, 1996). This material was deposited on Mýrdalssandur, the path of the jökulhlaup to the ocean, and formed a 3-4 km long peninsula where the main flood channel reached the sea (Jóhannsson, 1919; Tómasson, 1996).

Some information on the eruption site and conditions on the glacier exist (Sveinsson, 1919; Sveinsson, 1992), including photographs of the floodpath and from the ice cap in June and September 1919 (Þórarinsson, 1959). The eruption site has been located on basis of these records (Sveinsson, 1919; Larsen, 2000; Björnsson *et al.*, 2000) at a place where the ice thickness is presently about 400 m. The records and photos show that the surface of the glacier in June 1919 resembled that of Gjálp in the summer of 1997: a heavily crevassed, wide depression with piles and mounds of tephra protruding through the annual snow layer in places.

If the earthquakes at 1 pm on 12 October mark the start of the eruption, the subglacial phase was very short, no more that two hours long. This implies a penetration rate an order of magnitude faster than observed in the Gjálp eruption. It is probable that the initial melting rate was also an order of magnitude higher, or $50-100 \times 10^3$ m^3 s^{-1}. Alternatively, melting rates may have been slower if geothermal heating at the site of the eruption formed a subglacial lake before the eruption. However, the ice surface and bedrock topography is such that accumulation of large volumes of water at the base of the glacier is unlikely (Björnsson *et al.*, 2000).

6.2.3 Historic subglacial eruptions before 1900

Good descriptions exist of the impact of Katla eruptions back to 1625 (Safn til Sögu Íslands, 1907-1915) and a reliable tephrochronological record of eruptions shows that 20 eruptions broke the ice cover since AD 900 (Larsen, 2000). A similar record of eruptions that produced tephra layers in soil or ice exist for Vatnajökull back to about AD 1200 (Larsen *et al.*, 1998). Historical records and annals mention many of these eruptions but the tephrostratigraphical record extracted from the ablation areas of Vatnajökull revealed 27 previously unknown eruptions, mostly for the period before AD 1600 (Larsen *et al.*, 1998). Grímsvötn has been very active, with 60 known eruptions since AD 1200 accounting for almost 75% of the total number in Vatnajökull (Larsen *et al.*, 1998).

Almost all the eruptions included in the records had an emergent/subaerial phase. The total number of eruptions within these glaciers is therefore bound be an underestimate. Small to moderate size eruptions under thick ice may not penetrate the ice cover (as suggested here for Mýrdalsjökull in 1955) and therefore not register with a tephra layer. Such eruptions may in most cases have produced jökulhlaups but historical records of these are incomplete.

Öræfajökull and Eyjafjallajökull are relatively steep stratovolcanoes. Two eruptions have occurred in historical times in each volcano, in Öræfajökull in 1362 and 1727 and Eyjafjallajökull in 1612 and 1821-23 (Þórarinsson and Sæmundsson, 1979). Little is known about the 1612 eruption of Eyjafjallajökull but the one in 1821-23 was small, with the craters located within the summit caldera under relatively thin ice. The volume of tephra was small and associated jökulhlaups were not large. The Öræfajökull eruptions were more severe, causing major jökulhlaups. The 1362 eruption was one of the largest explosive eruptions in Iceland in historic times (Þórarinsson, 1958). Annals are very brief in their descriptions. However, it is clear that the eruption devastated the district of Litla Hérað to the southwest of the volcano and many of the inhabitants may have perished in the eruption. Öræfajökull has a 500 m deep, ice filled summit caldera

(Björnsson, 1988) and the potential for ice melting and catastropic jökulhlaups due to eruptions is probably little changed since the 14th and the 18th centuries.

6.3 PREHISTORIC HOLOCENE SUBGLACIAL VOLCANIC ACTIVITY

Data on subglacial volcanic activity in Iceland prior to the time of settlement are scarce. However, tephrocronological records provide evidence on the frequency of explosive basaltic eruptions. The common view is that during much of the Holocene, Iceland was mostly ice free, with only small ice caps on highest mountains, and that the present ice caps largely formed around 2500 ^{14}C yr BP during the onset of Subatlantic time (e.g. Eyþórsson, 1950; Björnsson, 1979).

The well-studied tephrocronology of the area around Mýrdalsjökull indicates that the ice cap has existed throughout most of the Holocene. Numerous prehistoric tephra layers in soil around Mýrdalsjökull suggest that water-magma interaction has been the dominant style of activity in this predominantly basaltic central volcano as far back as soil records exist (Larsen, 2000; Þórarinsson, 1975). This is a strong indicator of an ice cover. Moreover, studies of moraines formed by the outlet Sólheimajökull, show that it was more extensive in mid-Holocene than at present (Dugmore and Sudgen, 1991).

Analysis of the bedrock morphology of western Vatnajökull, which is a part of the volcanic zone, suggests that subglacial volcanic activity, producing ridges and mounds, has been the dominant force in forming the present subglacial landscape. Effusive subaerial eruptions, producing flat or gently sloping lava fields, have only had an impact on the presently subglacial landscape near the northwestern and northern margins (Langley, 2000). It is not clear whether this implies that an ice cap has existed for most of Holocene or, alternatively, that subaerial volcanism was not sufficiently vigorous during ice-free periods to reshape the landscape formed during the Pleistocene.

6.4 PLEISTOCENE SUBGLACIAL VOLCANIC ACTIVITY

6.4.1 The Moberg formation
The Pleistocene ice sheets in Iceland repeatedly extended beyond the present coast during their maximum extent (e.g. Hoppe, 1982; Norðdahl, 1990; Geirsdóttir and Eiríksson, 1994). Models relating glacier thickness with their size (e.g. Paterson, 1994) indicate that ice sheets of such extent may have been 1.5-2 km thick in the cental highlands. During glaciations eruptions under ice produced mainly pillow lavas, hyaloclastite tuffs and breccias (e.g. Kjartansson, 1959; Sæmundsson, 1979; Chapman *et al.*, 2000). A collective term for subglacially formed volcanics during the Upper Pleistocene (0.01-0.78 Ma) is the Moberg Formation (Kjartansson, 1959). The formation covers about 10-11 thousand km^2 within the volcanic zones, not counting presently ice-covered areas or subglacially-formed rhyolites (Chapman *et al.*, 2000; Jóhannesson and Sæmundsson, 1998). The widespread occurrence of hyaloclastites, pillow lavas and breccias in formations from Upper Pliocene and Lower Pleistocene (0.78-3.3 Ma) testifies to the intermittent existence of ice cover in Iceland during this period (Jóhannesson and Sæmundsson, 1998; Geirsdóttir and Eiríksson, 1994; Sæmundsson, 1979).

The contrasting environmental conditions between ice-covered and ice-free periods have led to the build up of a volcanic strata that contrasts with the regular subaerially-formed lava pile of the Tertiary (Sæmundsson, 1979). During the Pleistocene steep-sided mountains and other volcanic units of limited aerial extent were built, forming a landscape of high relief. During interglacials subaerial lavas piled up between the hyaloclastite mountains. At the same time glacial erosion has formed U-shaped valleys, fjords, and alpine landscapes in places (e.g. Sæmundsson, 1979).

Pjeturss (1900) was the first to suggest a Pleistocene age for the palagonites of the Moberg formation. Its volcanic origin was proposed by Peacock (1926) and further advanced by Noe-Nygaard (1940). Kjartansson (1943) provided a hypothesis explaining the formation of hyaloclastite ridges and tuyas (table mountains) in subglacial eruptions. Independently Mathews (1947) proposed the same explanation for the formation of the tuyas of British Columbia. Subsequent work (Bemmelen and Rutten, 1955; Sæmundsson, 1967; Jones, 1969, 1970) firmly established these ideas on the formation of hyaloclastite mountains.

6.4.2 Tuyas and ridges

The most conspicuous subglacial volcanic landforms in Iceland are the tuyas (table mountains) and hyaloclastite ridges (Figs. 6.9, 6.10). Figure 6.11 shows a cross section of an ideal monogenetic basaltic tuya composed of four major stratigraphic units. A mound of pillow lava (1) forms the lowermost unit, erupted subglacially under high hydrostatic pressure. Individual pillows may be 0.5-1 m in diameter (Jones, 1969). As the mound rises hydrostatic pressure is reduced and explosive activity causes fragmentation of the magma into glass shards with diameters in the submillimeter to millimeter range (e.g. Werner and Schmincke, 1999). The explosive eruption builds a tuff cone (2) that may melt its way through the ice sheet and emerge above water level. If the eruption continues beyond this stage access of water to the vent may stop and an effusive eruption commence leading to the formation of a lava cap (4). The lavas advance into the surrounding meltwater lake on top of a foreset breccia (3) made of hyaloclastite, pillows and pillow fragments. The horizon/contact between the lavas and the foreset breccias is the passage zone. Dykes and irregular minor intrusions are common in subglacial volcanoes although being a minor component volumetrically. These intrusions are probably mostly formed during the eruption itself and are commonly seen where erosion has exposed parts of the interior of tuyas or ridges.

Most subglacial eruptions of the Pleistocene did not last long enough or were of insufficient size to produce complete tuyas composed of all four main stratigraphic units. Most hyaloclastite ridges (termed tindar by Jones, 1969) do not have subaerially-erupted lavas or foreset breccias. Many ridges contain units (1) and (2) while some are solely composed of pillow lava and others may only have hyaloclastite tuffs. A complete tuya should therefore be regarded as an end member of a spectrum with formations composed exclusively of pillow lavas at the other end of the spectrum. The ridges are the subglacial counterparts of subaerial fissure eruptions and lava fields. The number of individual ridges within the Moberg formation may well exceed 1000-1200 (Chapman et al., 2000). The length of ridges may range from 1-2 to tens of kilometres, they are commonly 1-2 km wide and a few hundred metres high (e.g. Chapman et al., 2000;

result suggests that magma production is reduced during glaciations, while an order of magnitude increase in melt generation occurs during a relatively brief deglaciation period.

6.4.7 Tuya height and ice thickness

The lava caps on tuyas testify to subaerial effusive activity above the water level in an englacial lake within the Pleistocene ice sheets or glaciers. The height of this water level must be related to the total thickness of the ice sheets. When the water level in an englacial lake reaches about 90% of the ice thickness within a temperate glacier, the ice must be floated and water escape. However, evidence from draining of ice-dammed lakes in Iceland shows that such lakes are drained at a water level several tens of metres lower than this theoretical floating limit (Björnsson, 1988). At Gjálp the water level at the vents was 150-200 m below the original ice surface, reaching only 75-80% of the original ice thickness (Guðmundsson *et al.*, 2003). It is therefore likely that the level of englacial lakes in which tuyas formed is some 100-200 m below the pre-eruption ice surface. Tuyas therefore give an indication of ice thickness at the time of their formation, but reconstructions of ice sheet thicknesses must be regarded as very approximate.

Walker (1965) used the height of the largest tuyas in the Northern Volcanic Zone to reconstruct an ice thickness profile from south to north. In Fig. 6.13 the heights of 27 tuyas in the volcanic zones are plotted as a function of distance from the central east-west trending ice divide. This hypothetical ice divide is here placed in the centre of the highlands, approximating the main ice divide of the Weichselian ice sheet (Kaldal and Víkingsson, 1990). The graph shows that the highest tuyas rise some 1200 m above their surroundings and they all occur in the central highlands. The maximum height decreases with distance from the centre. However, tuyas of various heights exist in all areas, indicating widely different ice sheet thicknesses at the time of formation of individual tuyas. This may reflect differences in ages of tuyas. Further, this suggests that tuya heights are of limited value to reconstruct size and thickness of former ice sheets in the absence of absolute ages of individual tuyas.

6.5 DISCUSSION

6.5.1 Water drainage and jökulhlaups

Jökulhlaups appear to be an integral part of subglacial eruptions, occurring when the meltwater created in eruptions escapes from underneath the glacier. Guðmundsson *et al.* (2003) point out that conditions for ponding of water only exist where an eruption occurs within a pre-existing subglacial lake such as Grímsvötn. This is consistent with the 20th century record which shows that eruptions within the Grímsvötn caldera have not led to immediate drainage of the meltwater (Þórarinsson, 1974b; Björnsson, 1988). The effect of eruptions within Grímsvötn on lake level are small since melting of floating ice does not raise the lake level, the only rise that may happen occurs to make space for the volcanic material (Björnsson, 1988). In virtually all other cases initial conditions are such that meltwater will start to migrate along the base towards the glacier edge. Initial

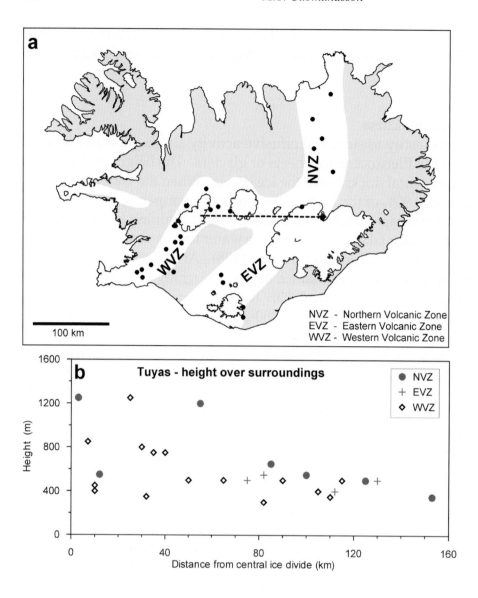

Fig. 6.13 Tuya height as a function of distance from the approximate location of a central ice divide of the Weichselian ice sheet (Kaldal and Víkingsson, 1990). The height is obtained by subtracting a low point in the elevation of the surroundings from the absolute height. This should minimise the effect of postglacial lavas that conceal the roots of the many tuyas.

temperature of the meltwater will greatly enhance the opening of pathways, as demonstrated in Gjálp.

In all reported cases the drainage away from the eruption site was subglacial (Fig. 6.14). In the cases where observations exist, the flow of water from the eruption site until a few kilometres away from the glacier snout, has been subglacial. It should be noted however, that details of glacial drainage in Katla 1918 are poorly known. Supraglacial flow occurred on Kötlujökull outlet glacier a few kilometers within the snout (Tómasson, 1996) but whether any water ever flowed along the surface right from the eruption site is unclear. Surface flow of meltwater occurred at the eruption site itself in Gjálp during the latter stages of the eruption. This supraglacial water flow was confined to the subsidence cauldrons, all drainage away from the eruption site itself remained subglacial.

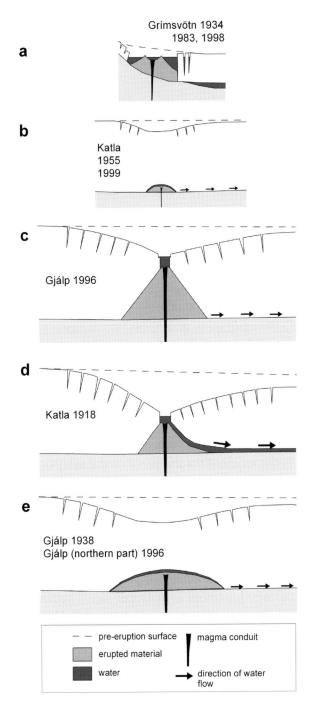

Fig. 6.14 Schematic drawings of conditions in the 20th century eruptions at Grímsvötn, Gjálp and Katla. The drawings are approximately to scale, vertical exaggeration 1.5-2.5. It is likely that an edifice formed in Katla in 1918 was proportionally smaller than in Gjálp 1996 due to very effective removal of erupted material by meltwater in 1918. The volume of meltwater present at the eruption sites may have been greater at some stages than shown, especially for the small subglacial Katla eruptions and the 1938 Gjálp eruption. The pathways of meltwater away from the eruption sites were subglacial in all casws.

The Pleistocene ridges, tuyas and pillow sheets provide limited evidence on style of drainage. If the glacier was warm-based, a more or less continuous drainage may have been the most common style of activity, at least during early stages of eruption. The continuity of the passage zone on many tuyas testifies to a semi-stable water level for considerable periods of time during construction of the volcanoes. How an englacial lake enclosing an erupting volcano can maintain such stable water levels has not been adequately explained.

6.5.2 Ice thickness and eruption behaviour

The effect of ice thickness at eruption sites suggests that subglacial eruptions may be broadly divided into two categories. Where ice thickness is relatively small such as in Grímsvötn (Fig. 6.14), the ice seems to react mainly in a brittle fashion, i.e. with fracturing and collapse of the ice associated with melting but ductile flow is relatively minor. Eruptions that start under thin ice quickly penetrate the ice and initial ice melting is largely confined to forming an opening in the ice cover and then a gradual widening of this opening. Total melting in such eruptions will be relatively small (Table 6.1). Another feature of these thin-ice eruptions is that the ice may act like any other earth material, with cones or crater rims being partly built on and against the ice walls, as observed in Grímsvötn (Fig. 6.14). The above description seems to apply to eruptions that start under ice that is at least 100-150 m thick.

For thicker ice (a few hundred metres) the response of the glacier to subglacial melting and drainage is fast flow of the ice by internal deformation into the developing depression. Brittle fracturing also occurs, as testified by intense crevassing of the top layer of the glacier. Estimates of the strain rates observed at Gjálp in 1996 suggest the this brittle top layer was some 150 m thick (Guðmundsson *et al.*, 2003). Below this brittle layer the ice deforms plastically. This fast ice flow towards the underlying volcanic vents delays penetration of the glacier by the eruption and the subsequent loss of heat to the atmosphere through an eruption plume. Thick-ice eruptions therefore are more effective at ice melting and produce more meltwater than do eruptions that quickly break through the ice cover.

There is probably no fixed ice thickness where a transition from thin ice to thick ice conditions occurs. It is likely that this boundary is dependent on the force of the eruption, a high magma discharge should lead to faster melting and higher strain rates in the ice which in turn increases the thickness of the brittle upper part. However, for the types of eruptions observed in the 20th century (Table 6.1) this transition probably requires ice that is at least 200 m thick.

Eruptions on steep stratovolcanoes where melting of glaciers and snow fields on the slopes occur would mainly belong to the category of thin-ice eruptions. However, they may fall into a different category to that of the ice cap/ice sheet eruptions, since surface melting by pyroclastic surges or flows may be a dominant factor, at least for Hekla eruptions. This is a common type of ice-volcano interaction in many volcanic areas throughout the world (e.g. Major and Newhall, 1989).

6.5.3 Pleistocene and present-day activity

The majority of recent eruptions have occurred within central volcanoes. The craters and other proximal deposits are initially unconsolidated tephra and they may be partly built on ice which makes the potential for preservation limited. These deposits may be removed by glacier erosion. The identification of such deposits in the Pleistocene strata may therefore be difficult. The hyaloclastite ridge formed in Gjálp appears to be similar to many Pleistocene ridges and is the best example linking present-day eruptions with a common type of Pleistocene activity. No tuyas are known to have been formed under present-day glaciers. Björnsson and Einarsson (1990) report a subglacial tuya-like mountain under Dyngjujökull in northern Vatnajökull but it is not known whether it is Pleistocene in age or younger.

6.5.4 Concluding remarks

Eruptions within glaciers and the associated jökulhlaups will in the near future continue to be a hazard in areas around Mýrdalsjökull and Vatnajökull. The need for a more complete understanding of this hazard and how to mitigate it, adds impetus to research in this field. Studies of subglacial volcanism have advanced in recent years both from direct observations of eruptions, mapping of deposits formed in subglacial eruptions during past glaciations, and by improved understanding of the physics of ice-volcano interaction. However, much remains to be done. Detailed geological mapping of the hyaloclastite formations will advance further our understanding of ice-volcano interaction. Radiometric dating of tuyas may become a useful tool in defining the extent and thickness of Pleistocene glaciers. Studies of volatiles in volcanic glass apparently hold great potential to infer confining pressure and hence ice thickness at the time of eruption (Dixon *et al.*, 2002). Further work is also required on the complex interaction of ice melting, ice rheology, heat transfer, subglacial hydrology and eruption mechanism. Laboratory studies of magma fragmentation (e.g. Zimanowski, 1998) should also provide useful constraints.

ACKNOWLEDGEMENTS

Helpful comments by Guðrun Larsen and the reviewers Fiona Tweed and Ármann Höskuldsson improved the quality of the paper. Fruitful discussions with Helgi Björnsson, Sveinn Jakobsson, Freysteinn Sigmundsson, Ian Skilling and John Smellie on various aspects of the interaction of volcanoes and glaciers are acknowledged. This work was supported by the University of Iceland Research Fund. Þórdís Högnadóttir helped preparing the figures.

7. Icelandic jökulhlaup impacts

Andrew J. Russell[1], Helen Fay[2], Philip M. Marren[3], Fiona S. Tweed[2], and Óskar Knudsen[4]

[1] School of Earth Sciences and Geography, Keele University, Keele, Staffordshire, ST5 5BG, UK (now School of Geography, Politics and Sociology, University of Newcastle-upon-Tyne, Newcastle-upon-Tyne, NE1 7RU, UK

[2] Department of Geography, Staffordshire University, College Road, Stoke-on-Trent, Staffordshire, ST4 2DE, UK

[3] School of Geosciences, University of the Witwatersrand, Private Bag 3, WITS 2050, South Africa

[4] Jarðfræðistofan ehf., Rauðagerði 31, Reykjavík, IS-108, Iceland

7.1 INTRODUCTION AND AIMS

Iceland contains the world's largest and best-documented active glacial outwash plains or 'sandar', which have been studied since the 19th century (Hjulström, 1954; Krigström, 1962, Boothroyd and Nummedal, 1978). Vigorous subglacial volcanic activity and the presence of numerous ice-dammed lakes, make Iceland the prime location for the study of glacier outburst floods or 'jökulhlaups' and their geomorphological and sedimentological impact (Þórarinsson, 1939; Björnsson, 1988, 1992; Tómasson, 2002). Previous reviews of Icelandic jökulhlaup impact have focussed on the relatively accessible 'sandar' or outwash plains of the country's southern coast (Maizels, 1991, 1997; Russell and Marren, 1999) and pre-date the spectacular November 1996 jökulhlaup on Skeiðarársandur, southern Iceland. However, increasing attention is being focussed on large jökulhlaup channels in other parts of Iceland, related both to modern and ancient processes. Jökulhlaup impact within Icelandic bedrock channels has so far received little attention despite the fact that such channels are abundant as sandar in Iceland (Fig. 7.1). Despite clear descriptions within Icelandic literature of jökulhlaup impact on glacier margins, only recently have there been attempts to link jökulhlaup feeder system dynamics with processes and products in the proglacial outwash system.

This review presents the latest research on the impacts of jökulhlaups generated by three main mechanisms: volcanic, ice-dammed lake drainage and glacier surge. As well as drawing attention to the most recent jökulhlaup literature, we identify and discuss the main advances in our understanding of jökulhlaup impact in Iceland.

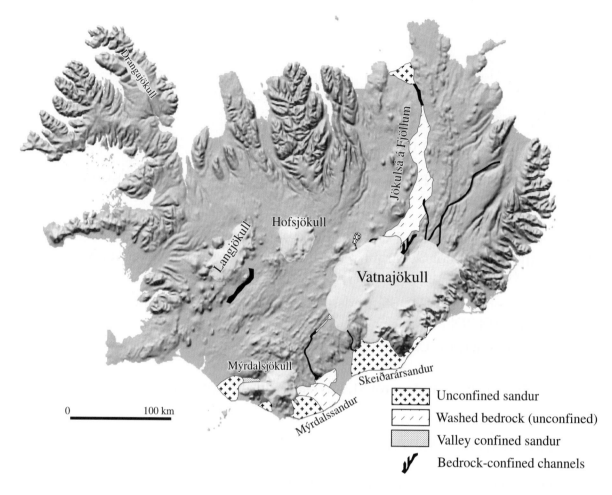

Fig. 7.1 Distribution of modern glaciers and meltwater rivers by morphological type: confined sandur, unconfined sandur, gorge, unconfined bedrock.

7.2 Morphology and distribution of Icelandic jökulhlaup systems

The distribution of jökulhlaups in Iceland can be considered over three time-frames 'modern', 'historic' and 'pre-historic' (Figs. 7.2, 7.3). 'Modern' jökulhlaups whose characteristics and impact have been observed or monitored in sufficient detail allow tight constraint of process-form relationships (Fig. 7.2). 'Historic' jökulhlaups encompass floods witnessed and subsequently documented, but do not deal specifically with either flood hydraulics or geomorphic and sedimentary impacts (Fig. 7.2). For these floods there is typically much less detailed information regarding specific hydraulic conditions or flood impacts (geomorphological or sedimentological), which instead have to be inferred from the written record. Nevertheless, these, often vivid, accounts provide a qualitative feel for former flood processes. There is good geomorphological and sedimentary evidence of 'prehistoric' jökulhlaups during the Holocene from ~10 ka ^{14}C yr BP (Fig. 7.3). Evidence of prehistoric jökulhlaups is associated with both significantly lesser and greater glacier extents than at present (Fig. 7.3).

We use the existing practice of distinguishing jökulhlaups resulting from volcanic

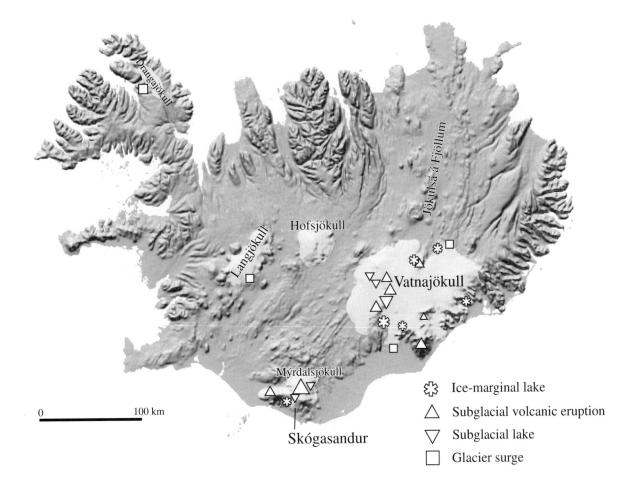

Fig. 7.2 Distribution of modern and historic jökulhlaups and flood generation mechanisms (subglacial volcanic eruption, subglacial lake, proglacial lake and glacier surge).

activity (volcanoglacial) from those generated by ice-dammed lake drainage (limnoglacial) (Þórarinsson, 1957; Björnsson, 1992; Maizels, 1991, 1997).

7.2.1 Proglacial jökulhlaup channel characteristics

Most Icelandic jökulhlaups reported in the literature are associated with relatively unconfined, distributary flow across the active sandar of the south coast. Mýrdalssandur, Skógasandur and Skeiðarársandur are type-sites for both limnoglacial and volcanogenic jökulhlaup deposits and are thought mainly to have formed by repeated jökulhlaups. The role of glacier fluctuation-generated topography on sandur morphology and sedimentology has been highlighted by many studies on Skeiðarársandur (Klimek, 1972; 1973, Galon, 1973a,b; Russell and Knudsen, 1999a, 2002a, b; Gomez *et al.*, 2000, 2002; Knudsen *et al.*, 2001a; Russell *et al.*, 2001a; Magilligan *et al.*, 2002; Marren, 2002b; van Dijk, 2002). In some cases, sandur are associated with bedrock feeder channels. At Skógasandur, jökulhlaups have been directed by a number of large bedrock gorges, which at present contain only misfit streams suggesting that the gorges were created or

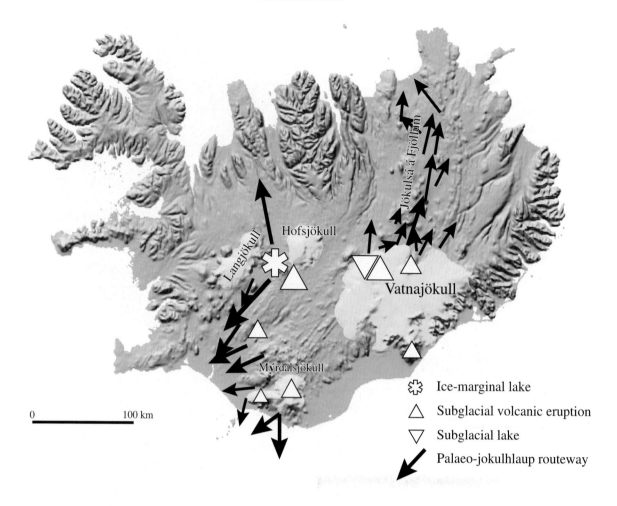

Fig. 7.3 Palaeo-jökulhlaup routeways and flood sources.

enlarged by jökulhlaups. Many modern proglacial rivers subject to recent jökulhlaups are actively incising older jökulhlaup fans. The Jökulsá á Sólheimasandi cuts into the Skógasandur fan, confining modern jökulhlaups to a relatively narrow high gradient course. Similarly, modern jökulhlaup channels in front of Kverkjökull are actively incising an older jökulhlaup fan.

Jökulhlaups have also been noted to flow across lava flows. On Mýrdalssandur, Kötluhlaups are known to have travelled across extensive lava flows (Maizels, 1993a; Larsen, 2000). Jökulhlaups in the Skaftá flow across the 1783 Laki lavas which, trap large volumes of flood transported sediment. On the northern margin of Vatnajökull, there is widespread evidence of jökulhlaups having washed over lava flows (Tómasson, 1973; Björnsson, 1992; Waitt, 2002; Russell *et al.*, 2000) (Fig. 7.4). Where valleys are plugged with lava, many jökulhlaups create and occupy channels along the valley sides where the lava margins are more susceptible to erosion (Russell *et al.*, 2000; Carrivick et al., 2002). Jökulhlaups descending the steep flanks of Kverkfjöll and Öræfajökull glaciated volcanoes have occupied both steep hillsides and complex valley networks resulting in the deposition of high gradient sediment fans upon reduction in slope gradient (Þórarinsson, 1958; Russell *et al.*, 2000). Jökulhlaups in the middle and lower

Fig. 7.4 (a, top) Jökulhlaup washed lavas in Hraundalur, Kverkfjöll. Palaeoflows of at least 50,000 m³s⁻¹ flowed directly towards the viewer. (b, bottom) High gradient, 2 km wide, jökulhlaup channel at Kistufell on the north-western flank of Dyngjujökull. Water scoured lava flows show widespread evidence of plucking and localised evidence of bedrock abrasion.

Fig. 7.5 (a, top) Cataract system in the Kistufell channel on the north-western flank of Dyngjujökull. Jökulhlaups have undermined the more competent surface layer. (b, bottom) Channel cut partially into the hillside composed of hyaloclastic rocks (top left) and younger valley filling lava flows (bottom right). Lava surfaces show evidence of both hydraulic plucking and fluvial abrasion.

reaches of the Jökulsa á Fjöllum are conveyed through deep bedrock gorges (Tómasson, 1973, 2002; Waitt, 2002) (Figs. 7.5, 7.6). Jökulhlaups have also been conveyed through many other bedrock channels of rivers such as the Skaftá, Djúpá, Hvítá, and the upper reaches of the Markarfljót (Björnsson, 1992, 2002; Tómasson, 2002). Study of the impacts of Icelandic jökulhlaups on bedrock channels has so far been restricted to the development of large canyons in the Jökulsá á Fjöllum and Hvítá river systems (Þórarinsson, 1950; Sæmundsson, 1973; Tómasson, 1973, 2002).

Fig. 7.6 (a, top) Large-scale flow expansion from a bedrock constriction 1 km downstream of Upptyppingar bridge on the Jökulsá á Fjöllum river. Note fracture control of 'normal' flows which turn through 90°. Jökulhlaup flows exiting the bedrock constriction (bottom left) expand into a 4 km wide jökulhlaup channel with flows locally upstream. (b, centre) Imbricated and clustered boulder bar downstream of the bedrock constriction. These clasts were plucked from a series of stacked lava flow units exposed in the stoss-facing cliff seen in (a). (c, bottom) Large boulder cluster in the Upptyppingar area.

7.2.2 Jökulhlaup sources

7.2.2.1 Volcanically generated jökulhlaups

Subglacial volcanic and hydrothermal activity in Iceland is frequently responsible for the direct release of jökulhlaups. There were several well-documented volcanoglacial jökulhlaups during the twentieth century from Mýrdalsjökull, Eyjafjallajökull, Hekla and Vatnajökull (Björnsson, 1992) (Fig. 7.2) and many reliable accounts made on the basis of historic floods. During volcanically-generated jökulhlaups rapid glacier melting results in the direct release of jökulhlaup water, without significant water storage. Given that one unit of lava melts roughly thirteen times the same volume of ice (Björnsson, 1988) subglacial volcanic eruptions have extraordinary potential for flood generation and impact.

Historic eruptions in ice-covered stratovolcanoes Öræfajökull, Eyjafjallajökull and Hekla have generated very rapid onset debris-laden floods (Þórarinsson, 1958; Kjartansson, 1951). The combination of steep slopes and melting ice cover during these eruptions stimulates rapid meltwater efflux and large-scale debris entrainment, often resulting in lahar-like jökulhlaup flows. The 1362 eruption in Öræfajökull generated a devastating flood that lasted less than a day, with water draining from the caldera via outlet glaciers and generating a peak flow that may have exceeded $10^5 m^3 s^{-1}$ (Þórarinsson, 1958). Less dangerous, but similarly rapid, debris-laden floods were also generated by an eruption in Öræfajökull in 1727 (Þórarinsson, 1958), in Hekla in 1845, and eruptions in Eyjafjallajökull in 1621 and 1821 (Kjartansson, 1951).

The Katla volcanic system lies under the centre of Mýrdalsjökull. Katla subglacial eruptions swiftly melt large volumes of ice producing some of the largest floods in Iceland (Þórarinsson, 1957; 1974b; Tómasson, 1996). Rapid heat transfer during Katla eruptions may be due to temporary storage of water at the start of volcanic activity, into which magma continues to erupt (Björnsson, 2002). The routing of floodwaters from subglacial eruptions in Katla has varied according to the position of the eruptive centre, ice thickness, local topography, and the magnitude of the jökulhlaup (Larsen, 2000). Kötluhlaups have mainly directed flow through Kötlujökull onto Mýrdalssandur, but floods have been known to exit Sólheimajökull to the south-west (Þórarinsson, 1975; Björnsson et al., 2000) (Fig. 7.8). Large hyperconcentrated jökulhlaups are released almost immediately following Katla eruptions, demonstrating the potential of subglacial volcanicity for the instantaneous supply of both water and sediment to these floods (Jónsson, 1982; Maizels, 1993a). In 1918, an eruption in Katla released a huge jökulhlaup on Mýrdalssandur with a peak flow at least ten times larger than the November 1996 jökulhlaup at Skeiðarárjökull, (Tómasson, 1996). The flood resulted in large-scale fracturing of Kötlujökull (see later sections), flows 60-70 m deep in places, and the entrainment of 0.5 km³ of ice blocks (Tómasson, 1996).

7.2.2.2 Ice-dammed lake drainage

Hydrothermal activity in Iceland generates a relatively large volume of meltwater. This meltwater, sometimes in combination with sub-aerial water inputs, accumulates in numerous ice-dammed lakes (Þórarinsson, 1939; Björnsson, 1992, 2002). Meltwater reservoirs form due to a combination of local topography, water supply and geothermal heat (Björnsson, 1974). Ice-dammed lakes are located in ice-marginal, supraglacial and

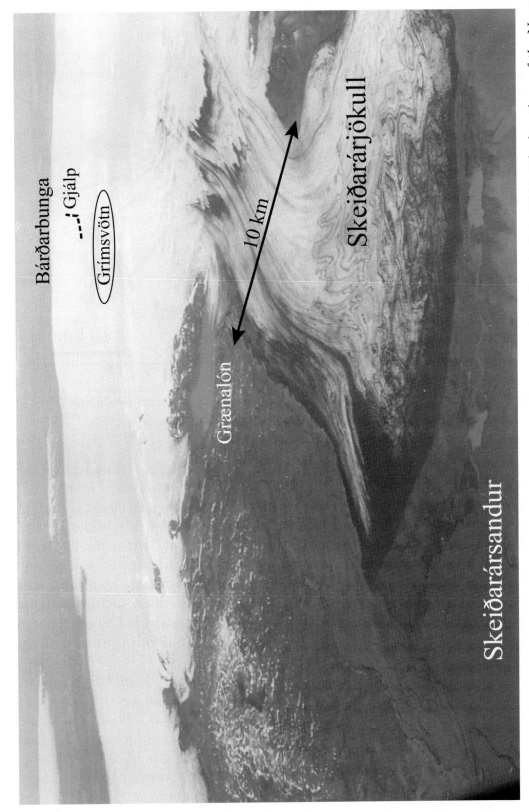

Fig. 7.7 Oblique aerial photo of Vatnajökull, Skeiðarárjökull and Skeiðarársandur taken in July 1988. The locations of the November 1996 eruption site, Grímsvötn subglacial lake and Grænalón ice-dammed lake are indicated. Bárðarbunga subglacial caldera on the northern side of Vatnajökull is believed to be the source of some of the largest Icelandic jökulhlaups during the Holocene, draining into the Jökulsá á Fjöllum river.

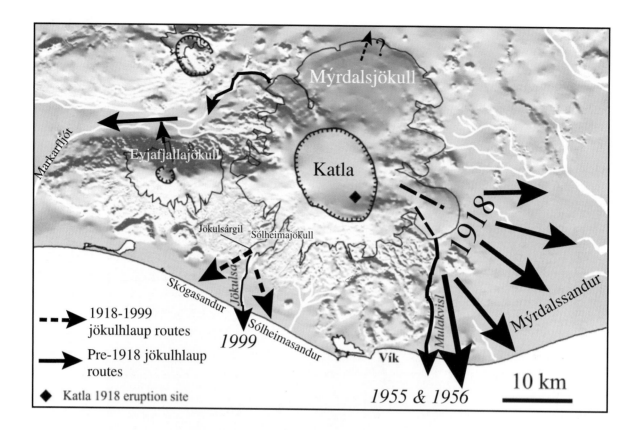

Fig. 7.8 Mýrdalsjökull and surrounding area showing historic and recent jökulhlaup routeways.

subglacial positions, with most producing periodic sudden releases of stored meltwater as jökulhlaups. Although meltwater usually builds-up steadily within many ice-dammed lake basins, large volumes of meltwater can also be supplied from nearby hydrothermal and volcanic activity (Björnsson, 1992; Tweed and Russell, 1999; Guðmundsson *et al.*, 1997). Under Vatnajökull, subglacial hydrothermal lakes exist at Grímsvötn, the Skaftá cauldrons, Pálsfjall and Kverkfjöll (Björnsson, 2002) (Fig. 7.7). Although these lakes owe their existence to subglacial hydrothermal activity, the floods that are associated with them are usually classified as 'storage-release' events. This is because such jökulhlaups tend not to be triggered by volcanic activity; their drainage is triggered by one or more of a suite of mechanisms associated with, for example, lake depth or the damming glacier's hydrological system (see Tweed and Russell, 1999).

Of all the subglacial lakes in Iceland, Grímsvötn, the largest, is probably the best documented (Björnsson, 1974, 1976, 1988, 1992, 1998, 2002). This lake is situated within the caldera floor of Grímsvötn volcano (Björnsson, 2002) and has a floating ice cover (Fig. 7.7). The magnitude and frequency (and thus, the impact) of jökulhlaups from Grímsvötn is dependent on the thickness of the ice dam (Þórarinsson, 1974b). Jökulhlaups from Grímsvötn are peculiar in that they are frequently triggered when the lake level is substantially lower than that required to float the ice dam (Björnsson, 1976, 1992), this may be due to the existence of Darcian flow near the seal (Fowler and Ng, 1996). Usually, floodwater drains through Skeiðarárjökull for a distance of approximately 50 km to showing a rheological transformation from hyperconcentrated to fluidal flow. High-pressure floodwater penetrated the glacier bed below the diamicton unit, propagating and infilling a series of hydrofractures within subglacial sediments. Hydrofracturing of diamicton was accompanied by extensive excavation of subglacial sediments within several rapidly enlarging sub glacier-bed channels. High-angled primary fluvial bedding suggests that sub glacier-bed fracture fills were deposited from a hydraulically supercooled flow. It is clear that jökulhlaup drainage circuits can extend to several metres depth within subglacial sediments. Due to excellent preservation potential, flood-filled fractures and channels within subglacial sediments may yield valuable data about ancient jökulhlaup dynamics (Russell *et al.*, 2003b).

During the 1996 and 1999 jökulhlaups, flow became progressively concentrated in larger conduits, exploiting areas of the glacier snout with lowest hydraulic potential. Meltwater elevated rapidly from great depths within the glacier during the November 1996 jökulhlaup exited the glacier ice at sub-zero temperatures (Roberts *et al.*, 2001; 2002a; Roberts 2002). Although this phenomenon known as 'glaciohydraulic supercooling' is known to occur under 'normal' conditions at a number of Icelandic glaciers (Roberts *et al.*, 2002a; Evenson *et al.*, 2001; Knudsen *et al.*, 2001b), increased water flux during jökulhlaups is thought to enhance this process. Jökulhlaups from the Skaftá cauldrons have been measured at or possibly below freezing (Jóhannesson *et al.*, 1985) and frazil ice was observed around a large fountain outlet during the 2002 Skaftárhlaup (Roberts *et al.*, 2002c). Accretion of ice and sediment under supercooled conditions during jökulhlaups is thought to throttle englacial passageways, thereby helping to maintain englacial water pressures (Roberts *et al.*, 2001, 2002a) and forcing the water to exploit new outlets. The accretion of large amounts of sediment within glacier snouts by supercooled water has major implications for glacial sediment flux and subsequent landscape development (Roberts et al., 2001, 2002a) (see later section).

Þórarinsson was the first researcher to systematically distinguish jökulhlaups from ice-dammed lakes from volcanically-derived jökulhlaups (Þórarinsson, 1939, 1953b, 1958). As well as proposing drainage trigger mechanisms for specific ice-dammed lakes (Þórarinsson, 1939, 1953b), Þórarinsson (1957) clearly identified the relationship between jökulhlaup generation mechanism and hydrograph shape, which will be explored later in this paper. Classic papers by Björnsson (1974, 1975) and Nye (1976) provided models both for the triggering of drainage from subglacial lake Grímsvötn and shape of resulting jökulhlaup hydrographs. Björnsson (1974) identified subglacial jökulhlaup routeways as well as mapping hydraulic potentials for the Grímsvötn area. Björnsson (1976) discussed mechanisms and characteristics of drainage for a number of supraglacial and ice-marginal lakes. Guðmundsson *et al.* (1995) presented new estimates of the volume of subglacial lake Grímsvötn which have allowed changes in jökulhlaup volume and peak discharge to be calculated between 1934 and 1991. Ice-dammed lake drainage mechanisms are generally well-understood, with the exception of jökulhlaups involving the drainage of hot water. Existing theories associated with exponential release of water through conduits cannot account for the rapidity of the jökulhlaup rising stage (Björnsson, 1992; Jóhannesson, 2002a,b; Roberts, 2002). Jökulhlaups with a rapid rise to peak discharge have been reported from the drainage of Grímsvötn (Þórarinsson, 1953b; Björnsson, 1974); the Skaftá cauldrons (Björnsson, 1976, 1992); a cauldron at Kverkfjöll

(Björnsson 1992; Sigurðsson *et al.*, 1992; Sigurðsson, 2002a); volcanically triggered jökulhlaups from Mýrdalsjökull e.g. 1918, 1955, 1999 (Tómasson, 1996; Þórarinsson, 1957; Rist, 1967; Russell *et al.*, 2000). Sudden increases in basal water pressure at the front of a rapidly propagating subglacial flood wave result in hydrofracturing, which in turn forces water to exit the glacier as a wave or flood surge front (Roberts *et al.*, 2000a,b, 2001, 2002a,b; Roberts 2002; Jóhannesson, 2002a,b).

7.3.2 Controls on jökulhlaup hydrograph shape

Jökulhlaup hydrograph shape has long been linked to flood generation mechanism (Þórarinsson, 1957; Rist, 1967; Nye, 1976; Maizels and Russell, 1992; Björnsson, 1992; Walder and Costa, 1996; Tweed and Russell, 1999). As outlined in the previous section, observations of several recent Icelandic jökulhlaups indicate that existing models of jökulhlaup drainage mechanism, glacier floodwater routing and conduit development are not universally applicable. Here we attempt to provide explanations for a number of distinctive jökulhlaup hydrograph shapes. We have divided jökulhlaup hydrographs into three categories, 'exponentially-rising', 'linearly-rising' and 'composite' (Fig. 7.13).

The exponential shape of the rising stage of many jökulhlaups from ice-dammed lakes can be realistically modelled by accounting for the rate of melting of a single conduit (Björnsson, 1974; Nye, 1976; Spring and Hutter, 1981; Clarke, 1982) (Fig. 7.13). Examples of exponentially-rising jökulhlaups include those from Grænalón (1935) and Grímsvötn (1934, 1954, 1972, 1986) (Björnsson, 1992). A positive feedback between conduit melting and increased discharge through the conduit results in the distinctive exponential jökulhlaup rising limb. The rate of exponential discharge increase is governed by the initial lake water temperature, conduit length, conduit roughness and hydraulic gradient. The waning stage of exponentially-rising jökulhlaups is usually more rapid, often reflecting complete lake drainage or drainage to a physical threshold. Nye (1976) explained that the sudden termination of jökulhlaups from Grímsvötn was due to the exponentially increasing rate of plastic deformation of ice as lake level drops. Gradual reduction of the waning stage of many jökulhlaups is also likely to reflect the more gradual release of jökulhlaup water from a number of temporary stores within the glacier. Fowler and Ng (1996) suggested that wide subglacial channels enlarging partly by the erosion of sediment provided better estimates of 1972 Grímsvötn jökulhlaup hydrograph shape than Nye's (1976) model, due to a better representation of conduit roughness. Jóhannesson (2002a,b) pointed out that existing models for exponentially-rising jökulhlaups cannot account for the rapid rates of tunnel enlargement witnessed recently during jökulhlaups involving the drainage of hot water.

Þórarinsson (1957) and Björnsson (1976, 1992) recognized that many jökulhlaups did not have an exponential rising stage. The November 1996 jökulhlaup from Grímsvötn focussed attention on these extraordinary 'linearly rising' jökulhlaups (Roberts, 2002; Roberts *et al.*, 2003). The November 1996 jökulhlaup from Grímsvötn melted a single large conduit immediately down-glacier of the ice-dammed lake seal, thereby imposing a subglacial flood wave on lower parts of the glacier (Björnsson, 1997; 1998; Jóhannesson, 2002a,b; Roberts 2002). Heat transfer from meltwater to ice within the first few km of the subglacial route was more efficient than could be accounted for by traditional models

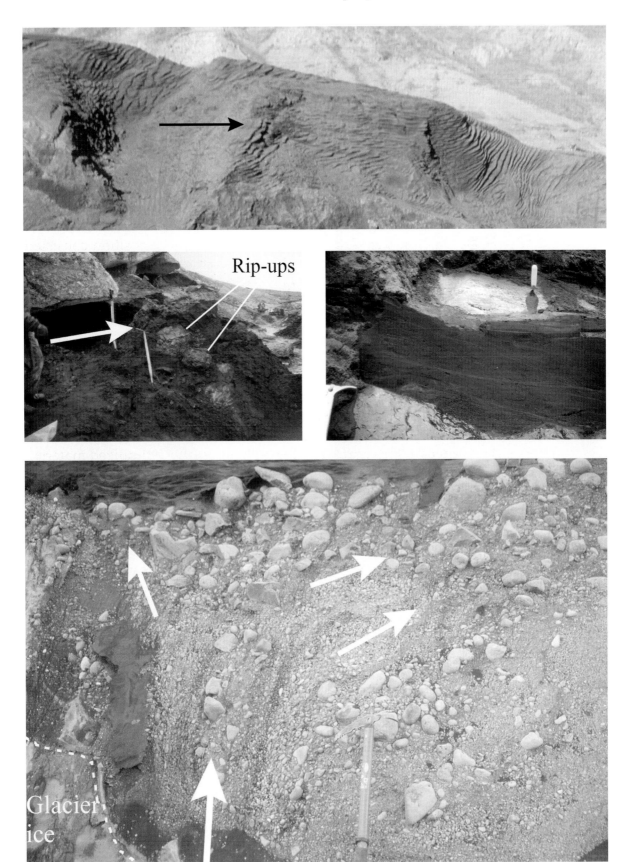

Fig. 7.15a top, 7.15b centre left, 7.15c centre right and 7.15d bottom, full caption opposite.

Fig. 7.16 (a, left) Rectilinear network of fracture fill ridges exposed by glacier margin down-wasting following the November 1996 jökulhlaup. Ridge pattern and location matches exactly those of ice fractures observed in the glacier immediately after the November 1996 jökulhlaup. (b, below) Longitudinal section through the ridge shown in (a). Upward fanning bedding within coarse-grained sediment is consistent with high energy jökulhlaup deposition within a fracture system generated during the early rising stage of the November 1996 jökulhlaup.

englacial jökulhlaup sediments were frozen within a temperate glacier near sea level, supports the hypothesis of freezing-on and deposition by supercooled meltwater flow during the flood. The potential of high magnitude jökulhlaups to freeze large volumes of sediment has major implications for the palaeo-glacier sediment flux and the sedimentary record of formerly glaciated areas (Roberts *et al.*, 2001,2002a; Roberts, 2002).

Englacial jökulhlaup deposits emplaced during the November 1996 jökulhlaup contain many rip-up clasts comprising sheared glacial diamicton and stratified glaciofluvial sediment (Roberts *et al.*, 2001; Waller *et al.*, 2001) (Fig. 7.15b). The rip-ups are contained within frozen englacial jökulhlaup deposits, often within fractures seemingly too narrow to have allowed their passage (Roberts *et al.*, 2001; Waller *et al.*, 2001) (Fig. 7.15b). Russell and Knudsen (1999a) suggested that the rip-up clasts were the product of mechanical excavation of the glacier bed by jökulhlaup waters. Indeed, the presence of frozen rip-up clasts composed of basal ice and diamicton indicate direct mechanical erosion of the glacier bed (Waller *et al.*, 2001). The fact that rip-ups are wider than the narrowest fracture aperture, suggests that fractures were dilated during the onset of the jökulhlaup by high basal water pressure and that the fractures closed as the jökulhlaup progressed (Roberts *et al.*, 2001; Roberts 2002). Rip-up or intra-clasts within englacial and ice proximal proglacial jökulhlaup deposits are diagnostic of subglacial mechanical excavation which may lead to the development of subglacial channels (Russell and Knudsen, 1999a; Russell *et al.*, 2001b; Roberts *et al.*, 2001).

Näslund and Hassinen (1996) identified fluvially-worked debris on the surface of Kötlujökull which they suggested was attributed to emplacement by flowing water during a previous Kötluhlaup. Despite vigorous objections by Krüger and Aber (1999) to the direct fluvial emplacement hypothesis, observations by Roberts *et al.* (2000a) of supraglacial outbursts at Skeiðarárjökull and Sólheimajökull suggest that jökulhlaups are a credible mechanism of emplacing fluvially-worked sediment to high elevations within glaciers. There is clear evidence that the 1918 and 1955 Kötluhlaups exited Kötlujökull initially as a supraglacial outflow, providing a mechanism for sediment to be carried to high elevations within the glacier (Rist, 1967; Tómasson, 1996). Water escape structures within subglacial sediment at Sléttjökull on the northern margin of Mýrdalsjökull are interpreted by van der Meer *et al.* (1999) as the product of a downward flux of high pressure meltwater supplied by the 1918 jökulhlaup. van der Meer *et al.* (1999) conclude that meltwater is highly capable of penetrating through a subglacial deforming bed layer into sediments below and is therefore unable to store large volumes of meltwater over prolonged periods.

7.3.6 Supraglacial embayments, supraglacial ice-walled channels, and ice block release

Icelandic jökulhlaups have long been known to result in considerable ice-margin disruption generating the release of large numbers of ice blocks (Þórarinsson, 1974b; Tómasson, 1996). In particular, jökulhlaups on both Skeiðarársandur (1913) and Mýrdalssandur (1918) were associated with large, m – km scale, ice blocks released primarily by the excavation of ice-walled channels or canyons (Þórarinsson, 1974b; Tómasson, 1996).

7.3.6.1 Ice embayments

During the November 1996 jökulhlaup arcuate ice-surface depressions or 'embayments' developed higher on the surface of Skeiðarárjökull at intervals along ice flow transverse fractures, created by the mechanical removal of fragments of glacier ice, already heavily fractured by high water pressure during the November 1996 jökulhlaup (Roberts *et al.*, 2001; Waller *et al.*, 2001). Although Waller *et al.* (2001) suggested that localized ice collapse above large up-glacier dipping fractures resulted in normal faulting, Roberts *et al.* (2001) reported that ice removal occurred instead in a zone of diffuse fractures which stem from a main or 'arterial' fracture. Roberts *et al.* (2000a,b) reported that flow was vigorous from these supraglacial outlets for only the first few hours of the jökulhlaup, after which flows concentrated at point sources along fractures and at progressively lower elevations along the glacier margin. The main Gígjukvísl outlet, located at the lowest elevation along the glacier snout and having lowest hydraulic potential received progressively greater amounts of jökulhlaup water. Roberts *et al.* (2000b) suggest that hydrofractures generated early in the jökulhlaup allowed higher stage flood flows to carve a large supraglacial ice-walled channel (Fig. 7.17). The glacier fractures developed and witnessed during initial rising stage of the 1996 jökulhlaup are believed to have provided a template for the highly irregular geometry of the ice-walled canyon system, which developed on the late rising flow stage during the hours of darkness during the night of November 5/6th (Roberts *et al.*, 2000b).

7.3.6.2 Ice-walled supraglacial jökulhlaup channels

The distinctive ice-walled channel that developed during the November 1996 jökulhlaup, provided the first opportunity for detailed analysis of processes governing the origin and post flood modification of such a feature. The ice-walled canyon extended supraglacially for over 700 m from the active pre-jökulhlaup glacier margin (Russell *et al.*, 2001b) (Fig. 7.17). The canyon, locally in excess of 50 m in depth and between 100-300 m wide, resulted from the removal of ~ 5 x 10^6 $m^3 s^{-1}$ of glacier ice by floodwaters (Russell *et al.*, 2001b). The unusual geometry of the ice-walled channel has played a major role in controlling the sedimentary architecture of channel-fill deposits (Russell and Knudsen 1999b; Russell *et al.* 2001b).

Much of the ice-walled channel fill consists of large-scale upstream dipping beds interpreted as the product of stoss-side accretion of a large channel-wide bar (Russell *et al.*, 2002; Cassidy *et al.*, 2003). The presence of two sets of large-scale stoss-side beds within the ice-walled channel in a stream-wise direction indicates progressive headward channel expansion. Polymodal matrix-supported backset deposits point to deposition almost directly from suspension. Excavation of the supraglacial ice-walled channel allowed the creation of new accommodation space for stoss-side bar accretion to take place.

Large-scale upstream dipping beds were overlain by a variety of lower angled up and downstream dipping beds. Upstream-dipping beds were arranged as trough-sets reflecting supercritical waning stage flows. Foreset beds deposited on the inner-side of a bend in the upper ice-walled channel reflect bar migration conditions under subcritical into deeper water (Russell *et al.*, 2001b).

The presence of a large 'chamber' off-set from the main ice-walled channel created ideal conditions for high energy slackwater sedimentation (Russell and Knudsen 1999a,b;

Fig. 7.17 Sequence of photographs showing down-wasting of ice surrounding the ice-walled channel feeding the Gígjukvísl river during the November 1996 jökulhlaup. Within 6 years the supraglacial ice-walled canyon has been transformed into a supraglacial esker ridge. (a, top) April (1997); (b, left) August (1998); (c, below) March 2001; & (d, bottom) July (2002). The ice walled canyon was created during the November 1996 jökulhlaup within less than 17 hours and conveyed a peak discharge in excess of 25,000 m³s⁻¹.

Fig. 7.18 *Rhythmically-bedded sands and gravels within a proximal slackwater location reflect deposition of sediment from suspension forming a large scale eddy bar. Note the presence of coarser sediment towards the top of the succession and the presence of large scale trough structures, which mark an increase in flow energy and progressive reduction of flow depths respectively.*

Russell *et al.*, 2001b). Sedimentation was dominated by a classic eddy bar at the chamber mouth sloping and fining away from the main channel (Fig. 7.18). Intra-clasts of up to 2 m in diameter were deposited rapidly from suspension with the proximal eddy bar and even distally, material of up to cobble size was deposited from suspension within a series of normally-graded units (Russell and Knudsen, 1999b). Each fining-upward unit was deposited by individual pulses of suspended sediment (Russell and Knudsen, 1999b) (Fig. 7.18). Estimates of stream powers and flow competence within the main ice-walled channel indicate that material in excess of boulder size was transported in suspension at flood peak (Russell *et al.*, 2001b) (Fig. 7.17). The highly angular geometry of the ice walled channel generated vigorous flow circulation cells, with local flow directions in an up-channel direction. Jökulhlaup bars deposited at high flow stage within flow separation cells were reworked by waning stage flows directed in a down channel direction (Russell *et al.*, 2001b).

Since the November 1996 jökulhlaup, the ice-walled supraglacial channel has been transformed into a supraglacial ridge (Russell et al., 2001b). In 2002, the double-headed ridge rose more than 30 m above the surrounding glacier surface taking the form of an esker. Progressive undercutting of the margins of the ridge reveal an irregular ice-contact, which was confirmed by the GPR surveys along the ridge axis (Cassidy *et al.*, 2003). The ice-walled supraglacial channel created in November 1996 is possibly the best documented modern analogue for large eskers found within the Quaternary landform and sedimentary record (e.g. Warren and Ashley, 1994).

An even larger ice-walled channel 'Rjúpnagil' developed during the 1918 Katlahlaup (Jóhannsson, 1919; Sveinsson, 1919; Tómasson, 1996). This channel was 1460-1830 m in length and 366-550 m in width, with walls 145 m in height (Jóhannsson, 1919; Sveinsson, 1919; Tómasson, 1996). Unpublished photographs taken on the waning stage of the 1918 jökulhlaup show vigorous flows filling Rjúpnagil completely, with only occasional bar surfaces emergent. It is highly probable that Rjúpnagil represents a former sub- or englacial conduit whose roof was subject to rapid failure and evacuation by the flow. Jóhannsson (1919) described the release of a huge mass of ice blocks from the Rjúpnagil area after 2 hours of the 1918 jökulhlaup. Little is known about the sedimentology of the Rjúpnagil channel-fill and whether subsequent waning stage flows or glacier advance have removed or over-ridden these features.

7.3.6.3 En- and subglacially deposited jökulhlaup eskers

> *"Eskers are typically a few tens of metres to hundreds of metres wide. Such dimensions are much greater than what is physically reasonable for an R-channel - even channels associated with gigantic Icelandic outburst floods have diameters no more than about 10 m (Nye, 1976). Thus eskers cannot form wholly within any single conduit, but must be built up over time as sediment melts out of channel walls."* Clark and Walder (1994)

Englacial eskers were emplaced at several locations within Skeiðarárjökull by November 1996 jökulhlaup flows (Russell and Knudsen, 1999a). Eskers are located on both reverse and normal slopes leading to the former glacier margin and show a direct connection to the proglacial jökulhlaup deposits. Esker locations match exactly the

Fig. 7.19 Esker ridge on the surface of Skeiðarárjökull in August 1998. This esker was deposited englacially during the November 1996 jökulhlaup and is clearly associated with proglacial jökulhlaup channels and the former ice-contact slope.

location of conduits observed on the waning flow stage. A prominent esker ridge was observed leading to the exact location of a conduit mouth clearly visible in Fig. 7.5a (Russell and Knudsen, 1999a) (Fig. 7.19). A November 1996 jökulhlaup englacial esker at the Sæluhusakvísl displays a coarse-grained, fine-grained matrix-supported core, suggesting deposition from a hyperconcentrated flow (Fig. 7.20). Price (1969, Figure 9) presents a model for the relationship between eskers, moraines and pitted sandur. Although not specifically acknowledging the role of jökulhlaups at Breiðamerkurjökull, Price's model highlights the process link between eskers and proglacial outwash for deposits attributed to jökulhlaups.

The 1999 jökulhlaup at Sólheimajökull, exited the glacier via a number of outlets, the largest of which was on the western margin of the glacier (Roberts *et al.*, 2000b, 2002b, 2003; Russell *et al.*, 2000, 2002; Sigurðsson *et al.*, 2000). Russell *et al.* (subm.) point out the presence of a wide subglacial esker leading to the western outlet of the July 1999 jökulhlaup. The esker ridge comprises a wide subglacial cavity filled with boulder-sized sediment. It is clear that recent Icelandic jökulhlaups have been responsible for the formation of both sub and englacial eskers extending commonly for 100s of metres in a down flow direction and 10s of metres in width. Clark and Walder's conclusions appear incompatible with Icelandic field observations. The presence of a 150-200 m deep linear collapse trough on the surface of Skeiðarárjökull leading from Grímsvötn subglacial lake points to the development of a conduit 100s of m in diameter. Esker formation by progressive melt out of englacial sediment is highly improbable in the case of the November 1996 jökulhlaup as meltwater at the glacier margin had little thermal energy available for conduit melt enlargement. Instead, widespread glacier hydrofracturing and mechanical erosion of sections of the glacier bed supplied large volumes of coarse-grained sediment to englacial jökulhlaup conduits. Deposition occurred simultaneously within the conduit over considerable distances exceeding 100m. It is clear that eskers do form during jökulhlaups and can be used to provide important information about englacial jökulhlaup flow conditions and sediment fluxes.

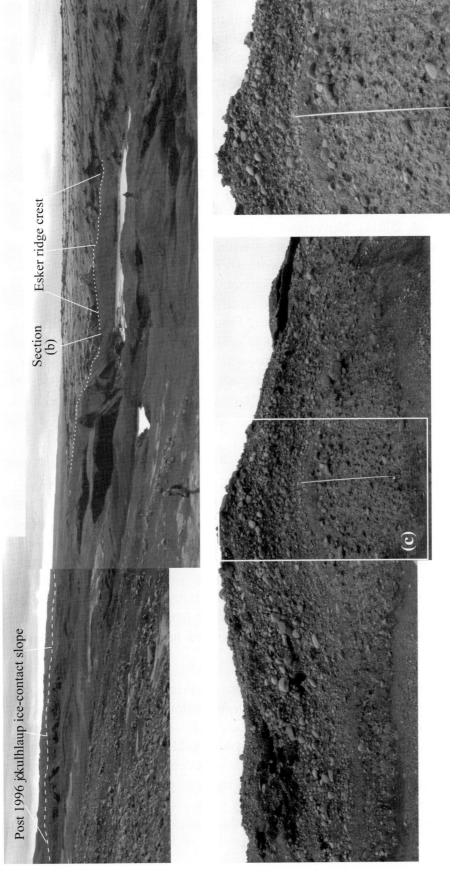

Fig. 7.20 (a, top) A steeply ascending November 1996 jökulhlaup esker ridge at the margin of Skeiðarárjökull. The esker ascends from right to left from the present ice-margin towards the November 1996 ice-margin position. (b, bottom) Section at right angles to the esker ridge axis reveals laterally extensive matrix and clast-supported units, which form a shallow anticline. (c, right) a lower, predominantly matrix-supported, unit is overlain by an open-framework unit containing cobble-sized clasts. The lower unit reflects deposition from a hyperconcentrated flow whilst the upper unit represents the product of deposition from a more fluidal flow, reflecting reduced sediment availability as the jökulhlaup progresses.

7.3.6.4 Controls on ice block release during jökulhlaups

Tómasson (1996) suggested that ice blocks constituted a significant volume of the total amount of material discharged during the 1918 Kötluhlaup. Ice blocks were released from 4 main sources: (i) through large-scale tunnel collapse, (ii) ice margin flotation or 'hydraulic jacking' where the pressure of basal flood water lifted sections of the ice margin causing ice blocks to break off, (iii) formation of supraglacial fracture outlets associated with hydraulic jacking (Roberts et al., 2000a, b); (iv) undercutting of the margin by flood water causing ice cliff collapse; and (v) from 'dead' ice in the proglacial zone. The location and degree of ice fracturing is therefore an important control on the size of ice blocks and amount of ice transported in a jökulhlaup (Fay, 2002a). Surprisingly large volumes of ice were removed from ice-cored moraines and ice covered outwash sediment during the November 1996 jökulhlaup (Russell and Knudsen, 1999a; Russell et al., 1999; Knudsen et al., 2001a). November 1996 jökulhlaup flows undercut buried glacier ice over a distance of several kilometres within the Gígjukvísl channel. The presence of large masses of 'dead' glacier ice, seemingly detached from the active glacier margin, has several implications for jökulhlaup impact. Undercutting of older ice-cored topography allows large blocks to be entrained and transported to relatively distal locations. Jökulhlaups with relatively sedate rising flow stages are associated with less glacier hydrofracturing and consequently the release of lower ice block volumes. Entrainment of dead-ice blocks during jökulhlaups displaying a gradual rising stage may significantly increase proglacial ice block impact. There will be more dead-ice entrainment to jökulhlaups flowing through proglacial trenches that have developed during periods of glacier retreat.

7.3.7 Proglacial ice block impacts

As early as 1918 it was acknowledged that large numbers of ice blocks were a frequent feature of outburst floods. Tómasson (1996) analysed eyewitness accounts of the jökulhlaup from Katla, Iceland, in 1918 and estimated that the total amount of ice moved was 0.5 km^3, which is about 10-15 % of the total floodwater. During the November 1996 jökulhlaup in southern Iceland 8.3 x 10^6 m^3 of ice was transported on to Skeiðarársandur including ice blocks as large as 55 m in diameter. Ice-block impact reflects glacial and topographical constraints, ice-block characteristics and jökulhlaup hydraulics. Since flow conditions of individual jökulhlaups can vary widely from one event to the next and during a single event, a wide range of ice-block related features are formed both during and after a jökulhlaup (Fay, 2001, 2002a,b).

7.3.7.1 Ice-block transport and grounding in Icelandic jökulhlaups

Ice blocks are grounded in locations where flow deceleration and expansion occurs, predominantly associated with zones of maximum elevation (Russell, 1993; Maizels, 1997; Fay, 2001) (Fig. 7.21). Zones include areas of: (1) pre-flood topographic high points; (2) backwater locations where fans are graded to lake levels and flow depths and velocities are low; (3) higher-level surfaces such as a roads or pre-existing bars. High rates of sediment deposition may partially or totally bury blocks further preventing their re-entrainment. During a jökulhlaup, the largest ice blocks will be transported in routeways characterised by high discharge, velocity, depth and stream power (Russell, 1993; Fay, 2001).

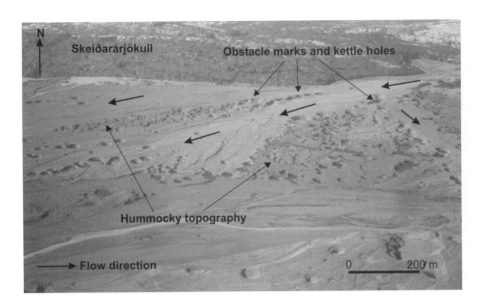

Fig. 7.21 Obstacle marks and kettle holes in front of the main flood outlet following the November 1996 jökulhlaup on Skeiðarársandur, southern Iceland.

In general, ice-block size decreases down sandur reflecting the progressive inability of the flow to move the larger ice blocks as slope gradient and stream power decrease (Russell, 1993; Maizels 1995; Fay, 2001). Within confined locations, ice-block size decrease is orientated parallel to flow direction (Russell and Knudsen, 1999a; Fay, 2001). At fan-scale, however, decrease in ice-block size is non-uniform. A non-uniform decrease in size of clasts down fan was noted by Maizels (1995) but was hitherto not reported for ice blocks. The non-uniform decrease in ice-block size down fan occurs because: (1) smaller ice blocks are arrested in the lower velocity wake of larger grounded ice blocks while large ice blocks bypass these zones to ground further downstream and (2) small ice blocks released on the waning limb of a flood ground in areas occupied by larger blocks grounded earlier during the flood.

Ice-blocks form clusters when a single, dominant grounded ice-block obstacle clast arrests floating ice blocks of a smaller or similar size causing them to ground in the stoss or lee of the obstacle. Although clusters formed from rock clasts have been widely recognised (e.g. Brayshaw, 1985; Whiting, 1996), similar formative processes had not been suggested for the formation of ice-block clusters until recently (Fay, 2001). The fact that ice blocks float means that flow depth and velocity are crucial controls on the formation of ice-block clusters. Within glacier proximal locations where jökulhlaup sediment flux is high enough ice blocks may be completely buried resulting in the subsequent formation of kettle holes.

7.3.7.2 Impact of grounded ice blocks during jökulhlaups

Where a high density of stranded ice blocks act as a focus for sediment deposition on the waning limb of a jökulhlaup and where there is a lack of waning stage reworking, topography with a hummocky morphology forms (Fay 2002b). Facies are poorly sorted and lack structure owing to the rapidity of sediment deposition. In topographically confined locations hummocky topography forms in areas subject to backwater

Fig. 7.22 Hummocky topography formed by numerous closely-spaced ice blocks on Skeiðarársandur, southern Iceland.

conditions. On unconfined outwash fans, hummocky topography forms on the lower velocity, distal lower portions of the fans (Fig. 7.22) (Fay, 2002b).

Chute channels develop where scour around ice-block clusters captures erosive waning stage flow. The amount of waning stage discharge and its rate of decrease primarily controls the amount of chute-channel incision. Development of chute channels around ice blocks was observed by Russell (1993). Where multiple ice blocks are grounded, scour around ice-block clusters not only leads to the formation of chute channels but consequently controls the within-channel routing of waning stage flow.

Although several models of obstacle-mark morphology and sedimentology have been proposed in the past (Karcz, 1968; Allen, 1984; Russell, 1993), Fay's 2001 study was the first to describe such a wide range of obstacle marks of large scale within coarse, up to boulder-sized gravel. Figure 7.23 lists 6 types of ice-block obstacle mark that can be formed during a jökulhlaup. The nature of ice-block obstacle scour is determined by (a) obstacle size, (b) flow velocity, (c) flow depth, (d) flow rheology, (e) nature of the substrate, (f) timing of ice-block release (i.e. scour time), (g) distribution of already grounded ice blocks (number and how closely spaced), (h) ice-block shape and (i) ice-block flotation. Controls a-h have been recognised as controls on scour occurring around bridge piers or foundations (e.g. Melville, 1997). Controls a, e, h and i have previously

Fig. 7.23 opposite. Morphology and composition of a range of ice-block obstacle marks formed during jökulhlaup flow. Type 1: kettle-scours with both proximal and lateral scour and tail formed by partial or total burial and subsequent partial exhumation. Type 2: semi-circular obstacle marks formed by partial exhumation of partially buried ice blocks on the margins of the main channels. Type 3 and Type 4: obstacle marks characterised by a proximal and lateral scour crescent, a ridge stoss-side of the scour crescent and a large aggradational tail whose structure is anticlinal. Truncation of anticlinal-shaped bedding produces the flat-topped morphology and armoured nature of Type 4 obstacle marks. Type 5: Obstacle marks <5 m in diameter with fine-grained gravel in the immediate lee of the scour hollow and coarse gravel lag characterising the distal part of the tail. Formed by waning stage deposition followed by late waning stage erosion. Type 6: entirely erosional obstacle marks. Formed by total exhumation of buried ice blocks or by scour around ice blocks grounded on the late waning stage (Fay, 2001).

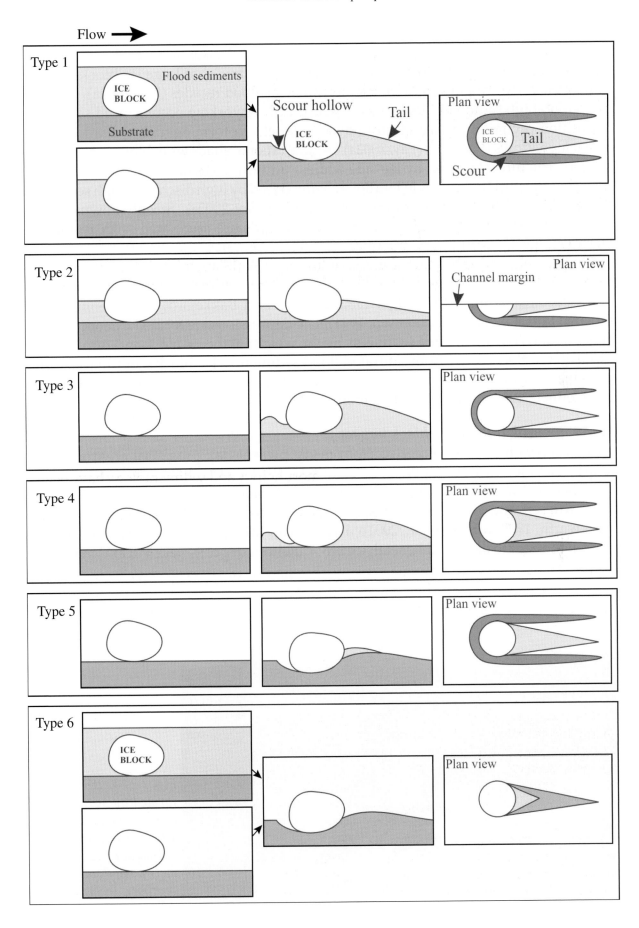

been recognised as controls on ice-block scour (Russell, 1993). Flow depth is an extremely important limiting factor as ice blocks are often not fully submerged. In such cases 'kettle-scours' (Figs. 7.21,24) are produced by partial burial of large ice blocks, scour during the flood, and *in situ* ice-block melt after the flood recedes (Fay, 2002a). In high discharge/high sediment flux locations where little reworking occurs, rapid sediment deposition occurs in the lee of grounded ice blocks to form aggradational ice-block obstacle shadows (Fig. 7.25) (Fay, 2002a). Aggradational shadows exhibit an anticlinal-shaped structure deposited by lee-side eddies. Antidune strata are deposited around ice blocks during jökulhlaups in locations characterised by high discharge, high sediment flux and very little waning stage erosion (Fig. 7.26) (Fay, 2002a). Steeply downstream-dipping beds and low-angle upstream-dipping beds are associated with antidune-washout and fill the antidune trough (Fig. 7.27).

The presence of ice-block tool marks following jökulhlaups has been observed by several authors. 'Iceberg gravity craters' or 'impact structures' are formed when an ice block becomes stranded (Dredge, 1982; Longva and Bakkejord, 1990; Bennett and Bullard, 1991; Longva and Thoresen, 1991). The depressions are formed by compaction of the underlying stream bed sediments by the weight of the ice block and through the expulsion of flood water from saturated flood sediments. Ice-block scour marks form as ice blocks are arrested in active movement (Collinson, 1971; Gustavson, 1974; Thomas and Connell, 1985; Longva and Thoresen, 1991).

7.3.7.3 Post-jökulhlaup ice-block impact

Kettle holes have been reported from many present day proglacial environments (Clark, 1969; Jewtuchowicz, 1971,1973; Churski, 1973; Galon, 1973 a, b; Klimek, 1973; Nummedal *et al.*, 1987; Maizels, 1992, 1995, 1997; Molewski, 1996; Ashley and Warren, 1997; Russell and Knudsen, 1999a; Russell and Knudsen, 2002a; Olszewski and Weckwerth, 1999; Russell *et al.*, 2001b; Fay, 2002b). Jökulhlaup-transported ice blocks are still thought to be buried beneath ice block meltout deposits near Vík and may date

Fig. 7.24 Kettle-scour formed by scour during the November 1996 jökulhlaup and subsequent ice-block melt post-flood, southern Iceland.

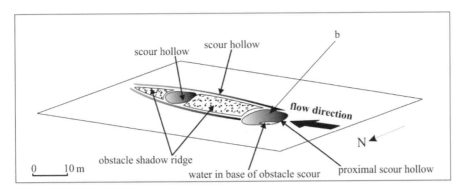

Fig. 7.25 (a, top) Morphology of an aggradational ice-block obstacle shadow.
(b, right) A transverse section of the obstacle shadow which consists of gravel beds producing an anticlinal-shaped feature (adapted from Fay 2002a).

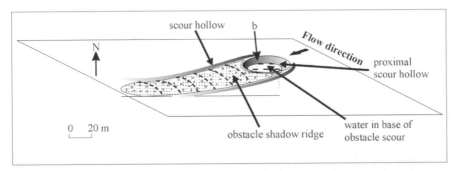

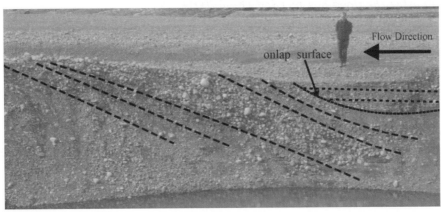

Fig. 7.26 (a, top) Morphology of obstacle mark formed during the November 1996 jökulhlaup, southern Iceland.
(b, right) Upstream-dipping gravel beds truncated by surface morphology (adapted from Fay, 2002a).

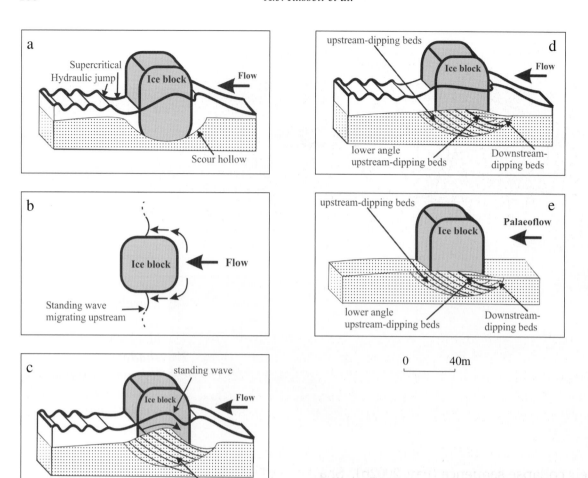

Fig. 7.27 Schematic model of the formation of antidune stoss-side strata around an ice block. (a) Flow becomes supercritical just down flow of the ice block, generating an upstream-migrating standing wave. (b) Plan view of the upstream-migrating standing wave. (c) Antidune stoss beds are deposited from the upstream-migrating surface wave. (d) 'Washout': with increasing bedform height, flow on the upstream side slows and deepens increasing the rate of sedimentation until the antidune is no longer stable and collapses. Downstream-dipping strata are produced by rapid migration of asymmetrical bedwaves immediately after the surface wave breaks. Associated with antidune collapse, water stored upstream of the antidune is released eroding the antidune strata and filling the antidune trough with low angle backset beds. (e) Resultant section after the flood has receded (Fay, 2001).

to the 1755 Kötluhlaup (Everest and Bradwell, 2003). Kettled or pitted sandur is glacial outwash in which numerous kettle holes have formed. The deposition of glaciofluvial sediments during a jökulhlaup may partially or completely bury ice blocks leaving hollows when the ice melts (Clarke, 1969; Maizels, 1977).

Kettle holes may be inverse-conical or steep-walled in shape (Fig. 7.28). Inverse-conical kettles form due the melt of partially or totally buried ice blocks and develop by sediment slumping and avalanching down the kettle walls. Steep-walled kettles form by collapse

Fig. 7.28.(a, left) An inverse-conical kettle hole. (b, right) A steep-walled kettle hole.

of overlying sediment into voids created by the in situ melt of completely buried ice blocks (Fay, 2002b). Ice blocks transported on to the sandur by water are often progressively buried by glaciofluvial sediments. However, during a jökulhlaup, sediment deposition can be so rapid that small ice blocks are incorporated within the flow and deposited simultaneously with flood sediments (Russell and Knudsen, 1999a; Fay, 2002b).

The physical properties of the deposited sediment, in which kettle holes form, control their collapse sequence (Fay, 2002b). Shallow kettles with vertical or inward-dipping walls, whose base is a coherent block of sediment, form in coarse, clast-supported sediments. In coarse sediments dominated by matrix-supported or in entirely fine-grained sediments, deeper kettles with steeply-dipping or overhanging walls form often through sudden roof collapse. Similar collapse features have been described by Branney and Gilbert (1995) following an eruption-induced jökulhlaup, and modelled experimentally by Maizels (1977). Steep-walled kettle holes may form over small buried blocks, or over larger buried ice blocks which melt irregularly. Steep-walled kettles can develop into inverse-conical kettle holes by slide or avalanche of sediment into the kettle hollow. The spatial distribution of steep-walled and inverse-conical kettle holes is controlled by the distribution of buried ice, thus kettle holes form in glacier proximal locations where flood sediment concentration was high.

Kettle holes and obstacle marks may possess raised diamict rims (rimmed kettles) (Fuller,1914; Thwaites, 1926, 1935; Maizels; 1992, Branney and Gilbert,1995; Olszewski and Weckwerth, 1999; Fay, 2002b) and/or diamict mounds in the base of the depression. Diamict mounds may also be deposited directly on the streambed (Maizels, 1992). Isolated conical sediment mounds form in slackwater locations where small ice blocks have been stranded on fine sediments and aeolian deflation processes are active (Fig. 7.29) (Fay, 2002b). Facies within conical sediment mounds consist of flood sediments in their lower parts and ice-block till in their upper parts. Ice-block till provides protection against erosion by the streambed beneath while wind deflation erodes around the till producing the mound.

Fig. 7.29 (a, left) An inverse-conical mound formed during the November 1996 jökulhlaup, southern Iceland. (b, right) The mound is composed of flood sediment overlain by ice-block diamict (Fay, 2002b).

The morphology of the hummocky topography has a very similar appearance to descriptions of pitted sandur formed by the melt of buried stagnant ice (e.g. Howarth 1968; Price, 1971), in particular kame and kettle topography (Price, 1969; Benn and Evans, 1998). Similarly, since the development of kettle holes is similar over both stagnant glacier ice and flood-related ice blocks, it may be difficult to differentiate between flood-related kettles and kettles produced by non-fluvially driven processes. However, on a palaeoflood surface, a flood origin for kettle holes is indicated by a distinct radial pattern of kettle holes on prominent outwash fans reflecting flow expansion and/or a decrease in kettle size down sandur relating to a progressive decrease in stream power. Since features of similar morphology such as hummocky topography and kettle holes may result from processes associated with ice transported in jökulhlaups and processes associated with the passive melt of stagnant glacier ice, it is important that the internal lithofacies of the landforms are analysed as well as their morphology. Evans and Twigg (2002) concluded that the pitted sandar of Breiðamerkurjökull were of jökulhlaup origin, due to their similarity to Icelandic sandur of known jökulhlaup origin.

The characteristics of anticlinal-shaped obstacle shadows can be very similar to those of eskers viewed in cross section (Fay, 2002a). Like, anticlinal-shaped ice-block obstacle shadows, the structure of eskers may also be anticlinal and consist of a variety of glaciofluvial facies, ranging from sorted silts, sands, gravels and boulders to matrix-supported gravels (Sharpe, 1953; Shreve, 1985; Syverson et al., 1994; Benn and Evans, 1998). Both eskers and ice-block related features may exist in the proglacial zone and may even both be produced within a single jökulhlaup. In contrast to eskers, ice block obstacle shadows display marked differences in morphology down their shadow ridge.

7.3.7.4 Palaeohydrological implications of ice block related landforms and deposits

Knowledge of processes responsible for the formation and pattern of ice-block related features within jökulhlaup channels, provides a remarkable insight into a process-form relationship that is truly diagnostic of jökulhlaups. Detailed knowledge of the associations between ice blocks, resultant bedforms and lithofacies enables the identification of jökulhlaups involved in the release and transport of ice blocks within

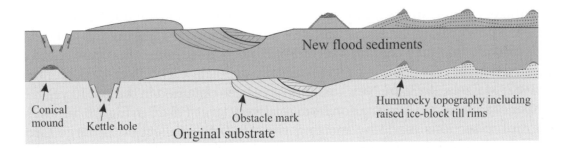

Fig. 7.30 Expected stacked sequence of conical mounds, kettle holes, obstacle marks and hummocky topography. The ice-block related features on the original substrate are buried by sediments deposited from a subsequent flood. New ice-block related features are formed on the new substrate (Fay, 2001).

both modern and ancient proglacial environments. Since obstacle marks, kettle holes, isolated conical mounds, ice-block till rims and mounds are essentially formed on the stream bed, their identification within the stratigraphic record can be used to recognise individual palaeofloods, thereby enabling reconstruction of jökulhlaup magnitude/ frequency regime (Fig. 7.30). Ice-block obstacle tails provide information on flow velocity and flood depth (Russell, 1993). Shadow sediments denote maximum grain size transported at the time of shadow deposition. Since ice-block obstacles are not fully submerged during jökulhlaups, obstacle tail height, whether depositional or erosional in nature, provides a minimum flow depth at the time of formation. Ice-block obstacle tails also record waning-limb flow direction (Russell, 1993). Fay (2001,2002a) interpreted large upstream dipping beds around the margins of kettle-scours as the product of supercritical flow conditions, with associated Froude numbers ranging between 0.8 and 1.2 (Allen, 1984). Upstream dipping beds associated with flow around large ice blocks can thus be used to reconstruct former flow depths and velocities. Following the *in situ* melt of the ice blocks, kettle scours may therefore provide excellent sections of jökulhlaup flood sediments. Perhaps, more importantly, the depth of kettle-scours provides a minimum thickness of a single jökulhlaup depositional unit. Ice blocks may either enhance or inhibit the preservation of jökulhlaup deposits. For example, the deposition of obstacle shadows and hummocky topography produces higher amplitude bedforms than would be formed in the absence of the ice blocks. However, localised scour around ice blocks may limit the thickness and lateral continuity of sediment sequences.

Knowledge of how ice blocks behave during transport and after deposition may aid jökulhlaup hazard identification and mitigation within active proglacial channel systems. Areas prone to ice block transport during jökulhlaups can be predicted, thus allowing mitigation of risk to existing and planned infrastructure. For example, areas in which damage to man-made structures such as roads and bridges may occur as a result of ice-block transport during a jökulhlaup can be identified. Information regarding the location of buried ice blocks following a flood and potential ice-block transport during a subsequent flood will be of importance when planning post flood reconstruction.

7.3.8 Controls on the stratigraphy and architecture of proglacial outwash deposits

Little is known about the stratigraphy of the large sandar flanking the Icelandic ice caps. It is likely that they contain a record of Holocene sedimentation, and it has been speculated that they are dominated by jökulhlaup deposits (e.g. Walker, 1965; Nummedal et al., 1987; Maizels, 1993a,b). New insights have been obtained in the last 10 years from a number of sources: (1) Information from relatively large (15-20 m high, 0.5-2 km long) stratigraphic sections made by incising rivers; (2) the collection of large Ground Penetrating Radar data sets mainly on Skeiðarársandur; (3) observations from recent floods and the reworking of flood deposits by between-flood flows; and (4) the role of glaciers in controlling aggradation and incision on sandar, and the role of glacier margin fluctuations in controlling flood impacts.

The role of jökulhlaups in controlling sandur architecture and stratigraphy has been discussed recently by Marren (2002b; in press), Marren et al. (2002; subm.), Cassidy et al. (2003) and Zielinski and van Loon (2002). Descriptions of exposures of sandur sediments containing the deposits of multiple events are provided in Russell and Marren (1999), Marren (2002b) and Marren et al. (subm.). Maizels (1993b) produced a predictive model for generating hypothetical sandur successions. From these studies it is apparent that a great number of factors can control the stratigraphy of a sandur deposit. The most important are whether the sandur is aggrading or incising, the size of jökulhlaups relative to the size of normal flows and the sandur size, the flow characteristics of the jökulhlaup, jökulhlaup frequency and base level change.

7.3.8.1 Role of glacier fluctuations in determining jökulhlaup impact

Glacier margin fluctuations are a direct control on the nature of aggradation or incision on a sandur, and consequently have a significant primary effect in controlling jökulhlaup impact. Glacier fluctuation history is also the main control on the overall stratigraphy of a sandur. The glacier margin position relative to surrounding moraine systems can also have an effect on the impact of a jökulhlaup, a factor that is discussed in the following section.

Advancing glaciers generally move over older sandar and moraines, and consequently sedimentation that occurs during the advance phase or at the maximum advance of the glacier tends to occur on top of pre-existing sediments. Glacier advance therefore causes aggradation (Maizels, 1979; Thompson and Jones, 1986; Marren, 2002a). Jökulhlaups that occur during an advancing phase or at maximum advance positions are likely to be deposited in an aggradational setting. In these circumstances, jökulhlaup deposits are likely to be more unconfined, and affect a greater proportion of the sandur. During glacier advance, steeper ice margin gradients allow meltwater to flow more independently of subglacial topography (Shreve, 1985). Such a situation would permit a greater variability of outlet position across the entire width of glacier snouts, and allow more even spatial distribution of sediment supply to and dispersal within the proglacial environment.

In contrast, glacier retreat involves moving away from moraine systems, and away from the sediments deposited during aggradational phases. In lowland settings, and where the glacier snout overlies thick accumulations of sediment, retreat therefore generally involves moving into a topographically lower position. The point of exit of rivers draining a glacier is therefore lowered, and incision is therefore likely to occur in the proximal part of the sandur (Thompson and Jones, 1986; Marren, 2002a). This may be accompanied by minor aggradation in more distal parts of the sandur, as eroded sediment is re-deposited (Germanoski and Schumm, 1993). In these circumstances, jökulhlaups are likely to produce incised channels with terraces and very confined depositional sequences (e.g. Maizels, 1987).

Retreat of a glacier from a prominent moraine system often produces a topographic low between the glacier and the moraine system. This topographic low is often occupied by proglacial lakes or minor fluvial systems that flow parallel to the glacier margin before breaking through the surrounding moraines (Klimek, 1972,1973; Bogacki, 1973; Galon, 1973a,b). Older sandur surfaces constitute progressively abandoned unconfined jökulhlaup fans and ice-marginal river courses, which develop during periods of glacier retreat and proglacial trench development (Maizels, 1991; Gomez *et al.* 2000). Molewski and Olszewski (2000) and Marren (2002b) provide sedimentary evidence in the Gígjukvísl channel of major changes depositional environment within the embryonic Skeiðarárjökull proglacial trench. Molewski and Olszewski (2000) suggest that three ice-advances are needed to explain the succession exposed in the Gígjukvísl area implying that there have been numerous phases of proglacial trench formation and destruction within the last few centuries. Marren (2002c) presents evidence of former proglacial lakes, which were subsequently in-filled rapidly by jökulhlaups. Similarly, the November 1996 jökulhlaup filled-in an ice-marginal lake occupying the proglacial trench within minutes (Knudsen *et al.*, 2001a; Russell and Knudsen, 2002a; Cassidy *et al.*, 2003).

Skeiðarársandur during the 20th Century provides an example of the differences in sedimentation patterns during the aggradation and incision phases. In descriptions of the 1934 and 1938 jökulhlaups, water is thought to have drained directly from the ice onto the sandur, and across most of the proglacial area of Skeiðarársandur. The withdrawal of the ice margin from the Little Ice Age moraines at Skeiðarársandur has caused drastic changes in the flow of jökulhlaups. During the 1996 jökulhlaup water was confined to the three main river systems, and only coalesced to occupy the entire width of the sandur in the most distal portion. This change is obvious by comparing the model of sedimentation of Boothroyd and Nummedal (1978) with the model put forward by Russell and Knudsen (1999a). The Boothroyd and Nummedal (1978) model is based upon data collected previous to the 1996 jökulhlaup when water flowed relatively freely across the sandur creating unconfined outwash fans. Boothroyd and Nummedal (1978) and Nummedal *et al.* (1987) although acknowledging the role of jökulhlaups for the transport of sediment and deposition of a limited number of sedimentary structures, attribute proximal-distal grain size and lithofacies variation to seasonal cycles of ablation dominated flows. Boothroyd and Nummedal's model does not consider the sedimentology or architecture of sandur deposits at depth, nor the role of topographically controlled sedimentation in ice proximal locations. Glacier margin fluctuations on large sandar can result in complex, variable responses across the sandur. Some of the channels occupied by floodwaters during the November 1996 jökulhlaup had long histories of flood impact. Other channels were younger, and had not been impacted by a flood previously. Russell and Knudsen (2002b) document the response of the various channels across Skeiðarársandur to the 1996 jökulhlaup pointing the fact that the greatest geomorphic impacts were associated with relatively new channels cutting through the moraine belt, such as the Gígjukvísl channel (Russell *et al.*, 1999).

An important point to realise is that a long history of glacier margin fluctuations can produce a complex sandur stratigraphy. This issue was raised recently by Marren (in press), in response to suggestions that, based on the Boothroyd and Nummedal (1978) model for Skeiðarársandur, outwash fans in general might be expected to show a simple 'alluvial fan' like structure and stratigraphy (Zielinski and van Loon, 2002). A number of problems were identified with this suggestion, the most important of which is that recent work, as highlighted by this review has shown that Boothroyd and Nummedal's model is largely inappropriate for Skeiðarársandur as a whole. The Boothroyd and Nummedal model (1978) only deals with shallow bar top sedimentology in the presently active rivers, rather than the more easily preserved jökulhlaup sediments that dominate the rest of the sandur. An 'alluvial fan' model is inappropriate for Skeiðarársandur and for all sandar in general, as glacier margins are dynamic. The single point dispersal system that occurs on an alluvial fan is not a feature of sandar over the medium to long term due to the impacts of the fluctuating ice margin. Consequently, no particular point on a sandur can be designated as 'proximal' or 'distal', except in the short term. Varying distance from the glacier, combined with changes from aggradational to incisional settings will result in complex stratigraphies on most sandur.

7.3.8.2 *Role of backwater effects for jökulhlaup channel and sandur morphology and sedimentology*

Implications of the development of a proglacial trench during a period of glacier retreat have been described by many researchers (see above section). Most of the above studies consider the role that proglacial trenches play in controlling the depositional environment between jökulhlaups as well as their routing. However, little attention has focussed on the interaction between the proglacial trench and jökulhlaup flow dynamics. Russell and Knudsen (1999a) presented a model highlighting the role of proglacial topography in generating within-jökulhlaup backwater effects with implications for downstream sediment and ice block concentrations (Fig. 7.31). At flood peak, $60\text{-}100 \times 10^6\,\text{m}^3$ of water was temporarily stored in a backwater lake upstream of the Gígjukvísl moraine constriction (Russell and Knudsen 1999a). Backwater ponding prevailed mainly within the first few hours of the jökulhlaup before the flood eroded various flow constrictions. The backwater effects acted to reduce the transport of coarse-grained sediment from the glacier margin (Russell and Knudsen, 1999a,2002a; Gomez *et al.*, 2002). Rising stage deposition into backwater lakes resulted in the formation of relatively flat-topped, delta-like, radial outwash fans showing rapid down-fan decrease in ice block size (Russell and Knudsen, 2002a). Ground Penetrating Radar (GPR) surveys of November 1996 jökulhlaup deposits within the main proglacial trench reveal progradational sedimentary architecture overlain by shallower angled structures indicative of aggradation (Cassidy *et al.*, 2003). Such a succession is compatible with a transition of deposition from backwater dominated to free flowing jökulhlaup flow. As local backwater controls were removed during the jökulhlaup, there was a greater potential for earlier rising stage sediments to be reworked and transported further down the fluvial system. Several cycles of storage and release can be identified for the Gígjukvísl channel system during the November 1996 jökulhlaup as various backwater lakes were created and destroyed.

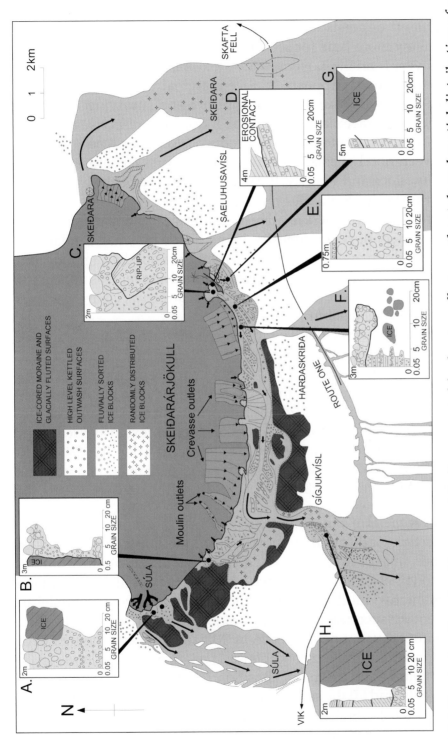

Fig. 7.31 Model illustrating the role of topographically generated backwater effects on the style and spatial distribution of proglacial jökulhlaup deposition (after Russell & Knudsen, 1999a). The model shows how vertical sedimentary characteristics might vary depending on whether jökulhlaup waters were confined by moraine ridges (A & B) or were free to spread out from the ice-margin (C-F). Note increased spatial variability of expected sedimentary successions associated with the moraine-confined conditions where backwater conditions act as a major control on sedimentation at different times during the flood. Unconfined jökulhlaup successions are divided into locations where waning-stage sediment flux decreases (A) and where sediment flux remains high even on the waning flow stage (B). In the moraine-confined scenario, backwater conditions result in much finer grained deposits downstream of the backwater zone (D). Sedimentary successions associated with ice-contact proglacial fans may also vary depending upon degree of backwater ponding or whether deposition is into shallow fast flows (C) or into sluggish deep flows (F).

7.3.8.3 Other factors controlling sandur stratigraphy and architecture

The other jökulhlaup-related factors controlling sandur formation are mostly involved in controlling the detailed facies architecture, rather than the overall sandur stratigraphy. The relative discharge of jökulhlaup events compared to non-flood flows will determine the rate at which jökulhlaup deposits are reworked. Similarly, a high jökulhlaup recurrence interval will allow more time for sediment reworking by non-flood flows. Monitoring of sandur recovery from the November 1996 jökulhlaup on Skeiðarársandur has been ongoing since the flood. It appears that ice-marginal deposits are quickly modified due to glacier retreat. Sandur surfaces appear to be surviving relatively unmodified at present, due to terrace surfaces being abandoned in response to glacier retreat. Low jökulhlaup recurrence intervals will produce 'flood hardened' channel systems that persist between flood events. This situation was first recognised for the Knik River in Alaska (Fahnestock and Bradley, 1973), and appears to be the case for the present day Skeiðará river (Marren et al., 2002).

The physical area covered by floodwaters relative to the physical extent of the sandur will determine the lateral extent of flood deposits. Large floods on small sandar will produce laterally extensive, sandur-wide flood units. Whether or not they aggrade and form a succession of flood units depends on the factors outlined above. This appears to have been the case at Kverkjökull, where stacked flood units can be traced in exposures for several kilometres (Marren et al., subm.).

Of all of the sedimentological factors involved in controlling the final appearance of a jökulhlaup deposit, the most significant appears to be the volume of sediment carried. Maizels (1989a,b, 1991, 1993a) provides details of the characteristics of Icelandic hyperconcentrated flow deposits. Examples from the November 1996 jökulhlaup are shown in Russell and Knudsen (1999a, 2002a). Marren et al. (in press) suggest that hyperconcentrated flow deposits tend to produce tabular, laterally extensive sediment bodies that thin down-sandur. In contrast, low sediment concentration floods tend to form large-scale channel systems, with recognisable bar and channel components (Russell and Marren, 1999; Marren et al., 2002).

7.3.8.4 Jökulhlaup impact modelling and prediction

Although the above sections indicate that knowledge of the long-term impacts of jökulhlaups on sandur is improving, there have been few attempts to produce quantitative predictive models of jökulhlaup impact. Maizels and Russell (1992) provide an early attempt at a qualitative model, and Maizels (1991) uses a similar model to predict the likely sedimentary succession in an area of jökulhlaup drainage. To date, the only attempt at a quantitative model has been by Maizels (1993b). The model of Maizels (1993b) uses four main controlling conditions: (1) relative changes in sea level; (2) fluctuations in glacier volume and ice-margin position; (3) the magnitude, frequency and duration of runoff events; and (4) the sediment concentrations of each runoff event. The model was initially tested using Sólheimasandur, for which the Holocene jökulhlaup history has been fairly well established (Maizels, 1989a,b, 1991). The model was then applied to Skeiðarársandur, for which much less is known about the pre-settlement jökulhlaup history, and which has only limited Holocene sedimentary exposure compared to Sólheimasandur. The limited sedimentary exposure at Skeiðarársandur makes it impossible to test the predictions of Maizels' (1993b) model. Nonetheless, this work provides the only framework for analysis of the Holocene evolution of Skeiðarársandur, and suggests the minimum amount of information that will be required for similar studies of other sandar.

7.3.9 Jökulhlaup sediment flux

Data concerning sediment flux during jökulhlaups have been very difficult to obtain mainly for logistical reasons and because most of the largest Icelandic jökulhlaups and have occurred before routine sediment sampling became common practice. Sediment flux estimates have been derived from within-event samples of suspended sediment load and from sediment budgets derived from volumes of sediment erosion and deposition, based on pre-and post event topographic data.

Tómasson (1974) and Tómasson *et al.* (1980) provided some of the first detailed measurements of jökulhlaup sediment load in proglacial rivers during the 1972 and 1976 Skeiðarárhlaups. The 1972 jökulhlaup transported 29.5×10^6 tons of sediment from Skeiðarárjökull (Tómasson, 1974) but the large 1938 jökulhlaup is thought to have transported 86×10^6 tons of sediment. During the 1972 jökulhlaup, large amounts of fine-grained sediment were frozen to proglacial channel margins whilst little fine-grained sediment was deposited during the 1976 jökulhlaup (Tómasson *et al.*, 1980). Tómasson *et al.* (1980) suggest that anchor ice formation trapped the fine-grained suspended sediment load during the 1972 jökulhlaup thereby generating drastically different suspended sediment dynamics in comparison with other events.

Nummedal *et al.* (1987) estimated that jökulhlaups account for 92% of all sediment transported across Skeiðarársandur. Smith *et al.* (2000) and Magilligan *et al.* (2002) use SAR interferometry to estimate net erosion and deposition during the November 1996 jökulhlaup to suggest a sandur wide net volume gain of $31\text{-}38 \times 10^6$ m^3. Net erosion and deposition does however vary considerably depending upon location. Most deposition ($50\text{-}96 \times 10^6$ m^3) occurs within the immediate proglacial area as defined by the belt of abandoned outwash surfaces and ice-cored moraines. By contrast there is considerable jökulhlaup channel erosion within the main outlet channels through the moraine belt and in the area immediately downstream of the moraine belt (Smith *et al.*, 2000). Smith *et al.* (2000) highlight the importance of 'detachment' or decoupling of the ice proximal zone from the rest of the sandur by erosion of a proglacial trench. Although this concept is appealing and very similar to the model presented earlier by Russell and Knudsen (1999a), glacier retreat accompanied by major down-wasting accounts for 'trench' formation rather than solely by channel incision. Snorrason *et al.* (1997, 2002) indicate that suspended sediment concentrations decreased progressively from 121 to 5 g l^{-1} within the Skeiðará River during the November 1996 jökulhlaup. Snorrason *et al.* (1997,2002) estimate that 180×10^6 tons of suspended sediment were transported by the 1996 jökulhlaup, stressing justifiably that this figure most probably severely underestimates the total amount of sediment transported during this event.

Using seismic reflection surveys, Guðmundsson *et al.* (2003) provide the first systematic attempt to calculate the thickness and volume of sediment on Skeiðarársandur. Based on a total reconstructed sediment volume of 100 km^3,

Guðmundsson *et al.* (2003) estimated Holocene sandur growth at a minimum of 1 km^3 100 yr^{-1} and equated this to five 1996-equivalent jökulhlaups per century. Guðmundsson *et al.*'s study provides clear evidence that jökulhlaups and glacier surges dominate sediment supply to the sandur. It is therefore highly probable that the majority of the Skeiðarársandur sedimentary succession comprises jökulhlaup and surge related units. Recent improvements in our understanding of both jökulhlaup and glacier surge-related fluvial processes and knowledge of the thickness of Skeiðarársandur, may allow testing and revision of Maizels' pioneering research on sandar modelling (Maizels, 1993b).

Tómasson (1996) used historic maps (1904 and 1946) and eyewitness descriptions of the 1918 Kötluhlaup to calculate net sediment deposition and probable sediment concentration. Total sediment and ash volume for the 1918 Kötluhlaup is estimated by Tomasson at 3.6 km^3 with a total meltwater volume of 8 km^3. Such high sediment concentrations are also suggested by the resultant deposits (Maizels, 1989a,b,1993a, 1997).

Jökulhlaup hydrograph shape may exert a strong influence sediment flux (Rushmer *et al.*, 2002). Jökulhlaups with a steep rising limb will be able to entrain relatively large volumes of sediment very quickly resulting in high rising stage sediment fluxes (Russell *et al.* 2000, subm.; Rushmer *et al.*, 2002). Assuming a finite volume of sediment available for transport, sediment concentrations on more prolonged rising flow stages are likely to be much lower. Exhaustion of sediment supply is more likely during prolonged jökulhlaup rising flow stages. This is illustrated by the marked reduction in suspended sediment loads measured during the November 1996 jökulhlaup (Snorrason *et al.*, 2002).

7.3.10 Glacier-surge related jökulhlaups and surge-jökulhlaup interactions

Subglacial drainage system reorganisation and release of stored water commonly generates unusual floods, which have been described as jökulhlaups. The occurrence of jökulhlaups during glacier surges has also been known (Björnsson *et al.*, 2003). High sediment and water discharge during the 1963-1964 Brúarjökull surge had a very high sediment concentration (Pálsson and Vigfússon, 1996; Björnsson *et al.*, 2003).

Skeiðarárjökull is known to have surged seven times since 1787 (Björnsson *et al.*, 2003), with the three most recent surges in 1929, 1985/86 and 1991 (Pálsson *et al.*, 1992; Björnsson and Guðmundsson, 1993b; Guðmundsson *et al.*, 1995; Björnsson, 1998; Björnsson *et al.*, 2003). At Skeiðarárjökull patterns of proglacial meltwater discharge were observed to alter radically during the 1991 surge (Björnsson, 1998). Meltwater discharge in the Skeiðará fell to less than quarter of normal, in marked contrast to the Gígjukvísl where discharges increased 3 or 4 fold (Pálsson *et al.*, 1992; Björnsson and Guðmundsson, 1993; Björnsson, 1998; Björnsson *et al.*, 2003). The discharge increase in the Gígjukvísl is consistent with observations of significant meltwater discharge from numerous regularly spaced new outlets in the central snout area of Skeiðarárjökull (Pálsson *et al.*, 1992; Björnsson and Guðmundsson, 1993; Björnsson, 1998). Truncation of the 1991 Skeiðarárjökull surge simultaneously with the jökulhlaup from Grímsvötn indicates the importance of surging for glacier hydrology and meltwater discharge (Björnsson and Guðmundsson, 1993b; Björnsson, 1998). Björnsson and Guðmundsson (1993) and Björnsson (1998) reported that 75% of the 1991 drainage of Grímsvötn, was halted by tunnel closure or tunnel inhibition associated with 1991 surge-induced enhanced ice-creep rates or the presence of a distributed drainage network respectively. Björnsson *et al.* (2003) describe how the 1994 surge of Síðujökull changed the ice-surface topography and consequent subglacial jökulhlaup routing to allow the 1994 and 1995 Skaftáhlaups to drain into the Hverfisfljót (Björnsson *et al.*, 2003). Such surge-controlled variations in proglacial meltwater discharge therefore need to be added to the present range of dominant flow magnitudes and frequencies controlling proglacial outwash in Iceland and elsewhere.

Russell *et al.* (2001a) and van Dijk and Sigurðsson (2002) present evidence of distinctive, proglacial outwash fans deposited by the 1991 Skeiðarárjökull surge. Kettled surge-related deposits on Skeiðarársandur closely resemble jökulhlaup deposits (van Dijk, 2001; van Dijk and Sigurðsson, 2002). Caution must therefore be exercised in distinguishing deposits of prolonged (weeks – months) from those of jökulhlaups occurring over shorter timescales (hours – weeks).

Surge-induced glacier margin fluctuations have major implications for proglacial jökulhlaup routing, flow characteristics and impacts. Glacier surges on Skeiðarárjökull allowed rapid proglacial trench infilling with ice thereby allowing unconfined proglacial jökulhlaup deposition to take place. The 1934 and 1938 Skeiðarárhlaups occurred just as the glacier margin began to retreat following the 1929 surge. Unconfined by proglacial topography, these jökulhlaups discharged radially from each outlet along the glacier terminus. Later jökulhlaups in 1945 and 1954 drained into the developing proglacial trench, initiating the present course of the Gígjukvísl River. Despite the fact that glacier surge cycles are superimposed upon climate-forced glacier margin fluctuations, surges have the potential to influence proglacial jökulhlaup routing and impact.

7.3.11 Jökulhlaup landscapes

7.3.11.1 Jökulsá á Fjöllum and Hvítá

Evidence for large pre-historic cataclysmic floods (peak discharges $0.2 - 1.0 \times 10^6$ m^3 s^{-1}) in the Jökulsá á Fjöllum has been presented by Sæmundsson (1973), Tómasson (1973, 2002), Elíasson (1977), Malin and Eppler (1981), Sigbjarnarson (1990) and Waitt (1998, 2002). Evidence of these floods includes large tracts of scoured 'scabland' topography, bedrock gorges, streamlined erosional hills, boulder fields, large bars and bedforms (Figs. 7.1,3-6,32). Several smaller historical jökulhlaups have been recorded in 1477, 1684, 1707, 1711, 1716, 1717, 1719, 1726 and 1730 (Þórarinsson, 1950; Ísaksson, 1985). These floods were confined to the gorge system created by the earlier, larger floods and only involved discharges of 10^4 m^3 s^{-1}.

So far, there is only a limited picture of the magnitude, frequency and geomorphic effectiveness of pre-historic cataclysmic floods. Controversy still exists regarding the number, age, size and sources of individual floods. Tómasson (1973, 2002) only presents evidence of a late Holocene flood (2500 ^{14}C yr BP). Sæmundsson (1973) and Waitt (1998) suggest only two great floods, one early Holocene ($<$7100 ^{14}C yr BP) and the other later in the Holocene (ca 2900-2000 ^{14}C yr BP). Elíasson (1977) and Sigbjarnarson (1990) propose 3 and 5 floods respectively in a period when Waitt (1998) stated there were none. Elíasson (1977) suggested a peak late Holocene flood discharge figure of 0.2×10^6 m^3 s^{-1}, similar to Malin and Eppler's figure of 0.3×10^6 m^3 s^{-1}. Tómasson (1973) calculated a late Holocene

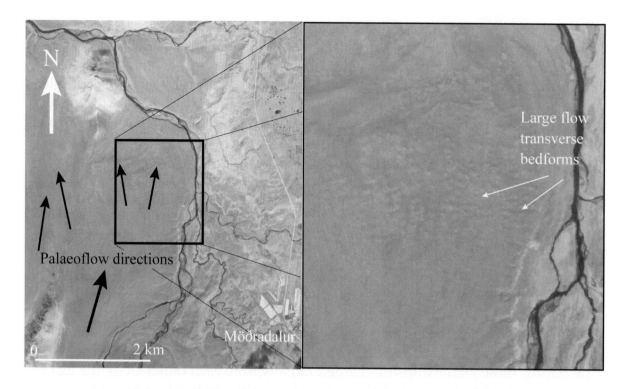

Fig. 7.32 Aerial photograph showing large-scale bar near Möðrudalur á Fjöllum. Palaeoflow directions are indicated by arrows. Note the presence of both flow transverse and parallel bedforms on the giant jökulhlaup bar surface.

average discharge of 0.5×10^6 m^{3}s^{-1}. Waitt (1998, 2002) suggests from step-backwater computation on the lower reaches of the Jökulsá á Fjöllum that the late Holocene flood reached a peak discharge of 1×10^6 m^{3}s^{-1}. Björnsson (1988,2002) and Björnsson and Einarsson (1990) suggested that volcanic activity in the Bárðarbunga subglacial caldera may be the source of the flows that created the deeply incised canyons of the Jökulsá á Fjöllum. Tómasson (1973) suggested that the source lies in the Kverkfjöll, Grímsvötn or Bárðarbunga calderas or even from an ice-dammed lake to the south of Kverkfjöll. Tómasson (2002) however favoured Bárðarbunga as suitable source of volcanically triggered jökulhlaups. Björnsson (2002) and Käyhkö *et al.* (2002) identify Kverkfjöll as potential sources of jökulhlaups. Waitt (1998, 2002), suggested that the Kverkfjöll caldera as the source of some of the floods, but considered the largest jökulhlaups to have drained from Dyngjujökull. Marren *et al.* (subm.) suggest at least 6 jökulhlaups of possible historical age have drained from the snout of Kverkjökull on the basis of sedimentary evidence. Russell *et al.* (2000, 2001) and Carrivick *et al.* (2002, subm.) suggest that at least two jökulhlaups with peak discharges of $10^4 - 10^5$ m^{3}s^{-1} drained from the flanks of Kverkfjöll during the Holocene. Alho (2003) used satellite remote sensing to map large-scale patterns of jökulhlaup erosion and deposition within the Jökulsá á Fjöllum river system in northern Iceland.

The Hvítá river is perhaps best known for Gullfoss waterfall and canyon system, which are believed to be formed originally by jökulhlaups (Tómasson, 1993, 2002). Tómasson (1993) presents evidence of extensive high-energy erosion of bedrock surfaces both aerially and as bedrock channels. Large bars located downstream of Gullfoss are also attributed to jökulhlaups (Tómasson, 1993, 2002). A series of over 11 ice-dammed lake shorelines in the Kjölur area are though to have formed over a 100-200 year period ~9500 years ago. Tómasson (1993,2002) calculated a maximum ice-dammed lake volume of 25 km³ resulting in peak discharges of over 200,000 m³s⁻¹.

7.3.11.2 Mýrdalssandur, Sólheimasandur, Skógasandur and Markarfljót

Jökulhlaups associated with eruptions of Katla subglacial volcano have had a profound impact on sediments and landforms on Mýrdalssandur, Sólheimasandur, Skógasandur and Markarfljót (Jónsson, 1982; Maizels, 1989a,b, 1991, 1992, 1993a,b; 1997; Olszewski and Weckwerth, 1999; Wisniewski, *et al.*, 1999). Maizels' 1997 model of sandur development in areas subject to volcano-glacial jökulhlaups was based upon sedimentological studies of Sólheimasandur, Skógasandur and Mýrdalssandur. There is also ample evidence of jökulhlaups in the Markarfljót (Haraldsson, 1981; Björnsson *et al.* 2000; Smith 2001, Tómasson, 2002) and possible routing of jökulhlaups to the northern margin of Mýrdalsjökull (van der Meer *et al.*, 1999).

Jónsson (1982) presented a model for Kötluhlaups as volcanogenic debris flows where ice blocks were transported on the surface of relatively dense sediment rich flows. Maizels (1989a,b,1991,1993a,1997) identified massive pumice granule lithofacies as the product of hyperconcentrated flows (Fig. 7.33). Her model of outwash development in areas of volcanoglacial jökulhlaup drainage (Type III sandar) is dominated by debris flow and hyperconcentrated flow deposits displaying lobate depositional morphology and channelised and hummocky erosional morphology (Maizels, 1997). Maizels (1997) points to a lack of well-defined aggradational bar morphology and instead identifies widespread surfaces supporting numerous kettle-holes. Tómasson (1996) reconstructed a hydrograph for the 1918 Kötluhlaup emphasising the occurrence of a series of smaller jökulhlaups (~10⁴ m³s⁻¹) after the main 2-3 x 10⁵ m³s⁻¹ main flood surge. Although the dynamics of the 1918 jökulhlaup have been reconstructed from both eyewitness accounts and reconstruction of the sedimentary record there is still considerable uncertainty regarding the rheology, relative magnitude and sediment preservation of earlier Kötluhlaups. The role of down sandur jökulhlaup flow transformation and within event reworking also need to be investigated. Although Maizels identified the sedimentary characteristics of sandar dominated by volcanogenic jökulhlaups, it is entirely feasible that sandar can be subjected to a mixed population of jökulhlaup sources and drainage mechanisms. Further sedimentological and geophysical research is required on the sandur radiating from Mýrdalsjökull in order to pinpoint the controls on the magnitude, dynamics and sedimentary impact of Kötluhlaups.

7.3.12 Implications of jökulhlaups for offshore sedimentation

Volcanogenic jökulhlaups are believed to play an important role in the supply and transport of sediment to southern Iceland shelf and transporting sediment via turbidity currents into the Iceland Basin (Jónsson, 1982; Nummedal *et al.*, 1985; Lacasse *et al.*, 1995, 1996; Maria *et al.*, 2000). Lacasse *et al.* (1996) attribute the source of three silicic ash zones within the Irminger and Iceland deep-sea basins to volcanic eruptions within Icelands' southeast volcanic zone during glacial or cold periods around 11,000; 55,000; and 305,000 BP. Maria *et al.* (2000) examined the dispersal of sediments offshore during the November 1996 jökulhlaup and concluded that the source of most of the sediment

*Fig. 7.33: a, top:
b,centre: c, bottom.*

Fig. 7.33 opposite. Sections of jökulhlaup deposits on Sólheimasandur and Mýrdalssandur. (a, top) coarsening upward unit within jökulhlaup deposits on Sólheimasandur. This unit reflects deposition of sediment during the rising flow stage of a fluidal jökulhlaup flow; (b, centre) Massive pumice unit is consistent with deposition from a hyperconcentrated flow, only very faint sub-horizontal layering can be picked out; (c, bottom) detailed view of hyperconcentrated flow deposits showing clasts 'floating' within a structureless matrix. The vertical structures on the left of the photograph are post-depositional clastic dykes.

reworked offshore by the 1996 event was sources originally from Katla. It appears therefore that large jökulhlaups from Katla dominate the offshore sedimentary record even in immediate proximity to the Skeiðarársandur sediment dispersal system.

Given the potential for oceanic triggers of abrupt climate change in the North Atlantic region, there is now a pressing need to identify the occurrence of, and controls on, potentially significant influxes of fresh water to the marine system from Iceland. Work is needed to understand the characteristics of and controls on large palaeo-jökulhlaups, which are known to have drained from a number of sources in Iceland. Jökulhlaups may have conspired with other factors to trigger abrupt climate change.

7.4 CONCLUSIONS AND OUTLOOK

The November 1996 jökulhlaup acted as a catalyst for greater understanding of various aspects of the jökulhlaup system. Since previous review papers there has been a greater awareness of the coupling of the subglacial and proglacial jökulhlaup processes. The ability to examine the deposits of a single high magnitude jökulhlaup has provided valuable 'control' on process-form relationships within a depositional environment where instrumentation is extremely difficult if not impossible. Observations of recent jökulhlaups have brought to life descriptions of historical jökulhlaups within the Icelandic literature and have reaffirmed our faith in the accuracy of historic observations made by the inhabitants of areas subject to some of the world's largest contemporary floods.

ACKNOWLEDGEMENTS

We acknowledge fieldwork grants from: The Earthwatch Institute (AJR, PMM, HF, FST and MJR), NERC (GR3/10960) (AJR), NERC (GR3/12969) (AJR and FST) and NERC (GT04/97/114/FS) (HF).

8. Environmental and climatic effects from atmospheric SO_2 mass-loading by Icelandic flood lava eruptions

Þorvaldur Þórðarson

Department of Geology and Geophysics, School of Ocean and Earth Sciences and Technology, University of Hawaii at Manoa, Hawaii 96822, USA, and Science Institute, University of Iceland, Dunhagi 6, IS-107, Reykjavík, Iceland

8.1 INTRODUCTION

Volcanic eruptions are dynamic and multifaceted events that affect the environment in various ways, both locally and remotely. In areas remote to the volcano (>1000 km) the effects of the eruptions are largely confined to the atmosphere and widespread volcanic aerosol plumes have resulted in significant perturbations on a hemispheric to global scale (e.g. Lamb, 1970; Pollock et al., 1976; Self et al., 1981; Rampino and Self, 1984; Hoffmann, 1987; Robock, 1991, 2000).

Flood lava eruptions were not considered much in early discussions on volcanism-climate interactions, despite existing knowledge of the widespread atmospheric effects of the 1783-84 Laki eruption in Iceland (e.g. Þórarinsson, 1979, 1981; Wood, 1984, 1992; Angell and Korshover, 1985; Sigurðsson, 1990; Stothers, 1996; Þórðarson and Self, 2003). The perception was that flood lava eruptions were short-lived cataclysmic events with low eruption columns (<5 km), a view that resulted in ambiguous assessments of their potential capability to produce lasting atmospheric perturbations (e.g. Stothers et al., 1986; Stothers, 1989; Woods, 1993). However, recent studies have appreciably advanced our understanding of the mechanics of flood lava eruptions (e.g. Mattox et al., 1993; Þórðarson and Self, 1993, 1998; Hon et al., 1994; Self et al., 1996, 1997, 1998; Keszthelyi and Self, 1998) and thus altered our views of their eruption dynamics and duration. The results show that they are prolonged eruptions (months to years) characterized by recurring eruption episodes. Each episode begins with a short-lived explosive phase that is followed by a longer-lasting phase of lava effusion.

Flood lava eruptions are capable of releasing huge amounts of sulphur into the atmosphere, because of the high sulphur yield of basaltic magmas coupled with extremely efficient vent degassing in the recurring explosive phases. Consequently, prolonged flood lava eruptions can maintain elevated atmospheric concentrations of H_2SO_4 aerosols for periods of months to years (e.g. Þórðarson and Self, 1996b; Þórðarson et al., 1996). Accordingly, recent studies have emphasized the significance of flood lava eruptions for short-term environmental and climate changes whilst others have pointed out the apparent association of flood basalt volcanism with major climatic excursions in the past (e.g. Rampino *et al.*, 1988; Rampino and Caldeira, 1992; Stothers 1993, 1996, 1998; Grattan and Pyatt, 1994; Grattan and Brayshay, 1995; Þórðarson, 1995; Courtillot *et al.*, 1999; Larsen, 2000; Þórðarson and Self, 2001, 2003).

Here we present an overview of the atmospheric sulphur mass loadings and effects of the three largest Holocene flood lava eruptions in Iceland; the AD 1783-84 Laki, AD 934-40 Eldgjá, and ~8500 cal yr BP Þjórsá events. The dialogue on the environmental-climatic effects is focused on Laki because it is by far the best documented event. This is followed by a short discussion on the potential impact of the Eldgjá and Þjórsá eruptions.

8.2 ICELANDIC FLOOD LAVA ERUPTIONS

8.2.1 Eruptive histories and eruption mechanics

The Laki, Eldgjá, and Þjórsá events represent the three largest Holocene flood lava eruptions in Iceland. These eruptions occurred on volcanic fissures within the Eastern Volcanic Zone in South Iceland (Fig. 8.1). The composition of the erupted magma ranges from olivine tholeiite (Þjórsá) through quartz tholeiite (Laki) to transitional alkali basalt (Eldgjá) (Vilmundardóttir, 1977; Jakobsson, 1979; Þórðarson *et al.*, 1996, 2001). In all three eruptions >93% of the magma was erupted as lava, the reminder being tephra and dispersed up to 100 km from the source (Table 8.1).

These were all prolonged eruptions. The Laki eruption lasted for eight months (8 June 1783-7 February 1784; Þórðarson and Self, 1993) and a 3-6 year duration is inferred for the Eldgjá eruption (Zielinski *et al.*, 1995; Þórðarson, *et al.*, 2001). The duration of the Þjórsá eruption is not known. However, inflated sheet lobes in the distal sector of the lava field indicate an emplacement time in excess of 5 months (Þórðarson and Self, 1998) and thus the eruption may have lasted for several years.

The Laki and Eldgjá eruptions featured a series of eruption episodes, each associated with rifting of the crust and sudden increase in the magma discharge (Table 8.1; Fig. 8.2). Each episode was distinguished by a short-lived explosive phase at the beginning,

Table 8.1. Summary table for the Laki, Eldgjá and Þjórsá flood lava eruptions showing the eruption duration; length of vent systems, LVS; total volume of erupted magma, EV; volume of lava, LV; maximum length of lava, MFL; volume of tephra, TV; area within the 0.5 cm isopach, TFA; and number of eruption episodes, nEE. All volume figures are given as dense rock equivalent

Eruption	Duration (years)	LVS (km)	EV (km³)	LV (km³)	MFL (km)	TV (km³)	TFA (km²)	nEE
Laki (1783-84)	0.67	27	15.1	14.7	65	0.4	>7200	10
Eldgjá (934-40)	3-5	75	19.6	18.3	75	1.3	>10,000	>20
Þjórsá (~8600BP)	?	?	>22	22	120	?	?	?

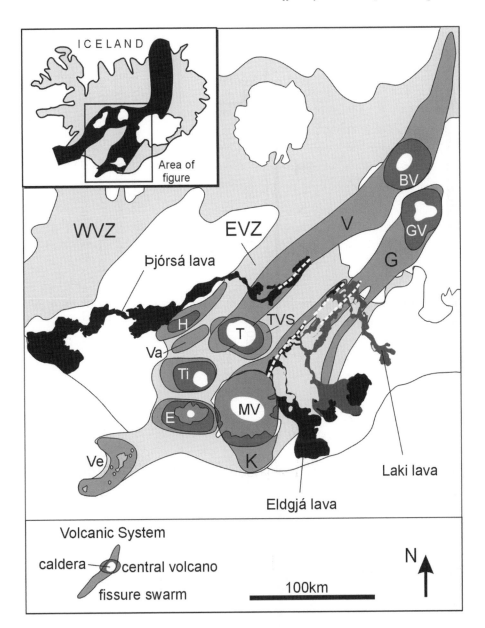

Fig. 8.1 Map of the Eastern Volcanic Zone (EVZ) in Iceland showing geologic setting of the Veiðivötn (V), Grímsvötn (G) and Katla (K) volcanic systems, the location of the corresponding Bárðarbunga (BV), Grímsvötn (GV) and Mýrdalsjökull (MV) central volcanoes, and distribution of the Laki, Eldgjá and Þjórsá vent systems and lava flows. The other six volcanic systems of the EVZ are indicated as follows: Vestmannaeyjar (Ve), Eyjafjöll (E), Tindfjöll (Ti), Torfajökull (T), Vatnafjöll (Va), and Hekla (H). Inset shows the volcanotectonic setting of the EVZ in relation to other volcanic zones in Iceland. Modified from Jakobsson (1979).

followed by a longer-lasting phase of lava emission. The explosive activity was typically of the sub-plinian type, producing eruption columns reaching to heights of 10-15 km (Miller, 1989; Þórðarson and Self, 1993; Larsen, 2000; Þórðarson *et al.*, 2001). Both eruptions also featured distinct phreatomagmatic phases and concurrent subglacial explosive eruptions at the corresponding central volcanoes (Figs. 8.1, 8.2).

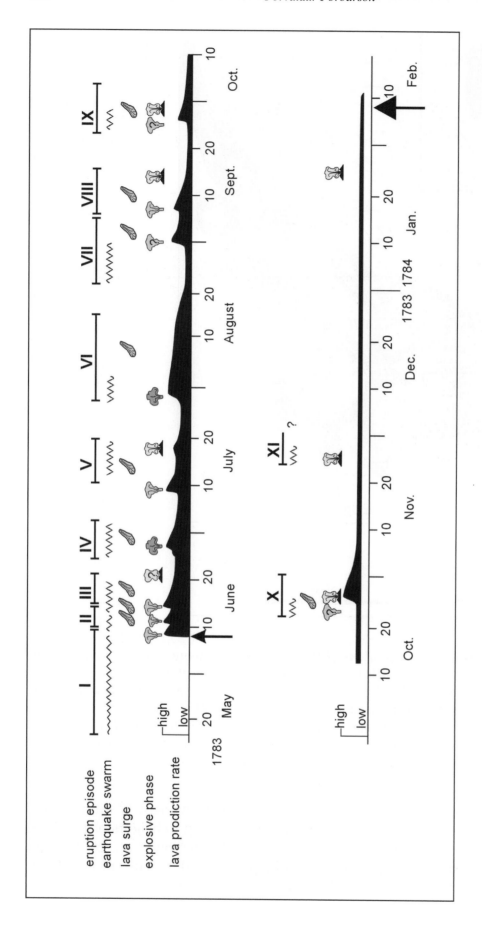

Fig. 8.2 Schematic illustration of sequence of events during the Laki-Grímsvötn eruptions. Extent of earthquake swarms is indicated by wiggly lines; fluctuations in lava discharge shown by shaded area (not to scale); eruption clouds denote explosive activity at Laki fissures; eruption clouds with a cone at the base denote explosive activity at Grímsvötn volcano; arrows indicate onset and termination of Laki eruption. The solid bars show the extent of each eruption episode, labelled I, II, III etc. Modified from Þórðarson and Self (1993).

Europe (July 1783 is the second hottest month on record in England after July 1995) and severe thunderstorms and hailstorms were a frequent occurrence. Inland regions of Eurasia (i.e. Poland to China) experienced unstable and cold weather, which among other things featured mid-summer snowfall. This summer is singled out as a particularly calamitous time in Japan. Stagnation of the polar front zone to the northeast of Japan resulted in unusually cool and wet weather conditions. As a consequence the country's rice harvest failed completely and brought about the most severe famine in the Japan's history.

Winter 1783-84 was one of the most severe in Europe and North America in the last 250 years. It began early (September to October) and was characterized by unusual and long-lasting frosts and heavy snowfall. In northern, western and central Europe the winter was unusually cold. The Icelandic lowlands and fjords were covered with thick ice until end of May 1784 and the ground was frozen at grass-root levels well into July. The Danish straits were completely frozen over and Jutland was still covered by ~1 m thick snow in mid April. In the Netherlands the severity was such that people travelled on ice across the Markersee and between coastal villages. In western and central Europe rivers were covered with ice due to long-lasting frost. This hindered travel and caused a severe shortage of firewood in major cities such as Paris and Vienna. The winter was harsh in Italy and the lemon crops were totally destroyed by intense frost. However, in the inland regions of Eurasia the severity appears to have been less. The "Long Winter" of 1783-84 is one of three landmark winters of the 18th century in the Eastern United States, the others being 1740-41 and 1779-80. It caused the longest ice-induced closure of the harbours and channels of Chesapeake Bay and in February 1784 ice floes covered the Mississippi River in the region of New Orleans.

Late 18th century temperature records from 26 stations in Europe and three in North America show that the 1783 summer temperatures were above average (Fig. 8.6) and that significant temperature variations existed between regions. In north, west, and central Europe the 1783 summer mean temperatures were 1-3°C above average, mainly because of the unusually hot July. In eastern and southern Europe, as well as North America, the summer temperatures were close to the long-term mean and no July anomaly is apparent. The winter mean temperature deviations indicate a very sharp and strong cooling in 1783-84 in Europe and eastern United States ($\Delta T = -3°C$) and the following two winters were also well below the mean but somewhat warmer. The summers 1784-86 were also characterized by cooler temperatures (Figs. 8.6a,b).

The annual mean temperature deviations show that the three years following the Laki eruption were by far the coldest years of the 31-year period centered on 1783 ($\Delta T = -1.3°C$). Analysis of the temporal distribution of the cold years and seasons within this 31-year period further emphasises the cooling that followed Laki and shows that the 1784, 1785, and 1786 have by far the highest counting frequency (i.e. register) of coldest years in this time period (Figs. 8.6c,d).

In the three years following the Laki eruption the surface temperatures in Europe and North America were well below average and they may have been the coldest years in the latter half of the 18th century (Fig. 8.6). The lower stratospheric H_2SO_4 aerosol burden from Laki translates into an excess optical depth of 0.7, a radiative forcing that has the capacity to force a 1.5°C drop in the annual mean surface temperature.

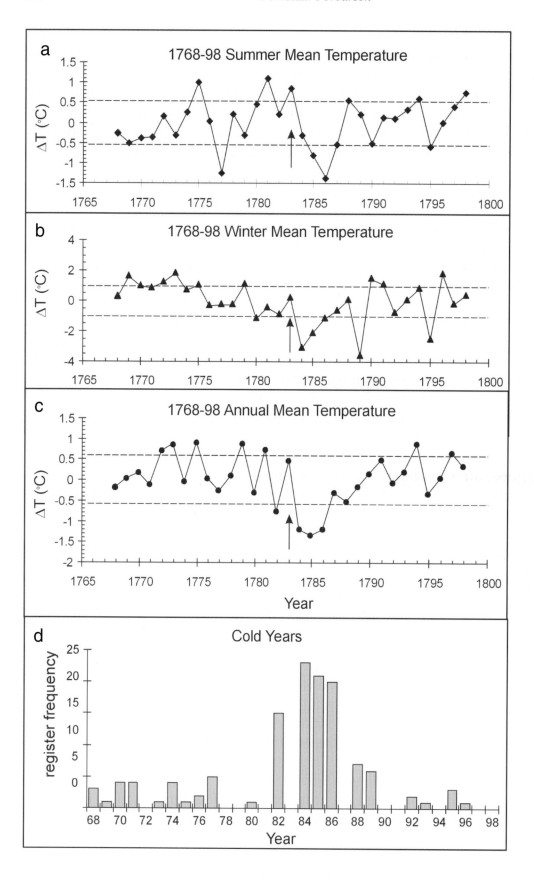

However, the evidence shows that bulk of the Laki haze was removed from the atmosphere by late summer of 1784. Consequently, the low annual temperatures in the years 1785 and 1786 cannot be directly attributed to radiative perturbations caused by the Laki haze. Nonetheless, back-to-back occurrence of cold years in Europe and North America implies a common source for this short-term climatic excursion and its close temporal association with Laki indicates that these two events are related. Is it possible that a one year-long-perturbation suppressed the climatic system to such a degree that it took two additional years for it to recover to a "normal state"?

The northerly location of the Laki fissures (~64°N) may be the key to this puzzle, because sulphur-rich eruptions at high latitudes are likely to cause high concentrations of volcanic aerosols in the Arctic atmosphere. Information from available Arctic sites (e.g. Fig. 8.4b) are consistent with a heavy H_2SO_4 aerosol burden in the Arctic and the GISP ice core data implies that it remained in the Arctic atmosphere through the summer of 1784. Accordingly, this aerosol burden may have caused a strong disruption in the Arctic thermal balance over two summer seasons when the incoming radiative flux is at its peak. The net effect of this type of perturbation is substantial heating of the Arctic atmosphere and subsequent reduction of the equator-pole thermal gradient. The consequence of this excess heating is a weaker westerly jet stream and increase in the frequency of mixed or meridional circulation.

Weakening of the westerly jet stream in the wake of the Laki eruption is consistent with the available historical climatic data. The frequency of progressive (westerly) weather over the British Isles during 1781-85 was well below the 1860-1978 five-year running mean, with a sharp reduction in 1784 and 1785. Similar reductions are observed in Central Europe, although the anomaly is less pronounced. Other evidence that indicate a weaker jet stream are flood and drought patterns in China from 1783 to 1785, severe drought in India caused by weaker summer monsoons, and the late summer stagnation of the polar front along the Pacific coast of Japan.

Graf (1992) used climate simulations to examine the effects of aerosol loading from a powerful volcanic eruption at high northern latitudes. The results indicate that the high latitude radiation deficit, similar to that caused by Laki, would have significant effects on the global climate. The model predicts a weakening of the westerly jet stream, prolonged winter monsoon conditions over India, and the development of a negative Walker anomaly in the Pacific. He also demonstrated that if the radiation anomaly were removed after 7 months, the forced weather and circulation pattern would still prevail for a few years.

Fig. 8.6 left. Late 18th century mean surface temperature deviations (°C) for a 31 year period centered on 1783, the eruption year of Laki: a) summer, b) winter, and c) annual. Data from 29 stations in Europe and northeastern United States were used in this reconstruction. The standard deviation (2s) of the 31-year-mean is shown by broken lines. Analysis by Þórðarson (1995) based on data sets from Jones et al., (1985), Landsberg et al. (1968), Groveman and Landsberg (1979), Reiss et al. (1980). (d) Frequency-distribution of cold summers, cold winters, and cold years in Europe and eastern United States during the period 1768 to 1798. The analyses are based on registering the 4 coldest years at each station and then adding the number of occurrences for each year from all of the stations involved.

The model results also indicate that the climate response to this type of forcing is not a uniform one, because it produces both negative and positive temperature anomalies and their distribution changes with the seasons. This may explain why the Laki signal is present in some dendrochronological records, but does not show up in others (e.g. Schove, 1954; Briffa *et al.*, 1988; and papers in Bradley and Jones, 1992).

8.4 POTENTIAL EFFECTS OF THE ELDGJÁ AND ÞJÓRSÁ ERUPTIONS

The very high H_2SO_4 yield of the Eldgjá and Þjórsá events strongly suggests that they were climatically significant eruptions. We also surmise that the eruption columns were relatively low (<16 km), implying an upper troposphere - lower stratosphere mass loading of aerosols, a dispersal by the westerly jet stream and a hemispheric coverage above 30° North (Lamb, 1970). Thus, it is reasonable to presuppose that the atmospheric effects of these two eruptions were of similar nature and magnitude to that of Laki.

Occurrences of volcanic haze (dry fog) over Europe in AD 934 and AD 939 that coincided with unusual winter weather in Europe and the Middle East support this view (Stothers, 1998; Þórðarson *et al.*, 2001). These records fit well with the notion of a prolonged eruption and episodic sulphur mass loading, indicating that the environmental and climatic effects of the Eldgjá eruption were at least equivalent to that of Laki and of longer duration. However, precise assessment of the impact of Eldgjá is not straightforward. The timing of the eruption in the Dark Ages of Europe precluded the detailed documentation of atmospheric and weather-related phenomenon that is available for the Laki eruption. In addition, the Eldgjá eruption was prolonged, possibly lasting for as much as 6 years, and featured a number of significant eruption episodes spaced at unknown intervals. Thus, despite much greater bulk mass loading of sulphur, the intensity of climatic effects induced by the Eldgjá event may have surpassed that of Laki because the sulphur emissions were drawn out over several years.

The magnitude of the atmospheric effects of the Þjórsá eruption is even more difficult to assess because relevant data are not available at present. However, it is noteworthy that the eruption occurred close to the onset of an early Holocene cold event (~8300 cal yr BP) of uncertain origin (e.g. Hu *et al.*, 1999). Whether these two events are linked is open to discussion and worth further consideration.

8.5 CONCLUSIONS

The available historic records give a good indication of the magnitude of the environmental and climatic effects of the Laki eruption. However, the fact that most of these records are from the North Atlantic and Europe restricts their effectiveness for assessing the exact magnitude and consequences of these effects on a hemispheric scale. A greater spatial coverage is desirable and may be obtainable because historic records from Asia (e.g. China and India) might contain further information on the effects of the Laki haze in these regions. Another attractive option is to analyse the effects of the Laki haze with global climate simulations, using the existing historical database as the ground truth. Such simulations would have to be able to account for volcanological factors, such as prolonged but episodic sulphur emissions and atmospheric factors, upper-troposphere/lower-stratosphere transport for the volcanic plumes and near-instantaneous atmospheric removal of a portion of the aerosols. Such a climate model would not only advance our understanding of the environmental and radiative effects of the Laki eruption, but would provide us with a tool for evaluating the impact of past flood lava eruptions, such as Eldgjá, Þjórsá and the much larger flood lava eruptions of continental flood basalt provinces. Also, such a model is critical for predicting the effects of future Laki-like eruptions, because eruptions of similar magnitude will occur again in Iceland and possibly in the near future. If such an eruption were to occur today, a presence of persistent haze at altitudes between 8-12 km would most likely result in halting and/or re-routing of air traffic over large portions of the Northern Hemisphere. This could also have serious economic ramifications. Another concern is how modern ecosystems, which are already over-stressed and weakened by human pollution (Park, 1987), would respond to an event of Laki-magnitude. Will they cope?

ACKNOWLEDGEMENTS

I thank Trausti Jónsson and an anonymous reviewer for constructive reviews and helpful comments.

9. Holocene glacier history

Maria Wastl and Johann Stötter

Institute of Geography, University of Innsbruck, Innrain 52, A-6020 Innsbruck, Austria

9.1 INTRODUCTION

This paper gives an overview of the present knowledge of Holocene glacier history in Iceland. The geomorphological evidence and the available dating control yielded by investigations of Holocene environmental and climatic history in Northern Iceland (Wastl, 2000; Wastl et al., 2003) are presented and compared to reconstructions of glacier history in other areas, with a discussion of the problems of establishing a reliable and high-resolution dating control for the Holocene glacier history in Iceland.

9.2 RECONSTRUCTION OF HOLOCENE GLACIER HISTORY IN NORTHERN ICELAND

9.2.1 Determination of Holocene glacier extents

Tephrochronological evidence from sections in Þverárdalur and Lambárdalur (Figs. 9.1, 9.2) (Stötter and Wastl, 1999; Wastl, 2000; Wastl et al., 2001) indicates that glaciers in Northern Iceland had reached extents at or within their Little Ice Age limits before ca 9200 ^{14}C yr BP. Post-Preboreal glacier extents were, at most, only in slightly advanced positions relative to those of the Little Ice Age.

The aerial photograph of the glacier in Þverárdalur (ca 65°54'N, 18°44'W) (Fig. 9.3) shows a distinct moraine ridge outside the forefield area. The greatest distance to the outermost Little Ice Age moraine is about 100 m (see Fig. 9.4). A section outside the pre-Little Ice Age moraine (Fig. 9.5, section G1998) contains a sequence of tephra layers back to the Saksunarvatn ash deposited ca 9200 ^{14}C yr BP (see Wastl, 2000). This moraine thus marks the maximum extent of the glacier since the late Preboreal. As there is hardly any difference in altitude between the outermost Little Ice Age and the older moraine the corresponding depression of the equilibrium line altitude (ELA) for the post-Preboreal Holocene lies within the margin of error of the reconstructed ELA for the Little Ice Age maximum extent.

On the aerial photograph of the glacier in Lambárdalur (ca 65°48'N, 18°34'W) (Fig. 9.6) two moraines can be distinguished outside the forefield area within a maximum distance of ca 200 m to the outermost Little Ice Age moraine. The record of tephra

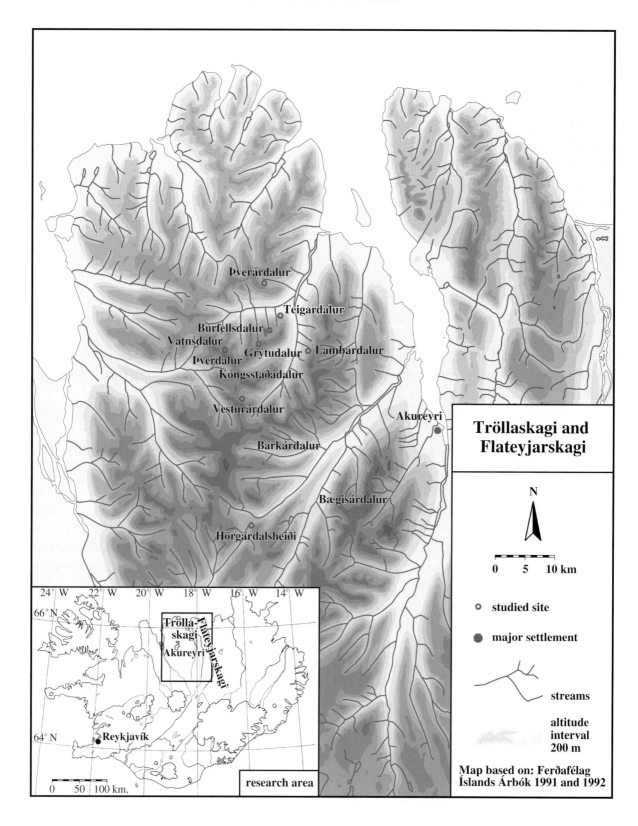

Fig. 9.1 Location of the sites studied for the reconstruction of Holocene glacier history in Northern Iceland.

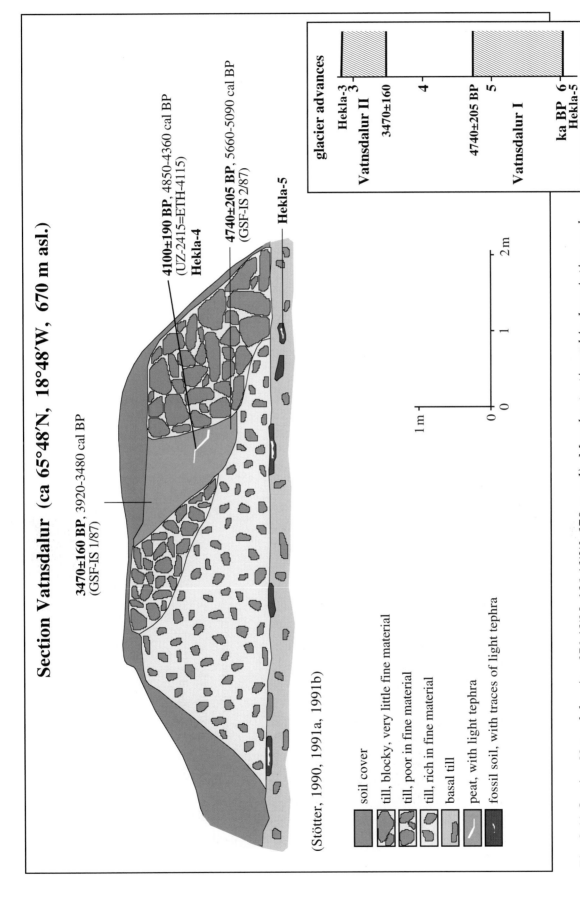

Section Vatnsdalur (ca 65°48'N, 18°48'W, 670 m asl.)

3470±160 BP, 3920-3480 cal BP
(GSF-IS 1/87)

4100±190 BP, 4850-4360 cal BP
(UZ-2415=ETH-4115)
Hekla-4

4740±205 BP, 5660-5090 cal BP
(GSF-IS 2/87)

Hekla-5

(Stötter, 1990, 1991a, 1991b)

soil cover

till, blocky, very little fine material

till, poor in fine material

till, rich in fine material

basal till

peat, with light tephra

fossil soil, with traces of light tephra

glacier advances

Hekla-3
3
Vatnsdalur II
3470±160
4
5
4740±205 BP
Vatnsdalur I
ka BP 6
Hekla-5

1m

0   1   2 m

Fig. 9.11 Section Vatnsdalur (ca 65°48'N, 18°48'W, 670 m asl). Morphostratigraphic description and dating control according to Stötter (1990, 1991a,b).

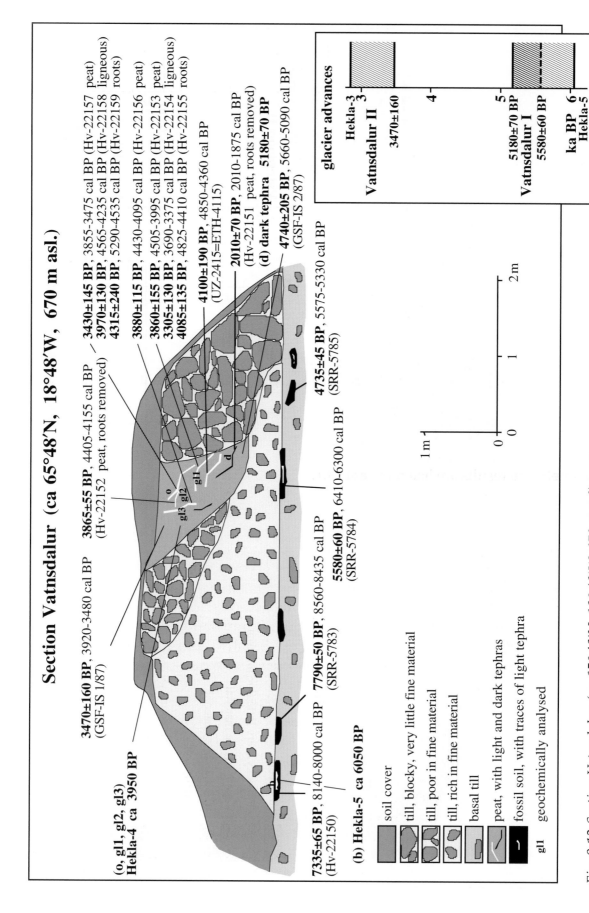

Fig. 9.12 Section Vatnsdalur (ca 65° 48'N, 18° 48'W, 670 m asl).

section Vatnsdalur (see Wastl, 2000). The dark tephra could be identified at two different sites between the forefield area and the inner pre-Little Ice Age moraine in Lambárdalur (Fig. 9.7, core B3 and section Ki). Neither core B3 nor section Ki, however, contain the Hekla-4 or Hekla-3 tephras, which form conspicuous marker horizons post-dating the dark layer. While it cannot be excluded that the upper part of the sediment column is missing in these profiles, as the sites lie close to a stream running from the glacier (see Fig. 9.6), additional evidence on the tephrostratigraphy of this area is clearly necessary. Before that, these findings should be regarded as preliminary.

If the present evidence can be confirmed, this would imply a minimum age of 5180±70 ^{14}C yr BP (AA-28979) for the formation of the inner moraine. The two glacier advances in Lambárdalur might thus correspond to the two phases of the Vatnsdalur I advance. As there are no traces of soil formation between the two till bodies of the Vatnsdalur I moraine (see Fig. 9.11), Stötter (1990, 1991a,b) assumed that the two advance phases occurred within a relatively short time. The proposed dating control for the Lambárdalur I and II advances would support these conclusions.

9.2.2.3 Site Þverárdalur

The sections on and inside the pre-Little Ice Age moraine in Þverárdalur contain several tephra layers (see Fig. 9.5). The geochemical analyses of the light tephras lying over the till material allow identification of the Hekla-4 tephra in sections G3, G5 and G1996/2 (see Wastl, 2000), which gives a minimum age of ca 3950 ^{14}C yr BP for the deposition of the moraine. In addition to the geochemical analyses, samples of brown silt in contact with light tephra layers in sections G4, G5 and G1996/2 and G1998 were radiocarbon dated. The humin and humic fractions of these samples were analysed separately, the results are listed in Table 9.1.

The idea underlying humic/humin extractions is to separate material which might have leached down through a soil profile, bringing younger organic components with it, from the material which is representative of growth at a given depth (C. Bryant and B. Miller, 1998 pers. comm.). The generally held view is that humic material is more likely to move in the profile, and is therefore often extracted and discarded, leaving the humin material, which consists of particulates and is thus less mobile and more likely to be representative of 'true age' at a given depth. In theory, this means that the humin fraction should either be the same age or older than the humic in a soil profile. According to C. Bryant (1998 per. comm.), the two standard deviation age ranges should be used to test if there is a statistically significant difference between the humic and humin dates for each sample. If the ages of the two fractions agree, they should represent the same material and the obtained date is considered reliable.

Table 9.1 gives the two standard deviation calibrated age ranges of the radiocarbon dates from Þveráradalur. The ages of the humic and humin fractions overlap for samples SRR-6217/SRR-6211, AA-29910/SRR-6212, SRR-6363/SRR-6364 and SRR-6365/SRR-6366 (in bold). However, all these dates are older that the ages of the tephra layers immediately over- or underlying the dated material (see Fig. 9.5). This might have been caused by extensive mobilization and redeposition of older soil material in the proglacial environment. Such processes would also explain why at this site, the humic fraction does not only produce younger (samples AA-29909/SRR-6209) but also significantly older dates than the corresponding humin material (samples SRR-6216/SRR-6210 and SRR-6215/SRR-6208).

Table 9.1. Radiocarbon dates from sections Þverárdalur
The calibrated ages are based on Stuiver and Reimer (1993).

sample	fraction	laboratory number	radiocarbon age	calibrated age (2 std.dev.)
G4, 22-24 cm	humic	SRR-6216	4380±50 BP	5210-4840 cal BP
	humin	SRR-6210	2840±40 BP	3065-2850 cal BP
G4, 35-40 cm	humic	SRR-6217	4250±55 BP	4870-4580 cal BP
	humin	SRR-6211	4405±45 BP	5245-4860 cal BP
G5, around 40 cm	humic	AA-29910	4890±55 BP	5730-5490 cal BP
	humin	SRR-6212	4690±45 BP	5575-5305 cal BP
G1996/2, 47.5-50 cm	humic	SRR-6215	2020±50 BP	2110-1840 cal BP
	humin	SRR-6208	1610±40 BP	1565-1400 cal BP
G1996/2, 66-71 cm	humic	AA-29909	3860±45 BP	4410-4095 cal PB
	humin	SRR-6209	4145±45 BP	4830-4460 cal BP
G1998, 30-33 cm	humic	SRR-6363	3765±50 BP	4270-3940 cal BP
	humin	SRR-6364	3945±45 BP	4520-4240 cal BP
G1998, 45-48 cm	humic	SRR-6365	4825±55 BP	5660-5340 cal BP
	humin	SRR-6366	5050±45 BP	5910-5660 cal BP

The minimum radiocarbon age of 4690±45 ^{14}C yr BP (SRR-6212) given for the post-Preboreal glacier advance in Þverárdalur (see Fig. 9.5) is based on the oldest radiocabon dated humin fraction of soil inside the moraine, which assumes that there is no contamination by particulate organic material pre-dating the deposition of the moraine.

9.2.2.4 Site Kóngsstaðadalur

The aerial photograph of Kóngsstaðadalur (Fig. 9.13) shows two discontinuous arcs of moraines in front of Gloppujökull. The outer arc reaches down to a position across the stream at ca 440 m asl, the inner arc terminates on the right bank of the stream at ca 450 m asl. The Little Ice Age maximum extent of Gloppujökull is more difficult to reconstruct. Stötter (1991a) tentatively connected it with the debris-covered lobe extending down to ca 800 m asl in the corrie. Later investigations in the field showed, however, that the outer ridge of this lobe is still ice-cored. Furthermore, no pre-Little Ice Age tephras have been found in sections between this ridge and the corrie threshold. Therefore, it is assumed that the Little Ice Age maximum terminated on the slope below the corrie threshold. A terminal moraine which could be connected with this extent has not been found, however, either because the glacier did not leave a prominent terminal moraine on the steep slope below the corrie, or because such a form has subsequently been destroyed by slope processes. Thus the available geomorphological evidence does not allow determination of the ELA for the maximum Little Ice Age advance of Gloppujökull as a reference for reconstructed earlier extents. Due to the irregular longitudinal profile of the glacier, however, changes of the ELA generally provide a less accurate parameter for variations of glacier extent and mass balance at this site.

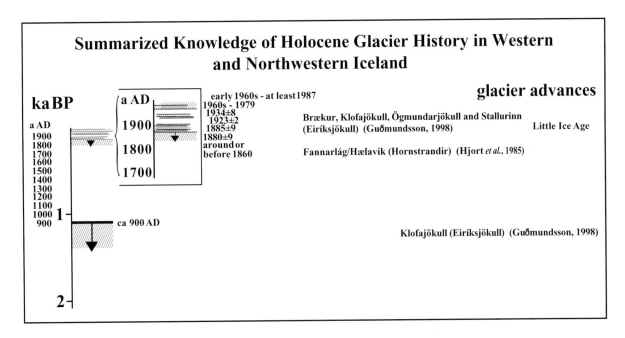

Fig. 9.17 Summary of knowledge of Holocene glacier history in western and northwestern Iceland

9.3 COMPARISON WITH RECONSTRUCTIONS OF HOLOCENE GLACIER HISTORY IN OTHER AREAS OF ICELAND

Apart from the record presented from northern Iceland (Fig. 9.8) there is dated evidence for Holocene glacier advances from southern and southeastern, western and northwestern Iceland. Most of these data come from the outlet glaciers of the ice caps in Southern and Southeastern Iceland. Figs. 9.16 and 9.17 summarize the dating control underlying the reconstructions of Holocene glacier history in these areas.

9.4 CONCLUSIONS

The data presented in Figs. 9. 8, 9.16 and 9.17 show that the reconstruction of the pre-19th century glacier history in Iceland still only defines relatively long time periods during which glaciers reached a maximum extent. The available dating control thus does not allow conclusions about possible synchroneity of glacier advances within Iceland. In addition to this, the chronological data from Northern Iceland discussed in this contribution underline that reconstructions which are purely based on (single) radiocarbon dates must be interpreted with caution. For the lichenometrically dated part of the record the comparison of results from different areas very much depends on the knowledge of the correct regionally specific lichen growth rates (see e.g. discussions in Hjort *et al.*, 1985; Evans *et al.*, 1999; Geitner *et al.*, 2003).

The record of glacier variations during the Little Ice Age also demonstrates that the reconstructed pattern of the pre-19th century glacier history may hide a considerable short-term variability. This should be kept in mind when the glacier history is compared to other records of environmental and climatic change both within Iceland and on a larger North Atlantic scale.

ACKNOWLEDGEMENTS

We thank the German Federal Ministry for Education and Research (BMBF), the German Research Council (DFG), the German Academic Exchange Service (DAAD) and the Austrian Science Fund (FWF), which funded the investigations on the Holocene glacier history in Northern Iceland. In this context we would also like to thank Professor Dr. Jörg-Friedhelm Venzke, Institute of Geography, University of Bremen. The majority of the radiocarbon dates were financed by the British Natural Environment Research Council (NERC) through a grant to Professor Chris Caseldine, Department of Geography, University of Exeter, whom we sincerely thank for his co-operation. We are most grateful to Professor Dr. Peter Mirwald and Mr Edgar Mersdorf for letting us use the electron microprobe at the Institute of Mineralogy and Petrography, University of Innsbruck. We thank the Carl Kühne KG for its support of the fieldwork in Northern Iceland. Permission for these investigations was kindly granted by the Icelandic National Research Council (RANNÍS). We thank the people in Northern Iceland for their interest in our work and the members of the Iceland Research Group at the Institutes of Geography in Munich and Innsbruck. These investigations could not have been made without their help in the field and in the laboratory. Finally we very much wish to thank Lene Zachariassen and Óskar Gunnarsson for their help and hospitality.

10. Variations of termini of glaciers in Iceland in recent centuries and their connection with climate

Oddur Sigurðsson

National Energy Authority, Grensásvegi 9, IS-108 Reykjavík, Iceland

Glaciers are intimately connected with climate. In Iceland the annual turnover in the glacier mass is relatively large so the consequences of mass-balance change are, in many cases, quite obvious. Glaciers increased in volume at a rapid pace in the 17th century and probably for some time before that. In general, glacier termini advanced slowly in the 18th and 19th centuries, reaching a maximum about 1890. The 20th century was a period of a great retreat of all glaciers in Iceland, especially during the warm period in the second third of the century; this left glaciers with less volume than in the mid-1600s. Most glaciers advanced in the years 1970-1995 primarily due to a colder climate. During the last five years of the century most glaciers in Iceland lost mass. Almost all variations of the termini of non-surging glaciers are closely related to changes in mass balance. Climate variations, large enough to cause noticeable change in the mass balance, lead to alterations in the position of the termini, usually with a time lapse of no more than 5 years from the shift in the climate.

10.1 INTRODUCTION

Glaciers are good indicators of climatic variability. In Iceland the primary climatic factors are summer temperature and winter precipitation. These two climatic factors are integrated over the entire area of a glacier and thus have a better correlation to regional climate than regular weather stations, because the latter represent observations only at a single point. The history of glacier oscillations is therefore a good indicator of the climate history of Iceland.

Geologists have reached varied conclusions about the oscillation of glaciers in Iceland in historical time (last eleven centuries). The two main authorities on glaciers in Iceland in the 18th and 19th centuries, Sveinn Pálsson (1945) and Þorvaldur Thoroddsen (1932), were of the opinion that there had been no trend in the changes of glacier fronts since the settlement of the country, only variations back and forth, which was contrary to the opinion of local people, who claimed that glaciers had expanded through time. During the 20th century, historical documents were re-examined and new geological evidence was gathered that indicated that there had been considerable changes indeed, and even a substantial trend towards larger glaciers at least through the 'Little Ice Age'. In the 20th century several scientists attempted to reconstruct oscillations of glaciers in Iceland (Bárðarson, 1934; Eyþórsson, 1935; Þórarinsson, 1943; Grove, 1988). Before the mid-1900s, the nature of surge-type glaciers was not well understood and theories on the response time of glaciers had not yet been developed. The present paper is an attempt at a new assessment of the recent history of glacier fluctuations in Iceland and the connection of glacier variation with climate changes.

Glaciers in Iceland have varied to a great extent over the past millennium. Climate variations have shifted between the extremes of the 'Little Ice Age', which was about the coldest period of postglacial time, and the middle part of the twentieth century that may have been the warmest 40 years during historical time in Iceland. During this period, the glaciers advanced considerably, in some places as much as 10 km or more. Many smaller glacier fluctuations during historical time cannot easily be reconstructed but for the last three centuries historical accounts help to determine the major trends in changes of Icelandic glaciers. In many cases, the glaciers had their Holocene maximum in the latter part of the 19th century. During the twentieth century, warming climate caused some glacier termini to retreat as much as 3-4 km.

10.2 HISTORICAL ACCOUNT

In his treatise on glaciers written in 1695 Þórður Þ. Vídalín (1754) states that glaciers in Iceland are in general advancing more than retreating. A substantial historical record on the extent and trend of glaciers in Iceland was compiled in the first decade of the eighteenth century by Árni Magnússon and Páll Vídalín (1980, information for 1702-1714). At that time, many farmlands were deteriorating because of glaciers advancing over grazing areas. From their account it is obvious that the glaciers were generally advancing over vegetated areas and had been for some time. Some farms were abandoned because the glaciers were getting uncomfortably close and the glacial rivers were constantly changing course and inundating more land. At least one prominent farmstead in southeast Iceland with ample grazing area and fertile land in the early part of the past millennium, was completely covered by a glacier at the turn of the 18th century. In the same area, woodland was lost in the 17th century due to advancing glaciers (Einarsson, 1918, written in 1712). Several other minor farms were also lost in the glacier advance. Eggert Ólafsson (1981), referring to the year 1754, stated that 20 years earlier a glacier in Kaldalón on the northwest peninsula was advancing over land that was green and fertile. Sveinn Pálsson (1945, written in 1793-1794) on the other hand was not convinced of the general increase of glaciers. He observed retreat of glaciers in the years 1792-1794 but states, however, that the general public considered the glaciers to be growing.

Thoroddsen (1932) said it was the opinion of some people that the glaciers had increased in the period AD900-1900, but he considered that there was no reason for reaching this conclusion. From Icelandic annals and other historical documents, it could be seen that the size and nature of glaciers had been the same in the Viking Age as it still was. Þórarinsson (1943) studied the oscillations of the glaciers in Iceland for the period 1690-1940 and concluded that the extent of glaciers had reached a temporary maximum in the middle of the 18th century and an even greater maximum in the middle of the 19th century. Furthermore, he stated that the glaciers were smaller in the 1940s than about 1680. At the time of Þórarinsson's paper, the nature of surge-type glaciers was not understood by scientists so his conclusion needs to be reassessed in that perspective.

10.3 RESPONSE OF GLACIERS IN ICELAND TO CLIMATE CHANGE

The termini of most non-surging glaciers in Iceland react to changes in the mass within 2-5 years. This applies both to medium size and large outlet glaciers with lengths in the range 3-50 km (Sigurðsson and Jónsson, 1995). The variations are easily verified at termini that extend far enough below the equilibrium line. The variations in the smallest glaciers are not as readily identified because their termini are commonly covered with snow even during the summer, especially in years of positive mass balance. The cirque glaciers of northern Iceland are at least in some cases surge-type (Hallgrímsson, 1972). The massive retreat of all non-surging glaciers in Iceland during the period of 1930-1960, recorded in the glacier variations database of the Iceland Glaciological Society (Sigurðsson, 1998), cannot be a response to anything but the warm climate that started no earlier than 1925 (Jóhannesson and Sigurðsson, 1998). Likewise, the general advance of Icelandic glaciers that started about 1970 is the result of increase in mass due to the cold period in the late 1960s as exemplified by Sólheimajökull (see Fig.10.4), southern Iceland (Sigurðsson and Jónsson, 1995). The negative mass balance of most Icelandic glaciers every year from 1995 to 2000 has resulted in the retreat of all non-surging glaciers at the turn of the millennium.

Most of Iceland's big, broad, lobate outlet glaciers are surge-type (Fig. 10.1). They have surge periods varying from less than 10 years for Múlajökull, an outlet glacier from Hofsjökull (Fig. 10.2), to about 70-80 years for Brúarjökull on the northern margin of Vatnajökull. The surge period of cirque glaciers in northern Iceland may be even longer. The periods can be quite irregular. It may be expected that the areal extent of surge-type glaciers at some specific point in the surge period, such as the termination of a surge, is a function of the total volume or mass of the glacier. In addition, some of the surge-type glaciers are composed of several branches that may surge independently making the interpretation of the variations of the terminus in terms of climate change, virtually impossible. An example of such a glacier is Breiðamerkurjökull, an outlet glacier on the southern margin of Vatnajökull with an area of about 910 km² (Björnsson, 1996). It is composed of four main ice streams. Some, possibly all, of these ice streams are surge-type, at least partly, with variable or unknown periods. The easternmost and the largest one is by far the most active. The ice streams come from a broad accumulation area and become constricted and constrained as the course narrows in the flat-bottom valley. When the glacier advances into the valley, the surface elevation increases greatly in the lowlands. This causes considerable feedback in mass balance that results in a

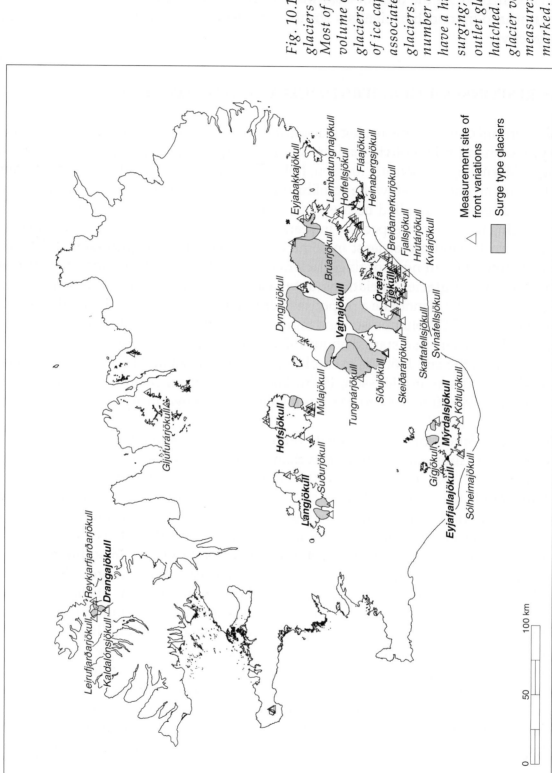

Fig. 10.1 Map of the glaciers of Iceland. Most of the area and volume of Iceland's glaciers is in the form of ice caps and associated outlet glaciers. A large number of glaciers have a history of surging; surge-type outlet glaciers are hatched. Sites of the glacier variations measurements marked.

slower reaction of the terminus to climate change. The eastern part of Breiðamerkurjökull started to retreat appreciably in the 1930s (Björnsson, 1998) which is about half a century after the last major climax of the 'Little Ice Age' and has since been a heavily calving tidewater glacier. Since 1932, Breiðamerkurjökull has retreated continuously except for a few minor surges, which have had very little effect on the front variations of the glacier (Fig.10.3). There is hardly any tendency of readvance to be seen in the terminus of Breiðamerkurjökull (except perhaps the westernmost ice stream) during the cold period in 1965-1985, although advance was characteristic of most other glaciers in Iceland. This is because of the tidewater nature of the glacier. Part of the easternmost ice stream is now calving off icebergs that are some centuries younger than the ice in the adjacent parts of the terminus.

Nevertheless, the areal extent of surge-type glaciers is only vaguely indicative of the time of maximum and minimum mass, because surges only rarely coincide with these occasions.

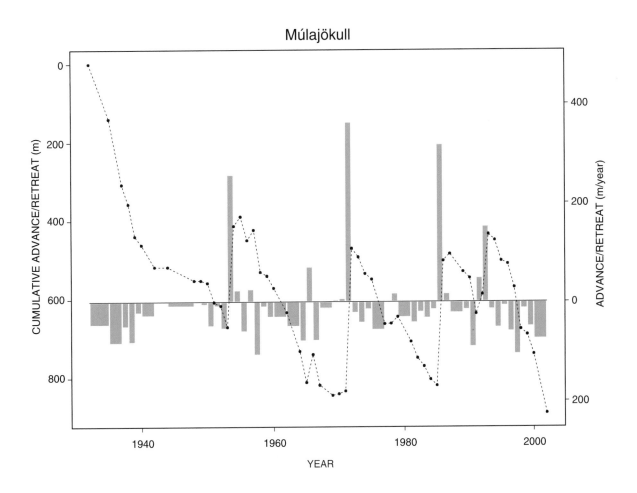

Fig. 10.2 A graph of the terminus variations of Múlajökull outlet glacier, Hofsjökull, central Iceland; 1932-2000.

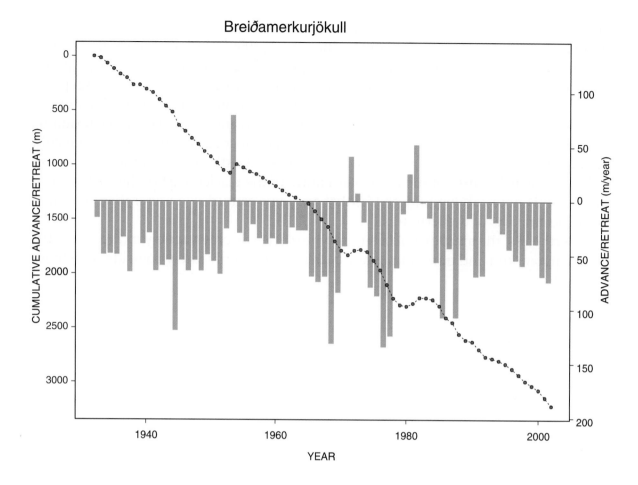

Fig. 10.3 A graph of the terminus variations of Breiðamerkurjökull outlet glacier, Vatnajökull, southeastern Iceland; 1932-2000.

10.4 EXPANDING GLACIERS DURING THE "LITTLE ICE AGE"

Ólafsson (1981, written in 1757-1766) stated that Skeiðarárjökull was 'formed recently, most likely during the 14th century'. This information is probably derived from local inhabitants acquired while he was traveling in the Öræfi district in the year 1756. It may very well be an indication from early in the 'Little Ice Age' of a great advance of this big outlet glacier, which now daily carries several million tons of ice through a mountain pass down to the lowlands. Jóhannesson (1985a) concluded that Skeiðarárjökull reached its maximum historical extent in the 18th century and fluctuated around that position until 1890. However, a surge in 1929 brought at least some parts of the terminus beyond the previous maximum extent (see below).

Breiðamörk was a prominent estate (farmstead) in the 11th century that was definitely not bothered by the closeness of glaciers (Þórarinsson, 1974a). About the turn of the 18th century Breiðamerkurjökull had engulfed the farmstead and its adjacent fields (Einarsson, 1918). Þórarinsson (1974a) estimates that the glacier had then advanced about 10 km since the 11th century. In 1756, Ólafsson (1981) estimated the length of the river Jökulsá á Breiðamerkursandi, which ran straight from the glacier terminus to the

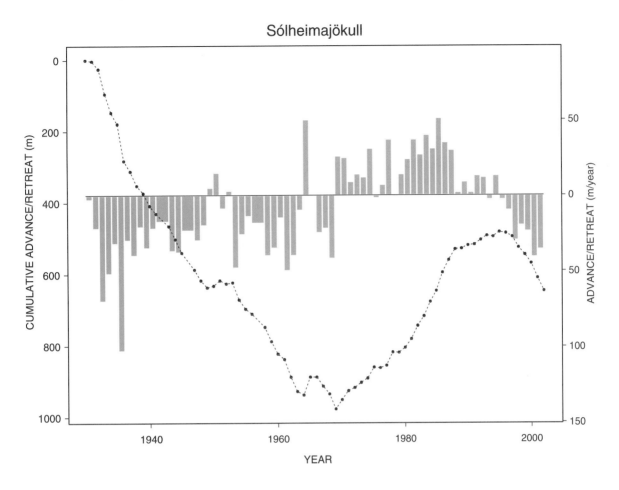

Fig. 10.4 A graph of the terminus variations of Sólheimajökull outlet glacier, Mýrdalsjökull, southern Iceland; 1930-2000.

sea, to be just short of one Danish mile (7.5 km). By 1878, the terminus was only about 250 m from the shoreline at the mouth of Jökulsá (Björnsson, 1998). Not long before 1700, the mountain Breiðamerkurfjall was surrounded by the outlet glaciers Breiðamerkurjökull and Fjallsjökull. These two glaciers were not separated again until 1946 (Björnsson, 1998). Likewise the outlet glaciers Skaftafellsjökull and Svínafellsjökull had joined termini in front of the mountain Hafrafell in the 17th century (Einarsson, 1918, written in 1712). These glacier tongues became separated again in 1935 (Þórarinsson, 1943).

In the *Land register* of Magnússon and Vídalín (1980, written 1702-1714), farmers complain about lost grazing fields in the district Eyjafjöll because of advancing outlet glaciers from Eyjafjallajökull. In the same register, 5 farms are listed that were lost to four different outlet glaciers from Drangajökull before 1700. This does not necessarily mean that the glacier itself advanced over the farm site. In at least some cases, perhaps all, the farmsteads were eroded or inundated by the glacial rivers.

Brúarjökull outlet glacier has an area of about 1500 km² and occupies a coherent basin and therefore acts, more or less, like a homogeneous mass. It is by far the greatest surge-type glacier in Iceland. In a single surge it may advance as much as 10 km. The area in front of it, Brúaröræfi, is vegetated and each time the glacier advances beyond its previous greatest extent it folds up the soil in a peculiar way (which is known as 'hraukar' to the local people). This was first described by Sveinn Pálsson (1945, written in 1794) from the verbal account of Pétur Brynjólfsson who stated that the distance from the terminus to the moraines was 300 fathoms (about 560 m). The folded soil was probably formed in an undocumented surge about 1775. Similar formations were noticed in surges in 1810 and 1890 showing that the glacier was then advancing over virgin land. Pétur Brynjólfsson also described to Sveinn Pálsson (1945) a turbid lake with floating icebergs at the northwest margin of Brúarjökull that was the source of the river Kverká. This ice-dammed lake, now called Hnútulón, burst in 1999 possibly for the last time until the next surge because the glacier has now retreated beyond the threshold of the previous ice dam. This indicates that the glacier had a more advanced position in 1794 than it has today.

Tómasson and Vilmundardóttir (1967) conclude that prior to 1500 and into the 16th century there was no glacier within the catchment area of river Tungná. They also argue that the southwestern part of Vatnajökull (Tungnárjökull outlet glacier) advanced on the order of 10 km from about 1600 to about 1890 when the glacier reached its greatest extent in historical time.

In descriptions of the counties and parishes in Iceland for the county of Skaftafellssýsla in the year 1839 it is stated that most outlet glaciers in Öræfi district are in forward motion and Heinabergsjökull and Fláajökull are slowly and continuously advancing (Bergsson, 1997; Thorarensen, 1997). In 1840 Hoffellsjökull is reported moving intermittently back and forth but more advancing than retreating (Erlendsson, 1997).

According to Dugmore and Sugden (1991) Sólheimajökull outlet glacier had its greatest historical extent relatively early. They suggest ice divide migration due to climate change as an explanation. Árni Magnússon (1955) described the shift in the course of Jökulsá á Sólheimasandi in the year 1690. This change must have been brought about by the retreat of Sólheimajökull, which has not since reached such an advanced position. This indicates that Sólheimajökull was at least for some time during the 'Little Ice Age' in anti-phase with most other glaciers in Iceland. Thoroddsen (1906) came to the conclusion that Sólheimajökull retreated distinctly in the period 1703-1783 but advanced after that. In the period 1860-1893 it retreated again which may very well be the result of loss of mass in the volcanic eruption of Katla in 1860 and consequent jökulhlaup. Reconstruction of the oscillations of Mýrdalsjökull ice cap as a result of climate changes will always be problematic due to huge mass loss during volcanic eruptions, which occur two times per century on average. The melting of the glacier induced by volcanic activity may have amounted to as much as 10% of the volume of the ice cap per century. The glacier has, in several cases, not recovered from the previous loss when the next one occurred. Possible topographic changes in the subglacial caldera and formation of subglacial ridges in the eruptions still ad to this confusion.

10.5 WHEN DID GLACIERS IN ICELAND REACH THEIR MAXIMUM EXTENT IN HISTORICAL TIME?

10.5.1 Non-surging glaciers

It is reported that in the early decades of the 20th century, it was the general opinion of the farmers in the districts in southeastern Iceland that all the steep, narrow outlet glaciers on the southeastern margin of Vatnajökull from Heinabergsjökull in the west to Lambatungnajökull in the east, had their maximum extent around 1890 (Þórarinsson, 1943). Fláajökull outlet glacier was overriding thick vegetated soil about 1880 (Benediktsson, 1972). None of these glaciers are known to surge.

Þórarinsson (1956) examined the variations of the outlet glaciers of Öræfajökull, southeast Iceland, and concluded that their maximum extent in recent times was about 1870. Björnsson (1998) states that Kvíárjökull spilled over the enormous moraine amphitheatre surrounding the terminus in the period 1886 to after 1890 thus correcting his previous statement of this happening in 1870 (Björnsson, 1956; Þórarinsson, 1956a). No information indicates that this had happened in recent centuries prior to the 1880s or later. Hrútárjökull and Fjallsjökull outlet glaciers on the eastern side of Öræfajökull were advancing across its outermost moraine in 1894 and Fláajökull was also advancing that same year after alternating 3 times between advance and retreat since 1882 (Thoroddsen 1906). In general Thoroddsen (1959) states, however, that most glaciers in southern Iceland were retreating a little in the years 1893 and 1894. All this and available information about the surge-type glaciers Breiðamerkurjökull, Skeiðarárjökull and Síðujökull, indicates that the maximum mass of the entire southern margin of Vatnajökull, including Öræfajökull, was reached late in the 19th century. The pattern of outlet glaciers from Öræfajökull may have changed through time due to volcanic activity and possible topographic changes associated with the eruptions. This does not affect the glacier variations postdating 1727 when the last eruption occurred.

Thoroddsen (1906) observed that Kötlujökull outlet glacier of Mýrdalsjökull ice cap was in a state of advance in 1893 and no terminal moraine was to be seen farther from the glacier. Of course terminal moraines might have been lost due to the enormous jökulhlaups that have inundated the proglacial area of Kötlujökull. Little information is to be had on the position of Mýrdalsjökull in the late 1800s and as noted before may be of little consequence as to the connection with climate because of volcanic activity.

On the southeastern side of Langjökull there is a proglacial lake, Hvítárvatn. Two non-surging outlet glaciers, Norðurjökull and Suðurjökull flow around the mountain Skriðufell at the head of the lake and Norðurjökull currently calves into the lake. A system of old cairns and artifacts from an early road building indicates that the main mountain path in the area in earlier times went between the glaciers and the lake (Matthíasson, 1980). Late in the 18th century, Suðurjökull did not extend quite to the shore of the lake, but it had reached far into the lake in 1888 (Thoroddsen, 1932). In the 1950s, Suðurjökull retreated out of the lake. This indicates that the southern side of Langjökull had its greatest historical extent in the latter part of the 19th century.

Very little is known about the oscillations of the small glaciers in northern Iceland before 1939 when Eyþórsson (1956, 1957) started the monitoring of glaciers in the area. However, Thoroddsen (1906) noted from an unidentified source that Gljúfurárjökull declined considerably in the years 1860-1894, which is surprising. Caseldine and Stötter (1993) determined from lichenometric dates that the age of the outermost moraines at 24 glaciers in northern Iceland falls within the 19th century and early 20th century, none earlier than 1813 and none later than 1921. Thus, this period is centred in the latter half of the 19th century. Kirkbride and Dugmore (2001), however, point out a discrepancy between lichenometric datings and tephrochronological datings of moraines in southern Iceland. An additional complicating fact is that at least some (maybe many) of the small cirque glaciers of northern Iceland are in fact surge-type (Hallgrímsson, 1972) with surges recurring at about a century's interval (Búrfellsjökull 1912 and 2002).

10.5.2 Surge-type glaciers

The best known recent maximum position of an Icelandic surge-type outlet glacier is that of the lobate Brúarjökull on the northern margin of Vatnajökull. It reached its greatest postglacial extent at the termination of a surge in the year 1890. It had previously surged in 1810 and again in 1963-1964, not counting the surge that only affected the eastern part of the glacier in 1938. The 1963/4 surge did not reach the same extent as the one in 1810 (Kaldal and Víkingsson, 2000a). The net advance of the terminus of Brúarjökull at the termination of the 1890 surge compared with the outermost position after the 1810 surge was about one kilometer, while the net retreat between the outermost position of the 1890 and 1963/4 surges was almost 2 km. These values are comparable and indicate the relative mass change of Brúarjökull between the years 1810, 1890, and 1964. They do not, however, allow a definite conclusion as to the exact time of the maximum mass of the glacier.

The more narrow outlet glacier Eyjabakkajökull, immediately to the east of Brúarjökull had its greatest Holocene extent, also following a surge (maybe a compound surge), in 1890-1894 (Thoroddsen, 1932). The surges of Eyjabakkajökull are much more frequent than is the case with Brúarjökull, and the quiescent period hardly exceeds 40 years. This indicates that Eyjabakkajökull was at its greatest mass no earlier than 1850. The first known surge of Eyjabakkajökull in the 20th century was in 1931 and the second one in 1938, coinciding with the surge in the eastern part of Brúarjökull. At the termination of the 1931 surge, the glacier terminus had retreated 500 m from its advance in 1890 and an additional 200 m in 1938. At the termination of the surge in 1972/73 the retreat from 1890 amounted to 1.5 km (Kaldal and Víkingsson, 2000b).

In 1889 Tungnárjökull was retreating and the outermost moraines were a short distance off the terminus (Thoroddsen, 1932) indicating a few years to a decade of retreat from the maximum extent. Tungnárjökull surged in 1945/46 (Þórarinsson, 1964) and 1994/95 and probably sometime during the period 1915-1920 (Freysteinsson, 1969). The surge period, although not well defined, indicates a maximum mass of Tungnárjökull in the period 1850-1910 approximately.

Síðujökull is a surge-type outlet glacier in the southwestern part of Vatnajökull. Its most advanced position was reached in 1893 which can be deduced from Thoroddsen's (1933) description, from his visit in 1893, of ice cliffs rising about 100 m over the surrounding terrain at its terminus. Such a description is typical of a surging glacier front. He also mentions a terminal moraine at the very edge of the glacier and no such feature farther away from the glacier which he would have been sure to describe if there had been one. Síðujökull has since then had three other known surges: in 1934, 1963/64, (Þórarinsson, 1964) and 1994. Assuming that the surge period for Síðujökull is no more than 30 years this information indicates that the southwestern part of Vatnajökull was at its maximum volume sometimes during the period from about 1863 to 1923.

During a surge in the year 1929, Skeiðarárjökull advanced about 400 m beyond the position it occupied in 1904 (Eyþórsson, 1963) when it was surveyed by the Danish General Staff (1905). The most advanced position of that glacier in the 19th century probably coincides with a terminal moraine shown on the map of the Danish General Staff. In the surge of 1929, the terminus of the glacier ran over a telephone line under construction (Björnsson, 1991). It is not reasonable to expect the telephone posts to have been founded on or within a recent terminal moraine so this indicates that the surge of 1929 was at least, along some of its arcuate margin, the greatest advance of Skeiðarárjökull in historical time.

Breiðamerkurjökull reached its maximum extent about 1880. Its easternmost part oscillated in a surge-type fashion close to that position for half a century before embarking on a fast retreat after 1932. Other glaciers in the vicinity of Breiðamerkurjökull, on the other hand, retreated slowly for about 3 decades after their maximum extent late in the 19th century (Björnsson, 1998). All these glaciers are non-surging except Breiðamerkurjökull.

Drangajökull has three main outlet glaciers, Kaldalónsjökull, Leirufjarðarjökull and Reykjarfjarðarjökull, running towards southwest, west and northeast respectively. Each of these is a distinct surge-type glacier, with surge periods of about 60-90 years. Their state of surge normally lasts 5-6 years while the termini of most other surge-type glaciers in Iceland move forward for only a few months to one year during each surge. Eyþórsson (1935), using information gathered from various sources, concluded that the termini of Drangajökull had very advanced positions in the first half of the 18th century and even more so about the year 1840. The positions of the 18th century advances are somewhat obscure. Each of the three above named outlet glaciers had a surge that started in the 1930s and 1990s (Fig.10.5). At the end of the surge in 1940, the terminus of Reykjarfjarðarjökull was about 1.5 km short of its 1846 terminal moraine. Likewise, at the end of the surge in 1942, the terminus of Leirufjarðarjökull was about 2 km from the mid-1800 position of the glacier tongue. As stated above this does not allow for an exact determination of the time of maximum volume of Drangajökull.

The beautiful Múlajökull piedmont outlet glacier on the southern margin of Hofsjökull is surge-type and surrounded by concentric moraine ridges. Schythe (1840) visited the terminal moraines in 1840 and described the outermost ridge as being a well vegetated gravel ridge at least 1000 years old about 2 cables (ca 350 m) away from the glacier terminus. In 1846, three sets of morainal ridges, of unknown age, were outside the terminus, the outermost estimated less than 1000 m from the glacier margin (Waltershausen, 1847). In 1902 Daniel Bruun (1925) passed by Múlajökull on the outer flanks of the moraines which were covered by grass, willow and angelica which indicates age of several centuries. In 1924, only one moraine ridge was to be seen 200 m beyond the glacier terminus, which was advancing at the time of observation (Nielsen, 1928). It is not known where this advance (surge?) of 1924 stopped, but a 200 m retreat would

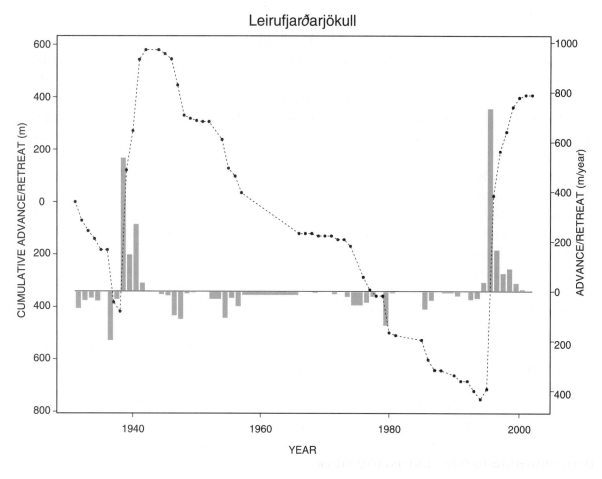

Fig. 10.5 A graph of the terminus variations of Leirufjarðarjökull outlet glacier, Drangajökull, northwestern Iceland; 1931-2000.

not be considered great in the period 1890-1924 if the glacier were assumed to have been at the outermost moraine at the end of the 19th century. Hjartarson (2000) concluded from tephrochronological and archaeological reasons that the ridge is no younger than from 1750 and not very much older than that. Small surge-type glaciers on northern Iceland are discussed above. Their record is obscure and it may be that the relative timing of maximum extent and maximum volume may differ by up to 50 years.

From the scanty meteorological information that we have from the 18th century and the early part of the 19th century (Ogilvie and Jónsson, 2001), there is no reason to believe that there was a major reduction in the mass of glaciers between the middle of the 18th century and the middle of the 19th century. The coldest 30 years on instrumental record (since about 1830) were 1860-1890. There is also very little indication of a decline to be found in the history of glacier variations between 1600 and 1890. Thus, it can be concluded that the maximum glaciation in historical time in most parts of Iceland was about 1890.

10.6 GLACIER VARIATIONS DURING THE TWENTIETH CENTURY

Glaciers in Iceland started to retreat in general just before the turn of the 20th century shortly after the last climax of the 'Little Ice Age' (Fig. 10.6). The retreat was slow during the first quarter of the century (Björnsson, 1998) but sped up dramatically in the warm period that started about 1925. In 1925, the snout of Gígjökull on the northern side of Eyjafjallajökull was still at the mouth of its amphitheatre moraines (unpublished information at the Iceland Glaciological Society), and little retreat is to be seen on photographs from 1928 (Ísólfsson, 1998), which indicates a slow retreat during the beginning of the century.

Since 1930 we have ample information on the variations of the glaciers in Iceland from the monitoring work started by Jón Eyþórsson (1931, 1963) and continued by the Iceland Glaciological Society to the present time (Sigurðsson, 1998). By 1960, all glaciers in Iceland, without exception, had retreated from their 1930 termini positions. Many steep glaciers such as the outlet glaciers from Öræfajökull had retreated to their pre-1700 position; however, Breiðamerkurjökull next to the east of Öræfajökull was still close to its mid-1700s position.

Although the retreat was massive, individual records from non-surging glaciers show that minor advances occurred intermittently (Fig. 10.4) (Sigurðsson, 1998). Such brief interruptions in the retreat may be interpreted as a reaction to short lived positive mass balance perturbations presumably within 5 years before the start of the advance. Indications are to be found of such short-lived retreats during the 'Little Ice Age' such as the one Pálsson (1945) observed on Öræfajökull in 1794. These incidences may have been numerous enough indeed to confuse the greatest authorities on glaciers in Iceland in previous centuries e.g. Pálsson (1945) and Thoroddsen (1932), as to the real trend in the oscillations of termini. However, their opinion was refuted by lay people in Iceland.

After the very fast retreat of glaciers in Iceland during the second third of the century there was a substantial advance of most of the glaciers. Most of the non-surging glaciers had their turning point about 1970 (Sigurðsson, 1998). The advance of some glaciers amounted to about half of the retreat during the previous four decades (Fig. 10.4).

Temperatures rose in Iceland in the mid-1980s following a 20 year long cold spell, and summer temperatures increased especially after 1995 (www.vedur.is/). Above average precipitation in the years 1988-1995 prevented the mass balance of the glaciers from becoming negative, but meagre snowfall in addition to a warmer climate after 1995 has caused a substantial loss of ice mass every year from 1996 to 2000. By the year 2000, all non-surging glaciers, recorded in the database of the Iceland Glaciological Society, were retreating.

10.7 SUMMARY

Glaciers in Iceland increased rapidly in volume and areal extent in the centuries preceding 1700. In the 17th century, outlet glaciers encompassed two prominent mountains in the Öræfi district, Breiðamerkurfjall and Hafrafell. Glaciers continued to advance, although more slowly, through most of the 18th and 19th centuries, most of

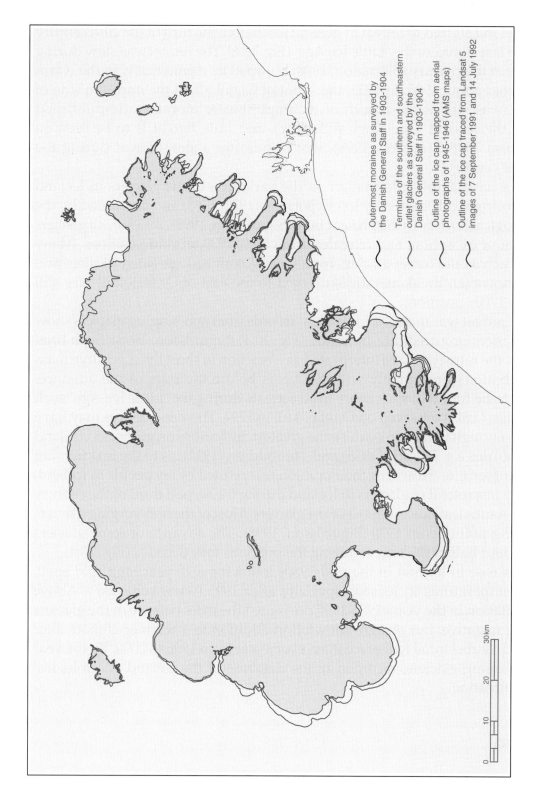

Outermost moraines as surveyed by
the Danish General Staff in 1903-1904

Terminus of the southern and southeastern
outlet glaciers as surveyed by the
Danish General Staff in 1903-1904

Outline of the ice cap mapped from aerial
photographs of 1945-1946 (AMS maps)

Outline of the ice cap traced from Landsat 5
images of 7 September 1991 and 14 July 1992

Fig. 10.6 Outline of Vatnajökull at four different times. The outermost one is based on the outermost moraines mapped by the Danish General Staff in 1903-1904 (only at the termini of the southern part). The second is the margin of Vatnajökull surveyed by the Danish General Staff in 1903-1904 (only the southern and southeastern part). The third is from U. S. Army Map Service (AMS/ Series C762, 1:50,000 scale) maps compiled from aerial photographs of 1945-1946. The fourth is traced from Landsat 5 images taken on 7 September 1991 and 14 July 1992.

them reaching their maximum volume and in most cases areal extent also, in the period in 1885-1894. A slow retreat started before the turn of the 20th century with dramatically increasing speed after 1930. Before 1950, the areal extent of glaciers in Öræfi had receded to their pre-1700 extent as indicated by the two above mentioned mountains being no longer encircled by outlet glaciers. A cold spell in 1965-1985 produced a distinct readvance of most glaciers in Iceland, but a warming and a decrease in precipitation after 1995 has caused all monitored glaciers in Iceland to retreat by the turn of the 21st century.

ACKNOWLEDGEMENTS

The Iceland Glaciological Society kindly made available data on the front variations of Icelandic glaciers. I thank Tómas Jóhannesson, Trausti Jónsson, Kristján Sæmundsson and Richard S. Williams, Jr. for useful comments.

11. Local knowledge and travellers' tales: a selection of climatic observations in Iceland

Astrid E.J. Ogilvie

INSTAAR, University of Colorado, Boulder, CO 80303, USA

Note: Unless otherwise stated, translations from Icelandic are by the author.

> It can be said with certainty that since Iceland was settled no significant changes in the weather or climate have occurred. Sagas and annals show quite clearly that good and bad years, in the past as now, have alternated with long and short intervals. Then, as now, the sea ice came to the coasts, the glaciers were the same, the desert areas the same and the vegetation the same (Translated from Thoroddsen, 1908-1922, vol. II, p.371).

Iceland is well-known for its rich literary tradition which includes a wealth of historical records containing accounts of climate and weather (Thoroddsen, 1916-17; Þórarinsson, 1956b; Bergþórsson, 1969a; Ogilvie, 1984, Ogilvie, 1991, 1992a; Ogilvie and Jónsson, 2001). In this paper, some of these sources will be described and evaluated and the information gathered from them will be used to cast light on variations in the climate of Iceland over the last 1000 years or so. The historical records discussed here consist primarily of information from native Icelanders, in other words, the products of local knowledge. Some accounts by foreign visitors to Iceland will also be considered. Of the many available sources, only a few can be highlighted here. These will include accounts taken from a number of different sources, from a variety of genres: the earliest historical works from Iceland; the Icelandic sagas; annals; geographical works; official reports, weather diaries and the accounts of travellers. The evidence from these sources for climatic variations in Iceland over the last 1000 or so years will be also be charted.

11.1 INTRODUCTION

The settlement of Iceland, which began around AD 871 (Vésteinsson, 1998) may be seen as part of the westward expansion of Norse peoples in the era of the Vikings. Approximately three-quarters of the original population were of Scandinavian stock, mainly Norwegians, and a large percentage of the rest has recently been demonstrated by mitochondrial DNA research to have originated from the British Isles, mainly from the Scottish islands and Ireland (Helgason, 2000; Helgason *et al.*, 2001) thus answering many questions in the long-debated topic regarding the exact origins of the Icelanders. The settlers undoubtedly tried to continue with the same agricultural practices that they brought from their homelands, but soon discovered that conditions on Iceland were harsher than they were used to. Although there is evidence that barley was cultivated in the first few centuries of settlement, the climate was, on the whole, unsuitable for this. The main crop, therefore, since early times, has been the grass upon which the livestock were dependent for winter fodder. Sheep and cattle have been the main domestic animals kept. Milk and meat products thus formed a major part of the daily diet from early times. Sea fishing was always important, but did not become a major industry until the late-nineteenth century. Throughout Iceland's history, difficulties due to the economic and political situation were often exacerbated by the impacts of unfavourable climatic variations (Ogilvie, 2001; Ogilvie and Jónsdóttir, 2000. See also Gunnarson, 1980). The variability of the climate may be attributed to the country's location in the North Atlantic in a region where contrasting air and ocean currents meet. Iceland is also affected by the Arctic sea ice which is brought to Iceland by the East Greenland current. These features of the climate of Iceland are illustrated in Fig. 11.1. Variations in sea-ice incidence off the coasts of Iceland provide an additional proxy climate indicator (Bergþórsson, 1969a; Ogilvie, 1984; Jónsdóttir, 1995; Ogilvie and Jónsson, 2001).

It is ironic that the pioneering compiler of so much weather and climate information in Iceland, Þorvaldur Thoroddsen (1855-1921) was of the opinion that the past climate of Iceland had not changed significantly, as may be seen in the epigraph of this chapter. In this paper, the climatic history of Iceland over the past 1000 or so years, which was indeed extremely variable, will be viewed through the eyes of those local and foreign observers who took the time to write down their perceptions and beliefs regarding climate change and climate impacts, and thus left us the legacy of the wisdom of their careful comments. Evidence from a wide variety of different types of works will be considered, from the *Sagas of Icelanders*, to naturalists of the Enlightenment, to meticulous recorders of weather diaries. A rough guide to the availability of data coverage on climatic matters may be found in Table 11.1. However, a detailed description of every single historical source which gives climate information for Iceland is beyond the scope of this paper. Furthermore, many of the available sources have already been discussed in Ogilvie, 1984, 1991 and 1992a. This paper focuses therefore on those sources which have received less attention in the literature, especially outside Iceland. Space constraints prevent a discussion here of the methodology used to derive climate and sea-ice indices from the historical evidence (for this see Ogilvie, 1991, 1992a and Ogilvie and Jónsson, 2001). It is also not possible to enter into a discussion of early meteorological observations. For an admirable account of this topic the reader is referred to Jónsson and Garðarsson, 2001. See also Garðarsson (1999) for a detailed discussion of the history of the meteorological office and observations in Iceland. Because of their importance, just two of the early observers will be mentioned here. These are Árni Thorlacius (1802-1891) who made the well-known observations at Stykkishólmur from 1846-1889 (Sigfúsdóttir, 1969) and Jón Þorsteinsson (1794-1855), who made observations in the vicinity of Reykjavík from 1820 to 1854. Together, these two observers have provided the foundation for continous systematic meteorological observations in the west of Iceland from 1820 onwards (Jónsson, 1989, 1998; Jónsson and Garðarsson, 2001). Despite their great value, however, these records do not negate the need for the historical observations of the nineteenth century which provide a wealth of detail on climate change and climate impacts. In the concluding section of the paper, climatic changes in Iceland over the past 1000 years will be charted.

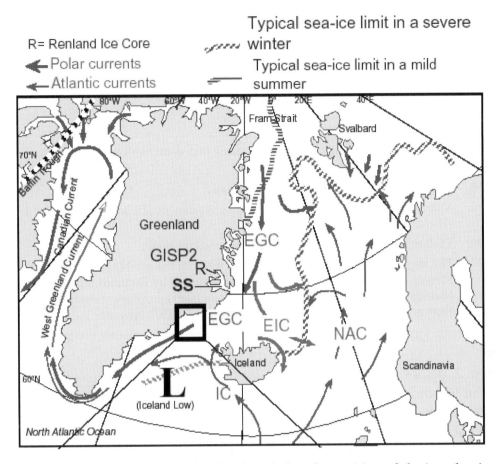

Fig 11.1. The surface currents around Iceland and also the position of the ice edge in a severe winter and a mild summer. 'R' is the site of the Renland ice core. The site of the GISP2 ice-record is shown and also the study site (Jennings and Weiner, 1996) of the Nansen fjord area (indicated by a square). SS - Scoresby Sund; EGC - East Greenland Current; IC - Irminger Current; EIC - East Irminger Current; NAC - North Atlantic Current; NAC - Norwegian Atlantic Current. The diagram has been modified after Hurdle, 1986.

11.2 EARLY HISTORICAL ACCOUNTS

The foundations of Icelandic historiography were laid in the early twelfth century, and by the end of that century the Icelanders had even established a reputation abroad as good historians (Bekker-Nielsen *et al.*, 1965, pp. 62-63). The two most well-known historians of this early period were Sæmundr Sigfússon (1056-1133) and Ari Þorgilsson (1067-1148). Sæmundr may have been the first Icelander to have studied abroad (in Paris), and he is believed to have written a Latin chronicle of the Norwegian kings. Unfortunately, none of his work has survived. Sæmundr seems to have exercised considerable influence on later historians (Turville-Petre, 1953). Ari Þorgilsson is frequently referred to as the father of Icelandic history, and is regarded as being extremely important for its development as a scholarly discipline in Iceland. The esteem in which he was held by later writers is evident from their praise of him (Foote and Simpson,

Table 11.1 Historical climate data for Iceland: type and availability from the Settlement of Iceland (*c.* AD 871) to the Present

c. 871-1145:	Sporadic non-contemporary accounts in early historical writings dating from the twelfth century onwards.
c. 1145-1430:	Medieval annals, certain sagas, early geographical descriptions. Mainly contemporary.
c. 1431-1490:	Virtually no historical information.
c. 1490-1560:	A few interesting historical works, mainly non-contemporary.
c. 1560-1600:	A few excellent accounts.
1600-1750:	Main sources are the later Icelandic annals.
1750-1894:	Main sources are the reports written by government officials (the Sheriffs' letters). Also diaries, annals, charts and maps.
From 1820:	Earliest continuous meteorological observations begin.
From *c.* 1872:	The beginning of modern-day observations with the establishment of the Danish Meteorological Office (Iceland was then governed by Denmark). In 1920 the Icelandic Meteorological Office was established. Data from aeroplanes began to be available from *c.* 1940 onwards, and from satellites from *c.* 1970 onwards.

1961-64, pt. 2, pp. 4-6). Ari was the compiler of a work known as *Íslendingabók* (The Book of the Icelanders) which is a brief history of Iceland from the beginning of the settlement (871) to 1120 and which was composed between 1122-33. In ten short chapters Ari tells of such matters as the discovery of Iceland by Vikings, its subsequent settlement, the settlement of Greenland, visits to the New World, to 'Vínland' (Ogilvie *et* al., 2000) and the adoption of Christianity in Iceland. In the past, *Íslendingabók* was regarded as a completely reliable source, but scholars now are more critical of this work (Nordal, 1957; Benediktsson, 1968. See also Sigurðsson, 2002). For our purposes here, of particular interest in *Íslendingabók* is one brief sentence which states: 'In those days, Iceland was wooded between the mountains and the sea' (*Íslendingabók*, ed. Benediktsson, 1968). There is no doubt that this reflects the environmental reality of a fragile ecosystem which was soon to be radically changed by the arrival of humans to a pristine landscape (Ogilvie and McGovern, 2001).

Another interesting early historical work is entitled *Landnámabók* (The Book of the Settlements). Its textual history is complex and hence its reilability as an historical source is, in many respects, uncertain. For a more detailed discussion of the work in the context of climate reconstruction see Ogilvie (1982, 1991). It is of interest here primarily because it includes a (non-contemporary) account of how Iceland acquired its name - reportedly from the presence of the sea ice off the coasts. The texts of *Íslendingabók* and *Landnámabók* are published in Benediktsson (1986).

Other well-known historians were Snorri Sturluson (1179-1241) to whom is attributed *Heimskringla*, the 'History of the Norwegian Kings' (Sturluson, 1979, 1999). He may also have written *Egil's Saga*, one of the best known of the *Sagas of Icelanders* (see the next section). Another was Sturla Þórðarson (1214-84) who composed *Íslendinga saga* which is part of the collection known as the *Sturlunga saga*. These, and other medieval Icelandic historians, were all highly regarded amongst their contemporaries for being wise, intelligent, possessed of good memories and for using their sources carefully. This shows that many of these early historians had an awareness of the problems involved in the writing of history, and the importance of reproducing a truthful account in a way that is thoroughly modern in attitude, and in sharp contrast to many medieval writers in the rest of Europe. Nevertheless, many difficulties still arise with regard to interpretation of the old Icelandic texts, mainly because of the way these have been preserved. We frequently do not have the original text, and many changes may have been made during the time since the author first wrote it. Such difficulties can only be solved by studying the individual texts. There is no space here to discuss the intricacies of historical source analysis which emphasises the need to carefully evaluate sources for reliability (instead, see Vilmundarson, 1969; Bell and Ogilvie, 1978; Ogilvie, 1984, 1991, 1992a).

11.3 THE ICELANDIC SAGAS

Iceland is renowned for its magnificent medieval prose narratives known collectively as "sagas". This term covers several literary forms, most of which are not strictly historical and which are therefore not suitable for use in climate reconstruction. The most well-known sagas are the *Sagas of Icelanders*, recently published in a new English translation in five volumes. There are many accounts of climate and weather events in these sagas. These descriptions are often used to create a mood or as a metaphor (Ogilvie and Pálsson, 2004) but some may well record a real weather event or climatic reality. Generally speaking, however, although many of the personages in the sagas are known to have been historical characters, these works are essentially literature rather than history and cannot be used to accurately reconstruct past climate. Furthermore, for the most part, they were written down some two to three hundred years after the events described. It is likely, therefore, that they reflect the climatic reality of the time in which they were written (the thirteenth and fourteenth centuries) rather than that of the events described (the tenth and eleventh centuries). However, in spite of their limitations as historical evidence, the sagas contain a rich reservoir of social information, and are hence a potentially valuable source for the understanding of Icelandic society and history during the time in which they were written (Pálsson 1995; Ólason 1998). In some cases, also, they may preserve memories of the past environment and landscape which are accurate. Thus, for example, it is possible that the description in *Egil's Saga* of the settlement of Skallagrím, the father of Egil, at Borg, in western Iceland, may indeed reflect an environmental reality of tenth-century Iceland:

> Skallagrím was an industrious man. He always kept many men with him and gathered all the resources that were available for subsistence, since at first they had little in the way of livestock to support such a large number of people. Such livestock as there was grazed free in the woodland all year round. Skallagrím was a great shipbuilder and there was no lack of driftwood west of Mýrar. He had a farmstead built on Álftanes and ran another farm there, and rowed out from it to catch fish and cull seals and gather eggs, all of which were there in great abundance. There was plenty of driftwood to take back to his farm. Whales beached there, too, in

great numbers, and there was wildlife there for the taking at this hunting post; the animals were not used to man and would never flee. He owned a third farm by the sea on the western part of Mýrar. This was an even better place to gather driftwood, and he planted crops there and named it Akrar (Fields)...When Skallagrím's livestock grew in number, it was allowed to roam mountain pastures for the whole summer. Noticing how much better and fatter the animals were that ranged on the heath, and also that the sheep which could not be brought down for the winter survived in the mountain valleys, he had a farmstead built up on the mountain, and ran a farm there where his sheep were kept (*Egil's Saga, The Sagas of Icelanders,* vol. I, p.66).

For further information on the *Sagas of Icelanders,* see, for example, Kristjánsson (1988) and Sigurðsson (2002).

Other works in the saga genre are the *Bishops' Sagas* and sagas from the *Sturlunga* collection. These concern twelfth and thirteenth century secular and religious leaders. These are known as 'contemporary sagas' because they were mostly written close in time to the events described, by men who had actually taken part in them, or who at least had access to eye witnesses and historical documents. It is relevant here to note that Iceland adopted Christianity in the year 999 or 1000 (Einarsdóttir, 1964; Vésteinsson, 2000). In 1056, Ísleifr Gizurarson was consecrated as the first Bishop of Iceland. Two bishoprics were established in Iceland during this early Christian period; at Skálholt, in the south, in 1056, and at Hólar, in the north, in 1106. Several sagas were written about some of the more notable of the medieval Icelandic bishops. These sagas are not all historically reliable, however, mainly because of their hagiographical element. During the Middle Ages in Iceland there was a strong religious movement to get some of the bishops canonized as saints. (After 1172, the Pope claimed that he alone had the right to select saints. See Jóhannesson, 1974, p. 192.) To this end, several sagas were written which stressed the miraculous and saintly qualities of the Bishops. However, the reliability of the different sagas, and even different passages in the same saga vary considerably, and each must be analysed separately. Lack of space prevents a discussion of all these individual works in a climate context. (For this see Ogilvie, 1982; 1991.) Just one Bishop's saga will be considered here. This is the *Guðmundar Saga Biskups Arasonar* (the 'Saga of Bishop Arason'). This Saga is a biography of Guðmundur Arason's life after he became Bishop of Hólar in 1203. Three versions of the Saga exist - the so-called 'Oldest Saga', the 'Middle Saga' and the saga by Abbot Arngrímur Brandsson (Vigfússon, 1858).

The 'Oldest Saga' and 'Middle Saga' both stem from a saga of Guðmundur that was put together about 1300 by an unknown author, but which is now lost (see Jóhannesson in *Sturlunga Saga* II, 1946, p. xxvii. See also Turville-Petre and Olszewska, 1942). The third version of the saga by Abbot Arngrímur was probably written around 1350 (Sigfússon, KL, 1956-), see *Guðmundar Saga Biskups Arasonar.* There was considerable agitation in Iceland to have Guðmundur canonized and the Saga must be seen as written in this context. It was for this reason that Arngrímur wrote in Latin as well as the more usual vernacular. However, only the Icelandic version has been preserved. Nevertheless, all versions of the Saga are of historical and cultural interest, not least because of their references to the poorer classes, an unusual feature in medieval literature.

It is presumably because his Saga was intended for foreign as well as Icelandic readers that Arngrímur included a short description of Iceland at the begining of the Saga. This gives us some interesting information on, for example, grain-growing, and the presence of sea ice off the coasts. Although there is a strong tendency in the Saga to stress the miraculous side of Guðmundur Arason's life, which means that it cannot be regarded as a strictly reliable historical work, there seems little reason to doubt the accuracy of his geographical description. It must, however, be regarded as relevant to the time when the Saga was written (ca1350), not for the time it was describing, 1203-1237, Guðmundur's term of office. Because it is so interesting, the text of this description is given below.

> This servant of God (Guðmundur Arason) was Bishop of the country that is in books called Thile but which northmen call Iceland. It must certainly be said that this is a suitable name for the country as there is plenty of ice both on land and sea. On the sea are great quantities of drift ice which fill up the northern harbours and in the high mountains there are permanently frozen glaciers of such an incredible height and breadth that they are immensly dense; something which is difficult for foreigners to comprehend. From under these mountains glaciers sometimes pour torrential streams in great floods. There is such a terrible stench from these streams that birds in the air die of it and also people and animals. There are other mountains in this country which produce terrible eruptions of fire and stones such that the crashing and crackling noise is heard all over the country... Such eruptions can be followed by such a darkening of the air that at noon in mid-summer one cannot see one's own hand clearly. It has happened that in the sea itself, about one sea-mile off the south coast, a great mountain was brought into being by a volcanic eruption but another sank into the sea again. There are plenty of bubbling hot springs and sulphur. There are no woods apart from birch and they are small. Grain grows in a few places in the south, but only barley. Fish from the sea, and milk, forms the basic diet of the ordinary people. This island lies so far to the north under the zodiac that in its lower part in some places there is sunlight all day and all night for a month or more at the end of the sign of Gemini and the beginning of Cancer. However, during the winter when the sun is in Capricorn, it is only visible for four hours in the day, even though there have been no eruptions and there is no cloud cover. The country is most widely settled along the coast but least in the east and west (Translated from the Icelandic *Guðmundar Saga Biskups Arasonar* in: *Biskupa Sögur* II, 1878, p.5).

This excellent description tells us much about the landscape and geological features of Iceland as well as the climate. Sea ice and the glaciers are noted as prominent features. Evidence is provided here of the growing of barley – something at this time (ca1350) apparently on the decline. We may compare the comment here on the lack of trees with the similar description in Ari Þorgilsson's *Íslendingabók*.

The name *Sturlunga Saga* is used to cover a compilation of different sagas relating to the history of Iceland from 1117 to 1264. The individual sagas were composed by different authors during the thirteenth century and put together by a compiler early in the fourteenth century (see Jóhannesson in *Sturlunga Saga* II, 1946, p.xix). The term 'Sturlunga Saga' is first found in the seventeenth century and has since become convention. The *Sturlunga Saga* covers the period which has come to be known as the 'Age of the Sturlungs', after one of the leading families of the time. Members of this family included Snorri Sturluson and Sturla Þórðarson (mentioned above in the section on 'Early Historical Accounts'). The major sagas in the collection are as follows: *Þorgils Saga ok Hafliða* (which covers the period 1117 to 1121); *Sturlu Saga* (1148 to 1200); *Prestssaga Guðmundar Góða* (1161 to 1202); *Guðmundar Saga Dýra* (1184 to 1200); *Hrafns Saga Sveinbjarnarsonar* (1203 to 1213); *Íslendingasaga* (1183 to 1262); *Þórðar Saga Kakala* (1242 to 1250); *Svínfellinga Saga* (1214 to 1252); and *Þorgils Saga Skarða* (1252 to 1258).

Vigfússon was the first to suggest that the compiler of *Sturlunga Saga* was one of the three sons of a priest, Narfi Snorrason, (1237-1284), all of whom pursued scholarly interests and were skilled lawyers and that, of the three sons, the most likely candidate was Þórður Narfason (d. 1308) of Skarð in the west (Vigfússon, in the Prolegomena to his edition of *Sturlunga Saga* I, 1878, pp.ciii-cv). Þórður knew Sturla Þórðarson and appears to have revered him as a mentor in whose footsteps as an historian he tried to follow (see, for example, what he has to say about Sturla in the prologue he added to *Sturlunga Saga*: '...we knew him to be a man of the greatest wisdom and moderation', *Sturlunga Saga* I, 1946, p.115). Later scholars have agreed with Vigfússon on this (see, for example, Jóhannesson, *Sturlunga Saga* II, 1946, p.xix).

The original manuscript of *Sturlunga Saga* has been lost, but we know that, at the beginning of the seventeenth century, there were two vellum copies in good condition (Jóhannesson, *Sturlunga Saga* II, 1946, p.xiii). These have been preserved, but are no longer complete. The difficulties of untangling the most accurate versions of the sagas from these manuscripts have perplexed scholars since the beginning of the last century. The *Sturlunga Saga* is the most important source we possess for the period it covers; the last phase of Iceland's commonwealth era and the events leading up to the acceptance of Norwegian rule in 1262. The sagas in the *Sturlunga* collection are known as 'contemporary' sagas because most of them were written close in time to the events described, by men who actually took part in these events or who got their information from reliable eyewitnesses, in addition to sometimes having access to written documents. Although all the sagas are now gathered into the *Sturlunga* compilation, they were originally conceived as separate works. Because of this, and because the individual sagas present a variety of different problems for the critic, they must also be evaluated separately. Just one of these sagas will be discussed here. This is *Þorgils Saga ok Hafliða* (the 'Saga of Þorgils and Hafliða'). For a discussion of the rest of the collection in the context of climate history, see Ogilvie, 1982.

This Saga tells the story of the feud and eventual reconciliation between two chieftains, Þorgils Oddason (d. 1151) and Hafliði Másson (d. 1130) and spans only a short period of time, from 1117 to 1121. The author is unknown, but he seems to have a good knowledge of ecclesiastical matters, and also of Saurbær and the surrounding area in the west of Iceland, where much of the saga takes place (Jóhannesson, *Sturlunga Saga* I, 1946, p.xxiii). Dates between 1160 and 1237 have been suggested by scholars for the writing of the Saga (see the discussion by Brown in the introduction to her edition of *Þorgils Saga ok Hafliða*, 1952, esp. pp.xi-xii). Several factors in the saga point to an early date; for example, the detail of the description suggests that the author had observed at first hand many of the events he relates, and a remark by the writer to the effect of 'as I remember' (*Þorgils Saga ok Hafliða*, 1952, chapter xxx, p.43) tends to confirm this. However, Brown (1952, p.xiv) has convincingly argued that the detailed narrative is more likely to be due to the saga-writer's mastery of his medium than his observation of a particular event, and points out that some of the most descriptive passages refer to situations which would have been very difficult for him, or indeed anyone else, to witness. In all likelihood, the saga was not composed until after 1237.

However, even if the Saga was not written as an eyewitness account, the author doubtless used information given to him by people who were close in time to the events described, and he does display a personal knowledge of certain places and traditions. This is also reflected in an extremely interesting reference to the fertility of the fields at Reykjahólar (in the south of the western fjords region). In 1119, a wedding was held at Reykjahólar. Describing the feasting and entertainments the saga-teller writes: 'At Reykjahólar in those times the lands were so fertile that the fields were never barren. It was quite usual to have new meal and good things to offer guests and there was a feast on St Olaf's day (3 August) every summer' *Þorgils Saga ok Hafliða* (1952, p.17). It seems unlikely that information such as this would be fabricated.

The cultivation of grain in medieval Iceland is recorded in a wide variety of documentary sources in addition to those cited above. These include ecclesiastical deeds, laws, sale contracts and many sagas. These sources are corroborated by the evidence provided by numerous place-names containing compounds meaning 'ploughed field', 'arable land' and so on. Archaeological evidence and the use of pollen analysis has furnished proof of grain-growing during the first few centuries of Iceland's history. The subsequent decline in grain cultivation has been documented by Olsen (1910) who lists all the written references to grain in medieval and later literature, and by Þórarinsson (1956b, 1974a) who argues strongly that the decline and final cessation of this practice was due principally (although not entirely) to a deterioration of the climate. While climate change was almost certainly an important element, it also seems likely that the depletion of the fertility of the soil due to erosion played a large part in the decline in grain-growing (Ogilvie and McGovern, 2000).

11.4 ANNALISTIC WORKS

The medieval Icelandic annals are collectively one of the most important sources of climate information in Iceland from the late twelfth century to 1430. Unlike the sagas, they do not form a continuous narrative but give brief notices for each year within the framework of a carefully constructed chronology. There are, in all, in this genre, eleven annals now extant. These are: I *Annales Reseniani* (*Resensannáll*); II *Annales vetustissimi* (*Forniannáll*); III *Henrik Høyers Annaler* (*Høyersannáll*); IV *Annales regii* (*Konungsannáll*); V *Skálholts-Annaler* (*Skáholtsannáll*); VI *Annalbrudstykke fra Skálholt* (*Annálsbrot frá Skálholti*); VII *Lögmanns-annáll* (*Lögmannsannáll*); VIII *Gottskalks Annáll* (*Gotttskálksannáll*); IX *Flatøbogens Annaler* (*Flateyjarannáll*); X *Oddverja Annáll* (*Oddverjaannáll*); XI *Nýiannáll*. These annals are all published in Storm (1888) except for *Nýiannáll* which is published with the later Icelandic annals in *Annálar 1400-1800 I*. The above list gives the traditional names and numbers of the annals used by their editor, Gustav Storm, and their Icelandic names (Pálsson, 1965). Here they are referred to by these Icelandic names or by the number given by Storm.

Much is uncertain regarding the origins of these annals, and scholars have disagreed both over their date of composition, and over their value as independent historical sources. Some scholars have suggested that early historians such as Sæmundr Sigfússon and Ari Þorgilsson wrote annalistic works, and that these may have formed the basis for the annals we know (Storm, 1888, pp.lxvii-lxx). Sæmundr in particular is sometimes given as the authority for a particular statement in the annals. Nevertheless, for the most part, we do not know who the authors of the medieval annals were. In common with annal writers in other parts of Europe they were, in all likelihood, employed in the service of the church. The annals have much information in common, but were probably compiled in different parts of Iceland; for example, it is sometimes possible to detect special knowledge of a particular district. It is known that in other countries in Europe, annal-writing developed from the practice of writing brief notes on the tables from which the date of Easter was calculated each year, and it is possible that a similar development occured in Iceland (Poole, 1926, pp.19-26). It has also been suggested that the annalists drew heavily on the sagas for their material (Einarsdóttir, 1964); an argument which suggests that annal-writing in Iceland developed fairly late, perhaps in the late thirteenth century. Certainly, the annals do record events from as early as at or around the birth of Christ, but up to the ninth century their entire contents relate to foreign matters. This early material is of little value. Information on Icelandic affairs begins in the late-ninth century when the country was first settled, and information on weather first appears in the late-eleventh century.

Many scholars have followed Storm in concluding that the annals do not become contemporary until ca 1300 (Storm, 1888, p.lxxii; Axelson, 1960; Einarsdóttir, 1964; Hauksson, 1972). This would fit with the theory that they drew on certain sagas, as by then many of the sagas, including the *Bishops' Sagas*, and the sagas in the *Sturlunga* collection, had already been written. However, there is much which testifies to the reliability of the information in the annals prior to this. We do know that there was a strong historical tradition in Iceland, and a particular interest in genealogy and chronology. This has led some scholars, notably Beckman, to suggest that annalistic writings were being made in Iceland as early as the first half of the twelfth century (Beckman, 1912). He bases this supposition mainly on the many reports of natural occurences within Iceland, such as volcanic eruptions, and maintains that such events in the annals can hardly have been drawn from foreign works. His comments are interesting: 'From our point of view eruptions of Hekla have the advantage that there is no question of non-Icelandic information', adding, 'but we lack all possibilities of verifying these events. We do not have, and we cannot find, any source with whose help we could verify the annals' information; it stands alone' (Beckman, 1912, p.2). Writing in 1912, he could not foresee that modern techniques of dating layers of volcanic dust and ice cores would go at least some way towards checking these statements (Þórarinsson, 1970; Dansgaard *et al.*, 1975; Dugmore and Buckland, 1991). Thus, for example, sections of Greenland ice cores which have been dated to ca AD 1104 preserve dust which is almost certainly from the Hekla eruption documented for that year in the annals (Hammer *et al.*, 1980). This is not to say that there are no difficulties associated with the use of the annals. That they contain errors can be demonstrated by comparison with other reliable sources, and they are frequently at variance with each other, both as regards chronology and general information. Because of this, even though they are not wholly independent works, it is vital to consider them individually for the purpose of evaluating their reliability. Space constraints mean that this cannot be done here. However, for further discussion of the medieval annals see Ogilvie (1982, 1991).

The compilation of the older Icelandic annals ceased in 1430 when the last entry was written in *Nýiannáll* and, as far as we know, no attempt at historiography was made in Iceland for well over a century. The reasons for this are complex, but are undoubtedly related to the general economic decline which affected Iceland after the end of the Commonwealth Period in 1262-64 and the beginning of the union with Norway and foreign domination. Also, Icelandic society was greatly disrupted by the devastating effects of epidemics which occurred in 1402 and 1492, presumably the bubonic plague, which had also affected other parts of Europe.

Although there are few sources covering the sixteenth century, some information is available from a specific annalistic work which is not part of any particular genre. This is a slim volume known as the *Biskupa-Annálar Jóns Egilssonar* (a history of Icelandic bishops) compiled at Skálholt in the south of Iceland in 1605 by Jón Egilsson (1548-1636) a priest and a learned man. Jón did not have many written sources at his disposal, but for the sixteenth century at least, relied heavily upon oral tradition, and notably on information passed on by his parents and grandparents:

> And my mother told me, after what she had heard from her grandmother, that a sign for this plague (in 1492) had been severe dearth years in the land. There were no fish in lakes and rivers so neither trout nor salmon was caught. This lasted for three or four years before the epidemic started (Translated from *Biskupa Annálar Jóns Egilssonar*, 1856. p.116).

In his introduction, Jón discusses his sources and states that he has tried to render a correct account. Jón's work contains information that concerns not only the individual bishops' lives but also events that occurred during their lifetimes, including severe seasons and other natural phenomena, particularly for the sixteenth century. It seems likely that Jón Egilsson was encouraged to write his annal by Bishop Oddur Einarsson, whose work is discussed below. The *Biskupa-Annálar* cannot be regarded as wholly reliable, but at a time when there are very few contemporary sources, it is of some value.

In the seventeenth century, annal writing resumed again, but in a rather different form. Presumably inspired by the medieval annals, seventeenth- and eighteenth-century observers took to writing, not just a few lines concerning a particular year, but much more detailed accounts, often covering several pages. Together, the some thirty extant annals form one of the most important sources of both the general history of Iceland and climatic variations for the period 1600-1800. Unlike the medieval annals, it is almost always known who the authors of the later annals were. They were invariably educated men, and usually combined a life as a farmer together with a professional position such as a minister or a teacher. As farmers and people living close to the land they took a keen interest in the weather and thus many of the later annals contain excellent contemporary descriptions of weather events in addition to more general information. However, the later annals do vary as to reliability and quality and all must be critically evaluated. Many of the annals are reliable only in part, with the first section deriving from earlier works, and only the latter part forming an independent and original source. It is not possible to discuss all the later annals and their authors here, but in the section below, some of the most interesting of these are discussed. A list of all these annals, together with their timespan, and the years during which they give contemporary information, may be found in Table 11.2. These later annals may be found in the references under *Annálar 1400-1800*.

One of the first to pioneer the resumption of annal-writing in the first half of the seventeenth century was Björn Jónsson (1574-1655). The annals were generally named according to the place where they were composed, and thus the annal that Björn compiled is known as *Skarðsárannáll* from Skarðsá in the Skagafjörður district in the north, where he lived. *Skarðsárannáll* covers the period 1400 to 1640 but, as most of the work on the annal was done during the years 1636 to 1639, its contemporary span is short. One of Björn's known sources is *Gottskálksannáll* (listed with the medieval annals above). This is contemporary for the period 1568 to 1578. On the whole, this annal is regarded as reliable and contains some useful weather information (Ogilvie, 1991). The importance of carefully analysing historical sources may be illustrated by the fact that Björn has taken weather information from *Gottskálksannáll* almost word for word for the 1570s. However, he accidentally missed the year 1570 and went straight on to 1571, with the result that, between 1570 and 1578, when *Gottskálksannáll* ends, the weather information that he has copied is one year out (Þorsteinsson, *Annálar* I, pps.33, 158). Apart from this unfortunate error and some other random errors in dating, there is evidence to suggest that Björn's work is reliable before 1636, at least from 1580. For example, apart from the information taken from *Gottskálksannáll*, Björn includes few weather references before the early 1600s, suggesting that he only included information that he considered to be accurate. His work must have been based partly on oral tradition, but he also had many documents at his disposal since he had access to the archives at the see of Hólar, and he often worked there for long periods (Þorkelsson, 1887; Benediktsson, 1948). Björn was a remarkable and learned man and his scholarly pursuits have left a legacy of many historical treasures.

Only three of the annals covering the seventeenth century are wholly independent, but these are all good sources of climate and other information. They are particularly useful because they provide long, continuous records. *Mælifellsannáll* was written by a father (Ari Guðmundsson) and son (Magnús Arason) at Mælifell, Skagafjörður, and spans sixty years, from 1678 to 1738. *Vallholtsannáll* is one of the most important and reliable annals for the seventeenth century and covers the period 1626 to 1666 (Þorsteinsson, *Annálar* I, p.320). It was written at Vallholt, also in the Skagafjörður district, by Gunnlaugur Þorsteinsson. *Ballarárannáll* covers the period 1597 to 1665 and is contemporary from 1620. It is one of the most original and entertaining of the annals and also gives very full details regarding the weather. Its compiler was Pétur Einarsson who lived at Ballará in the Dala district in the west of Iceland.

As may be seen in Table 11.2, in the eighteenth century the annals become more numerous. Examples of useful works from this time period are *Hítardalsannáll*, *Djáknaannáll* and *Sauðlauksdalsannáll*. These annals are also discussed in Ogilvie, 1992. Another valuable annal, not part of the collection *Annálar 1400-1800*, but relevant for the nineteenth century, is *Brandsstaðaannáll*. This was written by Björn Bjarnason (1789-1859) and covers the period 1783-1858. He lived mainly at Brandsstaðir in Blöndudalur in the Húnavatn district. He places great emphasis on recording the weather and states that one

Table 11.2. Seventeenth and Eighteenth Century Annals (*Annálar 1400-1800*).

Annal	Author	Timespan	Contemp
Skarðsárannáll	Björn Jónsson	1400-1640	*1636-1640*
Continuation	Brynjólfur Sveinsson	1641-1645	1641-1645
Seiluannáll	Halldór Þorbergsson	1641-1658	1652-1658
Vallholtsannáll	Gunnlaugur Þorsteinsson	1626-1666	1626-1666
Vantnsfjarðarannáll Elzti	Jón Arason	1395-1654	1645-1654
Continuation	Sigurður Jónsson	1655-1661	1655-1661
Vatnsfjarðarannáll Yngri	Guðbrandur Jónsson	1614-1672	1656-1672
Annálsgreinar frá Holti	Sigurður Jónsson	1673-1705	1673-1705
Ballarárannáll	Pétur Einarsson	1597-1665	ca 1620-1665
Vallaannáll	Eyjólfur Jónsson	1659-1737	1690-1737
Mælifellsannáll	Ari Guðmundsson	1678-1702	1678-1702
Continuation	Magnús Arason	1703-1738	1703-1738
Annáll Páls Vídalíns	Páll Vídalín	1700-1709	1700-1709
Fitjaannáll	Oddur Eiríksson	1400-1712	1700-1712
Continuation	Jón Halldórsson	1713-1719	1713-1719
Hestsannáll	Benedikt Pétursson	1665-1718	1694-1718
Hítardalsannáll	Jón Halldórsson	1724-1734	1724-1734
Continuation	Gísli Bjarnason?	1735-1740	1735-1740
Hvammsannáll	Þórður Þórðarson	1707-1738	1707-1738
Eyrarannáll	Magnús Magnússon	1551-1703	1673-1703
Annálsgreinar Árna á Hóli	Árni Magnússon	1632-1695	1662-1695
Grímsstaðaannáll	Jón Ólafsson	1402-1764	1734-1764
Setbergsannáll	Gísli Þorkelsson	1202-1713	-
Sjávarborgarannáll	Þorlákur Magnússon	1389-1729	1727-1729
Ölfusvantsannáll	Sæmundur Gissurarson	1717-1762	1740-1762
Ketilstaðaannáll	Pétur Þorgeirsson	1724-1784	-
Höskuldsstaðaannáll	Magnús Pétursson	1730-1784	1734-1784
Húnvetnskur annáll	unknown	1753-1776	1765?-1776
Hrafnagilsannáll	Þorsteinn Ketilsson	1717-1754	-
Íslands Árbók	Sveinn Sölvason	1740-1782	1740-1782
Continuation	Jón Sveinsson	1782-1792	1782-1792
Espihólsannáll	Jón Jakobsson	1768-1799	1768-1799
Þingmúlaannáll	Eiríkur Sölvason	1663-1729	1700-1729
Desjarmýrarannáll	Halldór Gíslason	1495-1766	-
Vatnsfjarðarannáll yngsti	Guðlaugur Sveinsson	1751-1793	-
Sauðlauksdalsannáll	Björn Halldórsson	1-1778	1740-1778
Annáll Eggerts Ólafssonar	Eggert Ólafsson	1700-1759	1752-1759
Djáknaannáll	Tómas Tómasson	1731-1794	1780-1794

of the models for his annal was the work by Hannes Finnsson *Mannfækkun af Hallærum* ('Loss of Life through Dearth Years') which is discussed below. Björn states that he feels that little is said in this work concerning good years but, from his point of view, it is important to document both good and bad weather, so that 'it is possible to see variations in the seasons' (translated from Bjarnason, *Brandsstaðaannáll*, edited by Jóhannesson, 1941). This concept of a changing climate did not become generally accepted until after the mid-twentieth century (Lamb, 1977). Clearly, Björn was far ahead of his time.

11.5 GEOGRAPHICAL WORKS

In this section may be placed two remarkable but little-used sources: Arngrímur Jónsson's *Brevis Commentarius de Islandia* (published in Benediktsson, 1950) and the *Qualiscunque descriptio Islandiæ* by Oddur Einarsson. Written in Latin in the early 1590s, they are the earliest Icelandic works to give detailed and accurate descriptions of sea ice. Both these works are descriptions of Iceland written within a few years of each other by native Icelanders. Both must also be seen in the context of popular geographical works of the sixteenth century written by foreign writers who had produced inaccurate and disparaging accounts of Iceland. Examples are Munster (1598) and Frisius (1582). (These accounts are discussed in some detail in Thoroddsen, 1892-1904, vol. I, especially chapters 10 and 11.)

The Bishop of Hólar, Guðbrandur Þorláksson (1541-1627) who was an outstandingly learned man, and one of Iceland's earliest cartographers (see e.g. Hermannsson, 1926) took the initiative in trying to combat these false accounts concerning Iceland, and eventually called upon a young relative, Arngrímur Jónsson, later called 'the learned', to produce a refutation of these works (Hermannson, 1926). Arngrímur's first attempt at this, *Brevis Commentarius de Islandia*, appeared in 1593. Unfortunately it did not have the desired effect at the time (people in the rest of Europe tended to go on reading the erroneous accounts of Iceland rather than Arngrímur's work), but it did arouse considerable interest in certain circles. Thus, for example, an English translation was included in Hakluyt's *The Principal Navigations* (1598). The main aim of Arngrímur's work was polemical, and he did not attempt a systematic description of Iceland. Nevertheless, his work includes much information on his own time. One example of this is his discussion of sea ice. In reply to the assertion by foreign writers that sea ice is nearly always to be found off the north coast of Iceland, Arngrímur has much to say. Because this interesting account has tended to be abbreviated in earlier discussions, it is included here in some detail:

> Regarding the notion that sea ice is always land-fast to Iceland or, as Munster asserts shortly after, that it is fast for eight consecutive months; neither of them are true. For the most part the ice melts in April or May and is driven towards the west. It does not then return before January or February and very often even later. It would be possible to count up many years in which this ice, the harsh scourge of our nation, has not been seen at all around Iceland. This was the case in this year, 1592. Thus we can see the truth of what Frisius wrote: 'navigation to this island is only possible for four months because the ice and cold close up the passage' when English ships come to us every year, sometimes in March, sometimes in April and some in May. German and Danish ships usually come in May and June and some do not leave again until August. Last year, 1591, a certain German ship laden with copper stayed in the port of Vopnafjörður in Iceland for about fourteen days in November after which it left without difficulty. Thus, since the ice of Iceland is neither perpetual, nor remains for eight months, Munster and Frisius can be seen to be wrong. (This quotation from *Brevis Commentarius* has been translated from the original Latin given in the edition by Benediktsson, 1950, p.16.)

The works by Munster and Frisius were published in 1598 and 1562, respectively. Clearly Arngrímur did not use Munster's edition of 1598, as his own work was written in 1592. However, an earlier edition of Munster, which is known to have existed (Thoroddsen, 1892-1904, vol. I) was published in 1544, and it is possible that it is this one that Arngrímur had seen. What is interesting in this passage is that it seems clear that despite Arngrímur's assertion that ice was not present off the coast of Iceland for seven to eight months in a year, it is evident that, around the time when he was writing, sea ice was a frequent and unwelcome visitor to Iceland. There is no question of the ice not being present. He says, for example, that 'for the most part the ice melts in April or May and is driven towards the west. It does not then return before January or February and very often even later'. The ice clearly came 'very often' to Iceland around this time (the late sixteenth century). Arngrímur is also aware of the variability of the presence and absence of the ice off Iceland, and states that the ice does not appear *every* year and furthermore, that this was the case during the year in which he wrote *Brevis Commentarius*, 1592. The fact that he specifically mentions this as an ice-free year suggests that this was the exception rather than the rule during his lifetime and up to the time of writing (he was born in 1568).

Arngrímur also seeks to refute the strange belief that hell itself is located in Iceland's Mount Hekla (a frequently erupting volcano) or on the ice, and in explaining that there are no 'tormented souls' on the ice, he gives us further information about sea ice. He tells us: "it is an ancient Icelandic custom that those who live by the sea go out onto the ice in the morning to catch seals, on the very same ice which the historiographers regard as hell, and in the evening return home unharmed" (Benediktsson, 1950, p.30). For the most part of his life Arngrímur lived in the north of Iceland where the ice most frequently occurs and, judging by his account, he must have seen it often.

Arngrímur's description of the ice fits well with the one written by his contemporary Oddur Einarsson in his work *Qualiscunque descriptio Islandiæ*. It was Oddur whom Guðbrandur Þorláksson first had in mind when he was seeking a young scholar to produce a refutation of the foreign accounts of Iceland. Oddur was at least as good a choice as Arngrímur. He was particularly interested in mathematics and astronomy and, whilst in Copenhagen, became a pupil of the Danish astronomer, Tycho Brahe (Ólason, *Íslenzkar Æviskrár*, 1948-, see Oddur Einarsson) who made meteorological observations on the island of Hven in the Sound between Denmark and Sweden during the years 1582-97. For some reason Oddur did not write the refutation that Guðbrandur asked for. What he did write, however, was an excellent description of Iceland both with regard to its people and to the country itself. As an early geographical work the *Qualiscunque descriptio Islandiæ* is outstanding for its detail and accuracy, and it has even been said of it that it is 'the foremost amongst all the learned writings on Iceland and the Icelanders' (Benediktsson, in his preface to the Icelandic translation of Oddur's work, *Íslandslýsing Qualiscunque descriptio Islandiæ*, 1971, pp.34-35). In addition to giving accounts of the Viking Age, the discovery of Iceland, Greenland and Vínland (North America) and information on such diverse subjects as birds, fish, food and amusements in Iceland and the Icelandic language, Oddur discusses Iceland's geographical position and climate, as well as natural phenomena such as volcanic eruptions and sea ice. Unfortunately, the *Qualiscunque* was not published until this century; this is partly the result of accident; for a long time all existing manuscripts of the work were believed lost. One manuscript had, however, found its way to Hamburg, and in 1928 was published by Fritz Burg, the librarian who discovered it. He did not attribute the authorship to Oddur, but this was later proved beyond all reasonable doubt by Benediktsson (1956). An Icelandic translation was published in 1971. Oddur has this to say regarding the ice. Like Arngrímur's account above, it is quoted in some detail because it is so interesting.

> If it so happens that someone drifts north past the island (Iceland) he will see before him the blue-green ice of immense breadth and horrible extent which reaches out over all the oceans surrounding the northern coasts of Greenland. And some people think that it was the proximity of the ice that caused those who first discovered Iceland to give it the name they did, either because they first encountered it near Iceland off the coast or because great quantities of it reached the shores of the land itself. The Icelanders who have settled on the northern coasts are never safe from this most terrible visitor. The ice is always to be found between Iceland and Greenland although sometimes it is absent from the shore of Iceland for many years at a time, by the grace of God. Certainly those who live in these regions suffer from great hardship caused by its presence when it has been evident for long periods of the time. In particular, on account of the barrenness of the fields which it caused. This occurs because the vitality goes out of the earth and the sap which gives fertility is wasted as soon as the ice has become land-fast and the damaging cold has touched the fields. This island could not be inhabited by men for long if such an unwelcome guest came to trouble it every year, but divine providence assures us that this ice is held within certain bounds so that it does not approach the island of Iceland, except when God has decided to punish our people, and this occurs at varying intervals. Sometimes it is scarcely to be seen for a whole decade or longer, sometimes it comes after five years' absence, sometimes it occurs almost every other year, sometimes twice or three times in the same year. It moves with such speed and strength... and although on one day the ice can scarcely be seen even from the high mountains, it has often occurred that on the next it fills up all harbours and fjords to such an extent that to observers it seems as if all the ice in the sea has suddenly arrived at the coast of Iceland...It makes a great deal of difference at what time of year the ice comes. In the autumn and at the time of the winter solstice when the frost has already got into the ground and there is snow cover, its presence does less damage but, during the spring and summer when the weather is becoming milder, the ice invariably brings disaster with it, because that is when it has the greatest power and the grass is most adversely affected. The northerners are therefore far worse off than the southerners who never see this ice as, whenever it starts to drift to east or west, it is swept back by the strong currents of the great sea (Translated into English from Burg, ed., *Qualiscunque descriptio Islandiæ*, 1928, pp. 4-5; and *Íslandslýsing*, 1971, pp.34-35.)

Oddur, like Arngrímur, is aware of the variability of the ice, but he is even more informative. It seems very likely from his descriptions that sea ice occurred often in his lifetime. He writes too authoritatively to be recording hearsay. He knew, for example, not just that the ice has a damaging effect, but that it makes a difference what time of year the ice comes, and that its presence is particularly disastrous during the spring and summer. It is noteworthy that Oddur says that the ice never reaches the south coast. This is not quite accurate. The ice does occasionally reach the south when it drifts around from the east coast, but only very rarely. It is also interesting that there is a description of sea ice in the work by Jón Egilsson, the *Biskupa-Annálar* mentioned above, which may apply to 1552. This account has actually been added by Oddur Einarsson. It seems that, as stated above, Jón was encouraged by Oddur to write his work. The two were at Skálholt at the same time and Oddur went through his friend's manuscript and added some comments of his own (see the introduction to Jón Egilsson's work by Jón Sigurðsson, 1856, pp. 20-21 and p. 100, note 2). The insertion regarding sea ice is interesting, not least because, in his own work, he states that sea ice does not come to the south. Presumably he changed his mind after he had become Bishop of Skálholt, and spent some time in the south, where he must have heard about this severe ice year. Before that he had always lived in the north, apart from time spent in Copenhagen. Furthermore, his work, the *Qualiscunque*, referred to above, was written in Copenhagen, in 1589, quite some time before he moved to the south of Iceland.

11.6 OFFICIAL REPORTS BY LOCAL OBSERVERS

Iceland had been a colony of the Danish crown since 1380, and, as such, was ruled over by the central authority in Copenhagen. However, from before that time, Iceland had been divided into a number of districts (*sýsla*, plural *sýslur*). From the seventeenth through the nineteenth centuries, these districts were administered by *sýslumenn* or 'sheriffs' who were representatives of the king. In the 1680s, a 'Governor' or 'Prefect' (*Stiftamtmand*) was appointed over Iceland. Up to 1770, the Governor was resident in Copenhagen, thereafter in Iceland. The Governors were all Danish until 1752 when it became more common for Icelanders to be appointed. The Governor was required to give an annual report on the situation in Iceland to the Danish government, and the lesser officials in Iceland, the *sýslumenn* or sheriffs in each district, were similarly required to send in annual reports on their district to the Governor. The Sheriffs were generally Icelanders and there were usually twenty-two in office at any one time (one for each district of Iceland). For information on the individual sheriffs see Benediktsson, 1881-1932, *Sýslumannaævir*).

The Sheriffs' letters were written in the form of a letter or report to the Governor. They contain information on, for example, legal, economic and social matters, issues with regard to trade, the state of the hay harvest, fishing, and also weather and climate. Table 11.3 shows a list of examples of reports on sea ice from sheriffs' letters for the year 1802. Occasionally, additional information is included in the form of an 'Assembly Testimony' (*Þingsvitni*), taken at the local courts, and detailing, for example, how many people died during a severe season, and how many farms were deserted. These letters are reliable contemporary records and they form the major source in recent reconstructions of climate and climate impact in Iceland during the eighteenth and nineteenth centuries (Ogilvie 1992a, 2001; Ogilvie and Jónsdóttir, 2000; Ogilvie and Jónsson, 2001). They are currently being further analysed by Ogilvie. The letters are unpublished and are written mainly in Danish, with some written in Icelandic, and are for the most part in Gothic handwriting. They are located in the National Archives in Reykjavík and cover the time period 1700 to 1896. They are especially detailed from

Table 11.3   Sea Ice information from Sheriffs' letters: 1802

Report from Governor of Iceland, Ólafur Stefansson, Viðey. 22 July 1802

The Greenland ice is preventing all trade in the north as the entire northern coast is hemmed in by ice and no ships can get through. Much of the western coast is also affected.

Report from Sheriff Magnús Ketilsson, Búðardalur, Dalasýsla. 29 July 1802

The situation in neighbouring Strandasýsla is far worse than here. There the Greenland drift ice is still lying off the coast.

Report from Governor of Iceland, Ólafur Stefánsson, Viðey. 20 August 1802

The Greenland ice is still keeping all northern harbours closed. According to the latest report, Vopnafjörður (in the east) is similarly affected. Thus, no trading ships can get to the northern harbours and, not, presumably, to Vopnafjörður either.

Report from Governor of Iceland, Ólafur Stefánsson, Viðey. 6 September 1802

The Greenland ice left the northern coasts in the middle of last month (August). On the first of this month (September) I also heard firm news that 7 merchant vessels have sailed into Eyjafjörður.

Report from Sheriff Jón Jakobsson, Espihóll, Eyjafjarðarsýsla. 14 October 1802

...thus when the animals are dead; the people must consequently die also, here in the north where there is no other food source apart from the animals, especially when all the northern coasts are spanned by the evil sea ice, as this year, when the ice first broke up on 23 August.

Report from Sheriff Jón Jónsson, Bær in Hrútafjörður, Strandasýsla. 28 December 1802

The ice spanned the whole of Strandasýsla's coast from January to late August.

Report from Sheriff Magnús Ketilsson, Búðardalur, Dalasýsla. 3 January 1803

The Greenland drift ice did not leave neighbouring Strandasýsla before the end of August. From long experience it is known that (the ice) has the same, or even colder, influence on the weather here in Dalasýsla.

---

around 1780 onwards. An example of part of a report may be given from the famously cold year of 1816 (Harington, 1992). This was also a difficult year in Iceland with moderate sea ice (Ogilvie, 1992b). The Governor, Stephen Thorarensen, who is mentioned below in connection with his description of Iceland, published in 1792, and who was based in the Eyjafjörður district, wrote a report for the north and east of Iceland on 26 September 1816. Part of this is translated below.

The winter was more than usually severe in several places here in the district with respect to much snow and severe frost and the frozen crust of snow on the ground. Some people lacked fodder, and thus lost quite a number of their outside livestock, especially horses...The spring also, in its way, was quite severe, on account of persistent northerly winds, frost and cold air which was a consequence of the sea ice which occurred off the northern coasts during the spring and enclosed a large part of its coasts. For a short while the ice hindered the arrival of some of the first ships to Eyjafjörður.

Famines and dearths which were due in large measure to climate impacts occurred many times during the history of Iceland. The last such famine occurred in 1882, and is discussed below in connection with William Morris and his interest in Iceland (see also Vasey, 2001). The severe winter of 1880-81, a precursor to the famine, is described in many sources of the time, and in all of the sheriffs' letters for this year. One example may be given. The extract cited here is translated from a report by Sheriff Lárus Blöndal written on 1 October 1881 in the Húnavatn district in the north.

> The extremely severe weather which began in earnest in the middle of November (1880) lasted until the beginning of April (1881). It was the general opinion, that no one now living had experienced such long-lasting and severe frost. This was frequently measured between 12 and 30 degrees Reamur and was often around 20 degrees...There was frequent fog due to the sea ice, and the bay of Húnaflói was full of sea ice...The spring was cold and dry... and the grass growth was of the poorest quality...The summer was also cold and dry and there was often night frost.

Although there is no doubt as to the reliability of these sources, they need to be examined carefully and evaluated, just as with all historical data. In the case of these examples, it is interesting to note that the Governor, Ólafur Stefánsson, based at Viðey, an island off Reykjavík in the south, has less accurate information on the sea ice than the northern observers who are giving eye-witness accounts. Thus, he states that 'The Greenland ice left the northern coasts in the middle of last month (August)'. However, Jón Jakobsson writing from Eyjafjörður in the north states more precisely: '...the ice first broke up on 23 August'.

11.7 WEATHER DIARIES

A large number of diaries which have weather and climate as their main focus were kept by Icelanders, particularly during the nineteenth century. A list comprising many of these diaries together with a classification according to the place where they were written and the time period covered, as well as the type of comments noted on the weather, may be found in Hávarðsson and Jónsson, (1997). There are a great number of personal diaries kept in archives and libraries in Iceland. Most of them are in the National/University Library in Reykjavík but many others are in smaller archives around the country. It became more common for farmers to write diaries during the second half of the nineteenth century, partly because paper became more readily available and partly because farmers were encouraged to keep a diary as part of an attempt by the authorities to document and improve farming conditions (Kristjánsdóttir, 1998). The most important factor, however, must have been the fact that diaries were useful in many ways. They served, for example, as an *aide memoire* to farmers who needed to note details concerning the farm: the harvest; the animals; and the weather; as well as other important elements of their daily life. Diaries have become an object of focus in the Icelandic research community in the last few years. Thus, for example, Magnússon has discussed the quality and nature of diaries as a tool in microhistory (1995, 1997).

One diary will be singled out for consideration here. This is the weather diary kept by a father and son, Jón Jónsson senior (1719-1795) and Jón Jónsson junior (1759-1846) who were both ministers (Kington and Kristjánsdóttir, 1978). This diary, which has not been published, was kept from 1747 to 1846 and thus spans virtually a whole century. The observations were all made in one locality, in Eyjafjörður, and although they were made at various sites, depending on where the Jón Jónssons lived, they were all fairly close together. From 1747 to 1758 they were made at Möðruvellir in Hörgárdalur and from 1759 to 1769 at Guðrúnarstaðir. Jón senior then moved to Grund and from there to Núpufell until 1798. He then lived at Möðrufell until 1839 and finally at Dunhaga in Hörgárdalur where he died in 1846, two weeks after he had recorded his last weather observation.

The observations in the Jón Jónsson's diaries are not instrumental but they are detailed, careful and very full. They also include daily reports on the wind, weather and state of sky, and also weekly weather summaries. There is also an annual summary of the weather during 1748 to 1768. The reports contain information on the fishing and the harvest. Other matters are described when the Jón Jónssons consider them relevant. For example, the summary for 1766 includes a detailed account of the sandfall in the Eyjafjörður region from the Hekla eruption of that year, and there is also an extensive description of the effects of the 1783 Lakagígar eruption on Eyjafjörður. Social phenomena are mentioned, as during the extremely severe year of 1756 when Jón senior writes, 'There was terrible hardship amongst people...Poor people left their homes in many places...and there were many people travelling around but few people had food to give them. As a consequence, people died of hunger'. In 1761 there is also hardship, and Jón describes how people dried seaweed for food. Details of the presence of sea ice, especially the times of its arrival and departure, are also included. Jón Jónsson's weather diary is thus an extremely valuable document, both because of the quality and extent of the observations, and also because of the length of time that it covers.

Another example of an excellent weather diary is the one kept by Magnús Ketilsson (1732-1803) at Búðardalur at Skarðsströnd in the west from 1779 to 1802. He noted the wind direction, precipitation and temperature and also made some instrumental meteorological observations. Magnús Ketilsson also recorded weather information in an early Icelandic newspaper, *Islandske Maanedstidender*, which he edited, covering the years 1773-1776. Magnús was a government official, a sheriff, and his reports while in office (1754-1803) have been incorporated into climate analyses by Ogilvie (1992a) and the present work. For more information on the observers of climate and the environment in Iceland, see Garðarsson, 1999; Garðarsson and Jónsson, 2001; Ogilvie and Jónsson, 2001.

11.8 TRAVELLERS' TALES: FURTHER ACCOUNTS BY LOCAL AND FOREIGN OBSERVERS

In the early eighteenth century, the Danish authorities commissioned a series of investigations into conditions in Iceland and the first systematic attempt to investigate the natural history of Iceland was begun in 1743. The published results of this are known as the *Sýslulýsingar 1744-1749* (Guðnason, 1957) and can perhaps best be translated as 'A Geographical Description of the Districts of Iceland 1744-1749'. This had its origin in the interest in science and nature awakened by the Enlightenment movement in Europe, and the recently formed (1742) Danish Scientific Society which, although initially set up specifically to investigate antiquities in Denmark and Norway, soon extended its interests to natural history and other branches of learning. In the nineteenth century, the Icelandic sagas were gaining a dedicated minority of devotees, and Iceland's seemingly exotic geology was also arousing interest. In the wake of these developments, a series of Icelandophiles, many of them British, visited Iceland and wrote down their impressions. A few of the many accounts and observations written by visitors to Iceland, as well as by native Icelanders, are highlighted below. An account of many of these travellers may also be found in Thoroddsen (1892-1904, vol. I, chapter 3). See also Ísleifsson, 1996; Garðarson, 1999; and Ogilvie and Jónsson, 2001.

From the point of view of the history of meteorological observations, the visit to Iceland during 1749 to 1751 by the Dane, Niels Horrebow (1712-1760) is of particular interest (see Horrebow, 1752, 1758). His sojourn was made in connection with the Danish Scientific Society mentioned above. During his stay in Iceland, Horrebow travelled around the country, investigating the economy, social life and natural phenomena. He also took instrumental meteorological observations at Bessastaðir. These are the earliest such observations for Iceland (Jónsson and Garðarsson, 2001).

After Horrebow's departure from Iceland in 1751, the task of undertaking further investigations in Iceland was given to the Icelandic students Eggert Ólafsson (1726-1768) and Bjarni Pálsson (1719-1779). They spent the years 1752 to 1757 travelling through each district of Iceland recording all aspects of life. They kept diaries and notes of all that they saw and these were later worked into a two-volume treatise by Eggert Ólafsson (Ólafsson, 1772). This contains a wealth of material and is also of interest because their travels coincided with the very severe years of the 1750s (Ogilvie, 1984, 1992a). When their travel book was published, it was the most important book that had so far been written on Iceland. It is still regarded as an outstanding work and an edition in English was published in 1975 (Ólafsson and Pálsson, 1975). Eggert Ólafsson also compiled an annal for the years 1700 to 1759. This may be found in the collection *Annálar 1400-1800*, volume VI (see the section on annals, above).

A similar, but far less ambitious study than Ólafsson and Pálsson's travel book, was later undertaken by Ólafur Ólafsson, known as Olavius (1741-1788) who travelled around the northwest, north and northeast of Iceland from 1755 to 1777 with the specific aim of investigating the natural resources, especially fishing, in these parts. His conclusions were published in 1780 (Olavius, 1780).

In 1780, the Faeroese naturalist, Nicolai Mohr (1742-90) was sent to Iceland specifically to investigate some of Olavius' discoveries, such as graphite in Siglufjörður in the north which, it was believed, might be of economic significance. Mohr spent two summers in Iceland and travelled around the northern and eastern parts. He later wrote a book about his travels: *Forsög til en islands Natur-historie med oekonomiske samt andre Anmærkninger*, ('Attempt at an Icelandic Natural History together with Economic and other Observations') published in Copenhagen in 1786. Mohr concentrates on the economy and natural history of Iceland, but his book is written in the form of a travelogue, and his observations on all that he saw help to build up a picture of conditions in Iceland at this time. He devotes much attention to Icelandic farming practices, particularly recent attempts at improvements, and he also records the weather when it strikes him as noteworthy. Thus, for example: 'On the third of September (1780) I travelled back to the trading station Reykjafjörður. On the seventh we saw outside the fjord a few small pieces of drift ice coming with the current from the sea' (Mohr, 1786, p.362).

Another major eighteenth-century description, this time by an Icelander, Magnús Stephensen (1762-1833) is the *Eftirmæli átjándu aldar* ('Iceland in the Eighteenth Century') first published in 1806 with a Danish translation in 1808. Here Stephensen concentrates on social and economic matters, with sections on, for example, farming, fishing, trade, population etc. There is also a section on natural history. This includes a discussion of 'good' and 'bad' years and their causes. Stephensen believed that the major cause of the 'bad' years was severe winters and sea ice. In this section, Stephensen concluded that out of the hundred years of the eighteenth century, forty-three of these had been dearth-years with severe seasons and that, in fourteen of these, there was considerable loss of human life, as well as of domestic animals. Stephensen's book is one of the most important contemporary accounts of eighteenth-century Iceland.

Several travel books written by individuals who visited Iceland give interesting and useful accounts of Iceland in the eighteenth century. These travellers came from all over Europe. One of the earliest was a Frenchman, Yves Joseph de Kerguelen Trémarec (1745-97). He sailed to the northwest coast of Iceland in 1767. He did not travel around the country but stayed almost exclusively in Ísafjörður. However, he was an observant visitor and his account, although quite short, contains a number of useful comments on the climate and environment. A Swede, Uno von Troil (1746-1803) together with Daniel Carl Solander (1735-82) also from Sweden, Sir Joseph Banks (1743-1820) from England, and several other distinguished companions, visited Iceland in 1772. Von Troil's travel book, in the form of a series of letters on different aspects of life in Iceland, was first published in Sweden in 1777, and translations into several languages soon followed (an English translation was published in 1808). It is a perceptive, entertaining and witty book. In the ninth letter, on Icelandic cuisine, von Troil describes all Icelandic food in detail and with enthusiasm. He also clearly enjoyed a dinner given to the travellers by Bjarni Pálsson (mentioned above), except for the pickled shark, which was 'so awful, that the tiny piece we tasted drove us from the table far sooner than intended' (translated from von Troil, 1961, p.73). Modern-day visitors to Iceland may empathise with this sentiment. The most recent edition of the book, published in Iceland in 1961, contains facsimiles of a number of fine drawings which were made on the expedition but which were not included in any of the earlier editions.

Another group of travellers, all from England or Scotland, visited Iceland in 1789. These were John Stanley (1766-1850), Isaac Benners (b. 1766), James Wright (1770-1794) and John Baine (dates not known). Stanley specifically asked that his companions keep diaries of their visit, and this they accordingly did. However, they were not published until comparatively recently (West, 1970-76). Other British Icelandophiles who visited Iceland were Sir George Mackenzie (1780-1848) and Henry Holland (1788-1873) a distinguished physician (who included Queen Victoria among his patients). When Mackenzie visited in 1810 he was accompanied by Richard Bright (1789-1858), subsequently known for his reseach into renal disease, and also by Henry Holland. The main purpose of this visit was to investigate Iceland's volcanic regions. Mackenzie's book, *Travels in the Island of Iceland during the summer of the year 1810*, published in 1811, secured Mackenzie fame amongst those interested in Iceland. Henry Holland also produced a manuscript which was meticulously prepared, but which was long neglected until its rescue from oblivion by the excellent edition of his work by Wawn (1987). The scholarly journal produced by Henry Holland after his 1810 visit to Iceland provides a detailed and fascinating account of Iceland at that time. It also includes a short weather journal for the year 1810 (Wawn, 1987, pp. 279-295).

Henry Holland returned to Iceland in 1871, now aged 84. In this year also, the multi-talented and philanthropic William Morris (1834-1896) visited Iceland for the first time (Morris, 1911). Morris was, amongst other things, a textile designer, printer, novelist and poet, as well as an early translator of the *Sagas of Icelanders*, together with his friend, Eiríkur Magnússon (1833-1913). In August of 1871, the two Icelandophiles chanced to meet at the house of mutual friends in Reykjavík. Most visitors to Iceland were, like Holland, inspired by the geology of the country. Morris, on the other hand, was an early pioneer in the appreciation of Icelandic literature, especially the *Sagas of Icelanders*. According to Wawn, the meeting between the two men was not a great success (Wawn, 1987, p.2). An incident at the site of the Great Geysir in Haukadalur prior to this meeting further illustrates the fact that Morris was more stirred by the spirit of the sagas than geological features (Wawn, 1987, p.2). Morris wrote :

'Let's go home to Haukadal,' quoth I, 'we can't camp in this beastly place.'

'What is he saying,' said Eyvindr to Gisli [the guides]:

'Why, I'm not going to camp here,' said I:

'You must,' said Eyvindr, 'all Englishmen do.'

'Blast all Englishmen!,' said I in the Icelandic tongue. (Morris, 1911, pp. 67-8. Quoted in Wawn, 1987, p.2).

Morris returned to Iceland in 1873. It is interesting to note that he was well acquainted with Árni Thorlacius (whose meteorogical observations for Stykkishólmur are mentioned above) and visited him at his home in Stykkishólmur (now preserved as a museum). In the early 1880s, Morris showed his genuine concern for the population of Iceland when he tried to arrange a relief committee, the so-called Mansion House Relief Committee, to alleviate the distress caused by difficult climatic conditions in the early 1880s. The hay crop failed in the summer of 1881 and many livestock had to be slaughtered. The next spring and summer were severe and it was a time of much sea ice (see the sheriff's letter for 1881 quoted above). Morris wrote to a friend: ' I am grieved indeed to hear that things are no better in Iceland...Please let me know what names you have got and tell me anything you want me to do: I think you should write to everybody you know and ask for help at once'. Unfortunately, despite all his efforts, Morris did not manage to rally the help which he had hoped for (Harris, 1978-9).

Another very important example of a work in the travel genre by a local observer is the travel book written by Sveinn Pálsson (1762-1840) a doctor, and one of Iceland's most brilliant naturalists (Jóhannesson, 1950; Hellund, 1882-3; Ingólfsson, 1991). Pálsson travelled around Iceland during the years 1791 to 1795 on behalf of the Icelandic Natural

History Society and subsequently submitted an account of his observations. These laid particular stress on the geology of Iceland. Pálsson was one of the first to understand many of the phenomena associated with glaciers and this makes it all the more regrettable that his account was not published until long after it was written. Sveinn Pálsson is also well-known for his careful meteorological observations, taken mainly at his residence in Vík, in the south of Iceland, from 1798 to 1840. For a detailed discussion of these and other observations, see Jónsson (1998) and Jónsson and Garðarsson (2001).

There are many other works by Icelanders which are on a smaller scale, in the form of pamphlets or essays, but which help to build up a picture of eighteenth-century life. One example of these is a description of the catching of birds on the island of Drangey in Skagafjörður in the north: *Beskrivelse over Fuglefangsten ved Drangøe udi Island* ('Description of Bird Catching at Drangey, Iceland'). This is by Ólafur Ólafsson (not the same person as Olavius, mentioned above). This gives a detailed account of the methods and equipment employed to catch sea birds such as puffins, guillemots and auks. The point of the article was said to be to: 'give the inhabitants of other districts of Iceland an account of how this is done, where it is not yet known, but where it could possibly be of benefit for them (Ólafsson, 1784)'.

The first scholarly treatise written in Iceland specifically in the genre of what would now be called 'climate impact-studies' or the 'human dimensions of climatic change' was written in 1786 by Bishop Hannes Finnsson (1739-96) and published in 1796. A new edition was published in 1970. Its title is *Mannfækkun af Hallærum* ('Loss of Life as a Result of Dearth Years'). By a dearth year Finnsson meant a year in which there was famine and deaths from hunger all over the country. The work is concerned not only with the effects of severe seasons, and Finnsson does point out that there may be many different causes of dearth and famine. However, it is clear that he did believe that the most important cause was climatic; specifically, poor grass-growth or failure of the hay harvest followed by a severe winter. He also suggests that famine is most likely to occur after a succession of unfavourable seasons, thus anticipating twentieth-century scholars by some two hundred years (see, for example, Parry, 1978).

Finnsson does not consider the possibility that the climate may have varied during the period of his survey, but he does briefly broach the subject of whether dearth years have increased or decreased. He says that it is difficult to make comparisons between the number of deaths from famine in the eighteenth century, and earlier times, because of the lack of data for the earlier period. However, he suggests that there were fewer dearth years before AD 1280 than after, and that the main reason for this was that trade was more favourable for Iceland at that time.

Finnsson wrote *Mannfækkun af Hallærum* in 1786, after a series of climatically severe years and the calamitous Lakagígar volcanic eruption in 1783 (Bjarnar, 1965; Þórarinsson, 1969; Ogilvie and Demarée, 2001). He was undoubtedly concerned by the distress he saw around him. Certainly, one of his main purposes in writing seems to have been to remind the Icelandic people that death and famine as a result of severe seasons had occurred many times throughout Iceland's history and that, as the country had survived these former vicissitudes, it could also survive the present hardship. Finnsson's work is of great value and interest; it is one of the earliest attempts to investigate climate-society interactions written in Iceland, or anywhere else, and it is, for its time, a most scholarly and well-written work.

Another interesting work from this period is a treatise written by Stephen Thorarensen (1754-1823), Governor of northern and eastern Iceland from 1783 to his death (see his report for 1816, above). A shortened translated title of his work is: 'Thoughts for Greater Consideration regarding Dearth Years and their Effects, in addition to the setting up of Food or Grain Reserves in Severe Years'. The full title may be found in the references. In the latter (and major) part of this treatise, Thorarensen considers the effects of dearth years and also offers suggestions on how the hardship they cause might be ameliorated. Although this was published before Finnson's work, in 1792, it was actually written after it, in 1790, and the earlier part, consisting of a list of dearth years between AD 976 and 1783, is acknowledged by Thorarensen to be an extract from Finnsson's as yet unpublished work.

Thorarensen suggests that there are four main causes of dearth years in Iceland: the failure of the grass or the hay harvest; the failure of the fisheries; severe and long-lasting winters, and the lack of necessities occasioned by the delay (for whatever reason) of the merchant ships from Denmark. He also mentions sea ice and volcanic eruptions as additional factors which contribute to hardship or dearth. While conceding that it is impossible to do anything to avert natural disasters or to change difficult environmental conditions, Thorarensen suggests that it would be possible for people to be much better prepared to cope with them. As a feasible measure, he suggests the setting up of grain stores which would provide insurance for years when food was scarce. He is aware that such a project could not easily be set in motion, and therefore submits a detailed plan on how this might be done. As such it must be seen as an early contribution to the concept of 'sustainable development'.

Many works of a similar nature which discuss severe seasons and dearth years and the possible effects of these were written in Iceland. However, Thorarensen's treatise is unique in that he suggested ways in which these difficulties might, at least in part, be overcome. Although other minor works on the general subject of the effects of severe years were written later on, throughout the nineteenth century, these are of little interest except as further examples of a particular genre. They consist mainly of lists of severe years, with little or no discussion, and most of them are no more than extracts from Finnsson's work. There were thus no new developments in the field of climate impact-studies and the theories of climatic change until the early decades of the twentieth century, when the concept of a changing climate, and the implications of this for societies, was first suggested and subsequently discussed. This latter chapter in the history of climate in Iceland may be read in Ogilvie and Jónsson, 2001.

11.9 CLIMATIC QUESTIONS: HOW COLD WAS THE MEDIEVAL WARM PERIOD AND HOW WARM WAS THE LITTLE ICE AGE?

Prior to 1820, when the first continuous meteorological observations begin in Iceland (see Jónsson, 1989) reconstructions of temperature and sea-ice variations are based on historical documentary evidence found in the kinds of sources discussed above. The climate information gleaned from these has been summed up in Ogilvie, 1984, 1991 and 1992a and in Ogilvie and Jónsson, 2001. In the section below, some of the more interesting questions regarding the past climate of Iceland over the last 1000 or so years are considered.

Much has been written concerning the issue of whether the climate was 'milder' when the early settlers first went to Iceland. 'Milder than what?' or 'milder than when?' should perhaps be our first concern. For we are here talking about the question of comparison. The heated debate which took place in the early- to mid-twentieth century concerning the nature of the climate of Iceland during medieval times has been documented in some detail in Ogilvie and Jónsson (2001) and will therefore not be repeated here. However, it is worth noting the comment of the Icelandic geologist Sigurður Þórarinsson who, in the 1950s, found it necessary to point out that the debate on medieval climate had gone on for so long that, during the time that had elapsed, the object of comparison, the climate of the present day, had changed to a large extent (Þórarinsson, 1956b). One difficulty with analyzing the exact nature of the climate of this time is that it is obfuscated by the fact that the Icelanders came to a pristine island and to an ecosystem which had developed without the interference of humans or their domestic animals. Early writers such as the authors of Íslendingabók and Egil's Saga, quoted above, noted that the landscape changed very rapidly. This view has been corroborated by modern research which shows that there is little doubt that when the settlers first came to Iceland they found a landscape which was some 30% covered by 'forest' - mainly of low-growing birch trees (Blöndal and Gunnarsson, 1991). Not realising that this was a finite resource, the settlers cut down the trees to clear land and allowed sheep and cattle to graze freely. The result was the very rapid loss of woodlands commented on by Ari Þorgilsson in Íslendingabók. Erosion thus set in very quickly and remains a major problem today. Currently the land mass of Iceland is 1.3 % covered by woodland (Blöndal and Gunnarsson, 1991). The 'natural capital' enjoyed by the original settlers was thus considerable but did not last long. However, evidence for a relatively mild climate with a low sea-ice incidence and prevailing relatively warm Atlantic water (as opposed to cold Polar water) around 730 to 1100 (perhaps similar to the present day) is provided by marine core evidence from the area of Nansen fjord off the east coast of Greenland (shown as a square on Figure 11.1 and documented in Jennings and Weiner, 1996; and Ogilvie, et al., 2000).

For the period 1200 to 1400 we have reliable historical data that are contemporary, and relatively cold periods are suggested during 1180 to 1210, and again during the 1280s and 1290s. It is likely that there were severe sea-ice years around, for example, 1233, 1261, 1306, 1320 and 1374. The cold period around 1300 is well documented in written sources (Bergþórsson, 1967; Ogilvie, 1991). Palaeoglacial work (see e.g. Guðmundsson, 1997) also indicates major glacial advances at that time. This time may perhaps be termed the 'late medieval cold period' in Iceland. The fourteenth century was markedly variable. The emphasis on sea ice and glaciers in the account from Guðmundar Saga Biskups quoted above which was probably written sometime around 1350 accords well with the cold latter part of the 1340s, 1360s, and 1370s, mentioned in other sources, but is at odds with the probably mild 1330s and the latter part of the 1350s (Ogilvie, 1991). The Nansen fjord marine record also suggests a colder climate regime around the time of the loss of the Norse Western Settlement in Greenland around 1360 (Ogilvie et al., 2000. See also Barlow et al., 1997). There is some evidence that temperatures in Iceland were relatively mild on the whole for the period 1395 to 1430 (Ogilvie, 1991).

Between 1430 and 1560 there are very few contemporary sources. However, circumstantial evidence may suggest a climatic regime that was not unduly harsh during the period 1412 to 1470. It was at this time that the English dominated the trade with Iceland and the major import item into Iceland was cloth – not grain or other food items – implying that food was not in short supply at this time due to dearths or failure of the grass crop (Carus-Wilson, 1933. See also Þorláksson, 1992). Some evidence is available from the *Biskupa Annálar*, written at Skálholt in the south of Iceland in 1605 by Jón Egilsson. He gives a synopsis of the 'best grass years' (1551, 1557, 1575, 1596), Egilsson, (1856, pp.116-117) and the 'heaviest falls of snow' (1532, 1562, 1571, 1581, 1595), (Egilsson, 1856, p.117). A mention of sea ice shortly after 1549 is likely to be an insertion (and an accurate one) by Jón's friend, Bishop Oddur Einarsson (see above). A reliable account (*Gottskálksannáll*) suggests that the 1560s were very cold with much sea ice while the 1570s were mild. The interesting accounts by Arngrímur Jónsson and Oddur Einarsson, discussed above, suggest that the 1590s were, on the whole, cold with much sea ice.

The general view of the past climate of Iceland has been that the period 1600 - 1900 was more or less uniformly bad, with heavy sea-ice incidence, except for a few short and insignificant warm spells. This point of view is expressed by Bergþórsson in an Icelandic version (Bergþórsson, 1969b) of a paper published in English in the same year (Bergþórsson, 1969a). Here he states: 'It is thus most likely that a nearly continuous cold period stretched from the late twelfth century until the early present century' (Bergþórsson, 1969b, p. 343). However, a scrutiny of Icelandic historical sources seem to indicate that there was, in fact, a great deal of climatic variability from early settlement times right up to the present day. From 1600, the sources become far more numerous and it is possible to make climatic indices. The major historical sources for the seventeenth century are later annals, and for the eighteenth century, the sherrifs' letters. Many different sources as well as systematic meteorological observations are available for the nineteenth century. These observations are not reviewed here as they are considered in detail in Jónsson and Garðarsson, 2001. For the reconstructions discussed here, however, the sheriffs' letters are also the main source used for the nineteenth century.

Fig. 11.2 shows sea-ice variations from 1600 to 1870. From this diagram it may be seen that the early and latter decades of the seventeenth century were years with much ice present. From 1640 to 1680 there appears to have been little sea ice off Iceland's coasts. During the period 1600 to 1850, the decades with most ice present were probably the 1780s, early 1800s and the 1830s. From 1840 to 1855 there was virtually no ice off the Icelandic coasts. From that time to 1860 there was frequent ice again, although the incidence does not seem to have been as heavy as in the earlier part of the century. Further clusters of sea-ice years occurred again from 1864 to 1872. Although not shown on this diagram, several very heavy sea-ice years occurred during the 1880s, including 1881 and 1882, the time when William Morris was trying to organize the relief fund for Iceland. Some sea-ice years occurred in the 1890s, but far less than in the 1880s. From 1900 onwards sea-ice incidence falls off dramatically.

As regards temperature variations, a cooling trend may be seen around the beginning and end of the seventeenth century. However, these periods are separated by a mild

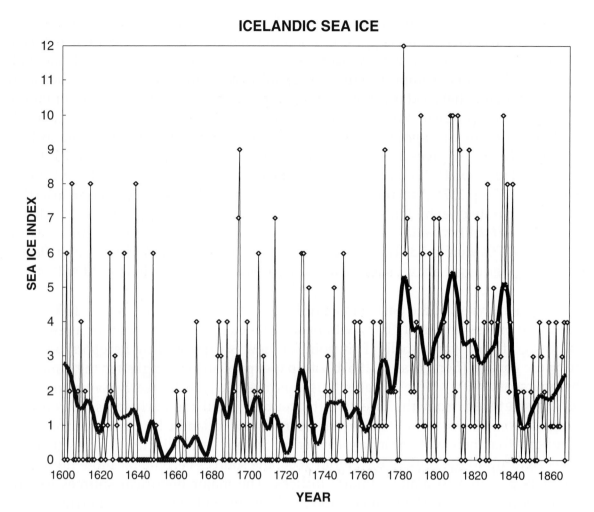

Fig. 11.2 A sea-ice index for Iceland showing variations in the incidence of ice off the coasts during the period AD 1600-1870. The data have also been smoothed to highlight the lower frequency variations using a 15-year low pass filter. The methods used to derive this index are described in Ogilvie, 1991.

period from 1640 to 1670. It is interesting to note that this coincides with a cold period in central Europe which is often regarded as being at the height of the so-called 'Little Ice Age'. (For a detailed discussion of the meaning of this term with regard to the climate of Iceland see Ogilvie and Jónsson, 2001.) This temperature pattern in Iceland correlates well with the sea-ice variations. The early decades of the 1700s were relatively mild in comparison with the very cold 1690s, 1730s, 1740s and 1750s. The 1760s and 1770s show a return to a milder regime in comparison. The 1780s are likely to have been the coldest decade of the century, but this was compounded by volcanic activity (Ogilvie, 1986). The 1801s, 1830s and 1880s were also comparatively cold.

A detailed account of temperature and sea-ice variations during the twentieth century is beyond the scope of this paper. However, some comments will be made. As background to this it may be noted that the Danish meteorological office was established in 1872, and and in 1873 observations in Stykkishólmur in Iceland were taken over by this office. In 1920, Iceland took over observing responsibilities for itself (Garðarson, 1999). In Fig. 11.3, annual mean temperature variations over the North Atlantic around

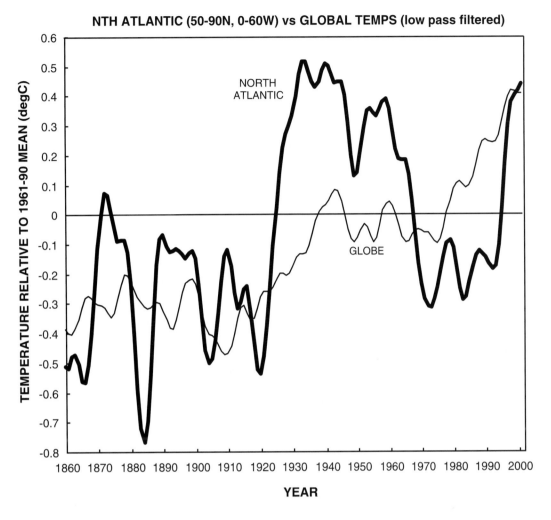

Fig. 11.3 Annual-mean temperature variations over the North Atlantic around Iceland (50-80° N, 0-60° W) compared with global-mean variations. The data have been filtered with a low pass filter to highlight changes on decadal and longer time scales.

Iceland are compared with global mean variations. As stated above, the North Atlantic area exhibits high climate variability. Its interdecadal variability during the past century is characterized by initially colder conditions between 1890 to the 1910s, an abrupt winter season warming around 1920 on Greenland and Iceland, a warm period into the 1940s and a much colder period in the 1960s, with attendant changes in sea ice and ocean salinity. For a more detailed discussion of twentieth-century climate in Iceland, see Ogilvie and Jónsson, 2001. As a reflection on the relative warmth of the twentieth century as compared with the cold years of the end of the nineteenth, it is interesting to note that, according to anecdotal evidence, older farmers in the mid-twentieth century reportedly said that by 1893, 'the worst was over' (Sigfúsdóttir, 1969, p.75). Certainly a warming trend was well under way by the late 1920s. It culminated in the 1930s and 1940s but ended abruptly in 1965 when the sea ice returned after being absent for many years. The period 1965 to 1971 is known in Iceland as the 'Ice Years'. The ice has been an infrequent visitor since the severe ice year of 1979. Cool years persisted to the mid 1980s. Fig. 11.4 shows average annual temperatures for Akureyri, northern Iceland, for

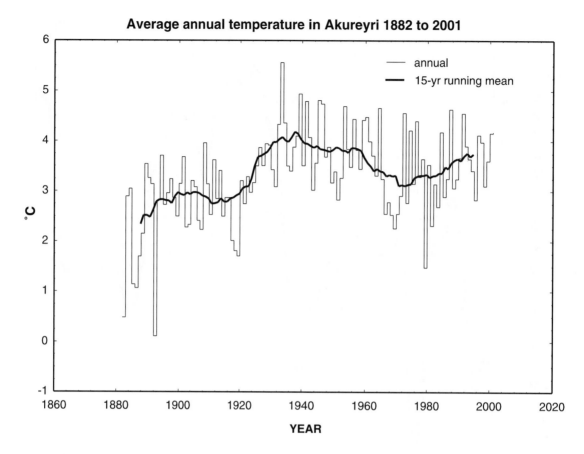

Fig. 11.4 *Average annual temperature and 15-year running means in Akureyri 1882-2001. Provided by Trausti Jónsson, Icelandic Meteorological Office.*

the period 1882 to 2001. Here the general pattern for Iceland described above can be seen, as well as the mainly warming trend after the 1980s.

To sum up, therefore, it is to be hoped that the discussion above of literary and historical sources has helped to demonstrate the great variability of the past climate of Iceland over the last thousand or so years. In particular, to answer, at least in part, the question posed above, we may note that the concept of a uniformly mild 'medieval warm period' and a uniformly cold 'little ice age' is very far from accurate.

11.10 CONCLUDING REMARKS

Whether apocryphal or not, the earliest accounts of the sighting and naming of Iceland have everything to do with climatic and environmental features. One early visitor apparently called it 'Snowland'. Another, Flóki Vilgerðarson, is said to have given Iceland its enduring name around AD 865 because of the sea ice he observed off the northwestern coasts (Landnámabók, Benediktsson, 1968; Ogilvie, 1991). Whether the country was really named for the ice, or for the glaciers that would have been clearly visible to travellers approaching by sea, the name has lasted. The name is indeed appropriate; Iceland's marginal location for agriculture has meant that the country's economy and history have been impacted by environmental and climatic change from early settlement times to the present day. This is reflected in the wealth of sources which document climate and weather events in Iceland. Although this correlation between a difficult climate and the recording of information about the climate and weather makes sense, it is in reality remarkable that the Icelandic nation has produced such an astonishing treasure trove of environmental literature. The small size of the population makes it even more remarkable. In 1703, when the first census was taken, the population was around 50,000. (The present-day population is around 300,000). The aim of this particular discussion has been to highlight the excellent local (and foreign) knowledge displayed in accounts of climate and weather in Iceland from the earliest times, and to focus on some of the sources which have received little attention heretofore. In many of the works discussed here, the dichotomy between 'official' reports and 'local' knowledge is blurred. Not until the twentieth century did these concepts become somewhat polarized (Pálsson and Helgason, 1998). However, throughout the twentieth century, the tradition of carefully maintained specific and detailed knowledge regarding local conditions has continued in tandem with the development of the scientific method. Whether this will survive the twenty-first century remains to be seen.

ACKNOWLEDGEMENTS

Funding from: the National Science Foundation (USA) grants OPP-9726510 and OPP-0002651; the Leverhulme Trust (UK); and RANNÍS (Iceland) is gratefully acknowledged. Trausti Jónsson kindly provided the diagram of Akureyri temperatures. Trausti also took the time to read the manuscript of this paper and afforded many useful and helpful comments. Figure 11.1 was provided by Anne Jennings. Figure 11.2 is based on data from the gridded land-plus-marine data set used by the Intergovernmental Panel on Climatic Change (Nicholls *et al.*, 1996) and were provided by P.D. Jones. I am grateful to Trond Woxen for his help in sorting the references.

Manuscript Sources
Landsbókasafn (National Library) Reykjavík

The Jón Jónssons Weather Diary, Lbs 332 8vo, ÍBR 82 8vo

Þjóðskjalasafn (National Archives) Reykjavík

Is. J. 10/2098 Ólafur Stefánsson, Viðey, 22 July 1802
Is. J. 10/2098 Magnús Ketilsson, Búðardalur, Dalasýsla, 29 July 1802
Is. J. 10/2098 Ólafur Stefánsson, Viðey, 20 August 1802
Is. J. 10/1848Ólafur Stefánsson, Viðey, 6 September 1802
Is. J. 10/2127 Jón Jakobsson, Espihóll, Eyjafjarðarsýsla, 14 October 1802
Is. J. 10/2098 Jón Jónsson, Bær, Strandasýsla, 28 December 1802
Is. J. 10/2098 Magnús Ketilsson, Búðardalur, Dalasýsla, 3 January 1803
Is. J. 12/2621 Stephen Thorarensen, Möðruvellir, Eyjafjörður, 26 September 1816
Is. J. 15/ 2194 Lárus Blöndal, Húnavatn, 1 October 1881

12. Chemical weathering, chemical denudation and the CO_2 budget for Iceland

Sigurður Reynir Gíslason

Science Institute, University of Iceland, Dunhagi 3, IS-107 Reykjavík, Iceland

The average relative mobility of elements in SW Iceland during chemical weathering and denudation is: S > F > Na > K >> Ca > Si> Mg > P > Sr >> Mn >Al > Ti > Fe. The sequence of mobility of the elements in the youngest rocks is different from that of the oldest rocks and it changes with increased vegetation cover. The secondary minerals that are left behind in soil are predominantly allophane, poorly crystalline ferrihydrite, and some imogolite. The total chemical denudation rate of the rock derived elements (TCDR) increases with increasing runoff, but decreases with increasing age of the rocks. Where the variation in the primary variables, runoff and age of rocks, is small the denudation rate for Ca, Mg and Si increases with increased vegetative cover. The TCDR for the major Icelandic catchments range from 19 to 146 t km^{-2} yr^{-1}. The calculated average for Iceland is 37 compared to the 26 t km^{-2} yr^{-1} for the continents. The high rate in Iceland is attributed to high runoff and high reactivity of glassy and crystalline basalt. The chemical denudation rate of relatively mobile elements like S, F, Na and Ca increased by a factor of 2 to 5 at constant runoff and vegetative cover but 0% to 100% glass content of the rocks. The chemical denudation rates of 5 m yr old crystalline basalt, including data on cation uptake by growing trees, indicate that the rate of release of Ca and HCO_3 to streams is about 3 times higher in tree-covered areas than in unvegetated areas. The variation in the chemical denudation rate in Iceland is dictated by the composition of the soil solutions and the saturation state of primary and secondary minerals as reflected in the dependence on runoff.

More atmospheric CO_2 is fixed during transient chemical weathering in Iceland than is released to the atmosphere from Icelandic volcanoes and geothermal systems. However, on a geological time scale, more CO_2 is released to the atmosphere than will eventually precipitate in the ocean as Ca and Mg carbonates, as the result of Ca and Mg released during chemical weathering. The overall net transient CO_2 budget for Iceland in 1990, including CO_2 fixation by chemical weathering and vegetation, and the anthropogenic release of CO_2 and the release from volcanoes, was 'positive'. That is there was a net CO_2 release to the atmosphere of the order 0.3–1.3 m t CO_2 yr^{-1}.

12.1 INTRODUCTION

The objective of this paper is to review and summarise the findings of recent studies of chemical weathering, chemical denudation (Fig. 12.1) and the CO_2 budget for Iceland and to discuss their implications.

Chemical weathering and denudation of Ca, Mg-silicates on land and the ocean floor governs the long term atmospheric CO_2 content and therefore climate (Urey, 1952; Holland, 1978; Walker *et al.*, 1981; Berner, 1992; Brady and Gíslason, 1997). Chemical weathering and denudation influence strongly the chemical composition of soil waters, rivers, and the oceans. Because of its abundance at the Earth's surface and its high reactivity, chemical weathering of both glassy and crystalline basalt plays a significant role in the global cycle of numerous elements (Gíslason *et al.*, 1996; Brady and Gíslason, 1997; Louvat, 1997; Moulton *et al.*, 2000; Dessert *et al.*, 2001; Frogner *et al.*, 2001; Oelkers and Gíslason, 2001; Stefánsson and Gíslason, 2001). Low temperature alteration by seawater of Ca-bearing silicates in basalt and basaltic glass in the younger part of the ocean floor (< 6 m yr) is estimated to release as much Ca to the world's oceans as continental weathering (Brady and Gíslason, 1997). Regions dominated by exposed reactive Mg-Ca silicate rocks, high rainfall, and high relief are the most important long term CO_2 sinks on land. Iceland provides an opportunity to study the weathering of Mg-Ca silicates under uniform lithology, at a constant average temperature, variable rainfall and rock age, high relief, and variable glacial/vegetative cover. More than 90 percent of the island is of basaltic composition.

Numerous studies have been carried out on the chemical weathering and denudation of basalt in Iceland (e.g. Sigvaldason, 1959; Cawley *et al.*,1969; Jakobsson, 1978; Raiswell and Thomas, 1984; Gíslason and Eugster, 1987b; Sigurðsson and Einarsson, 1988; Crovisier, 1989; Gíslason and Arnórsson, 1990; Arnalds, 1990; Wada *et al.*,1992; Gíslason and Arnórsson, 1993; Bluth and Kump, 1994; Gíslason *et al.*, 1996; Louvat, 1997; Louvat, *et al.*, 1999; Moulton *et al.*, 2000; Stefánsson *et al.*, 2001; Stefánsson and Gíslason, 2001; Arnórsson *et al.*, 2002) that define the changes in rock chemistry and mineralogy during weathering, the chemistry of the water involved in the weathering process, the relative mobility of the elements, the saturation state of the minerals, chemical fluxes, the effect of vegetation on weathering rates and the formation of soils.

Experimental studies on weathering and/or alteration of basalt and basaltic glass have underscored the importance of pH, water composition, temperature, glass content and composition, adsorbed salts on pristine volcanic glass, and the presence of bacteria on the rate of weathering (e.g. Hoppe, 1940; Furnes, 1975; Gíslason and Eugster, 1987a; Crovisier *et al.*, 1992; Guy and Schott, 1989; Gíslason *et al.*, 1993; Berger *et al.*, 1994; Thorseth *et al.*, 1995; Brady and Gíslason, 1997; Daux *et al.*, 1997; Oelkers and Gíslason, 2001; Frogner *et al.*, 2001; Wolff-Boenisch *et al.*, 2002; Gíslason and Oelkers, 2003).

Chemical weathering is often referred to as a process at a specific site, the weathering site. When water comes in contact with rocks and soil it reacts with the primary minerals of the rock. The minerals dissolve to a varying degree and some of the dissolved constituents react with one another to form new or secondary minerals. Chemical denudation is the transport of dissolved solids, released by chemical weathering, away from the weathering site via soil water, streams, and rivers, and often all the way to the oceans.

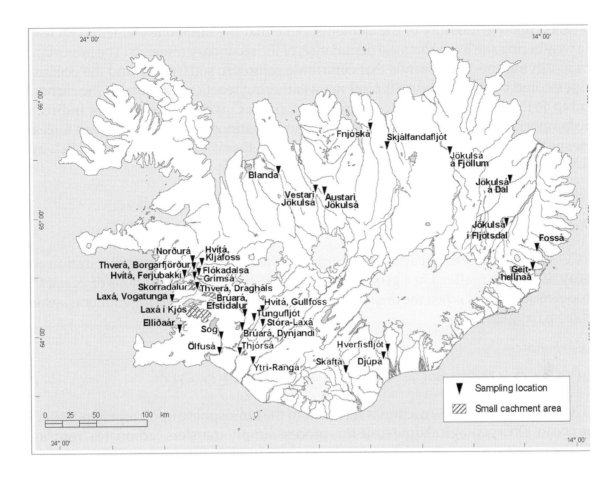

Fig. 12.1 *Location of rivers sampled for chemical weathering and denudation studies. Gíslason et al. (1996): all the rivers in the southwest, from Þjórsá to Norðurá in the north-west excluding Laxá í Kjós. Louvat (1997): all the rivers marked on the figure in the north, east and south, fro m Blanda in the north to Ytri-Rangá in the south, plus Þjórsá and Ölfusá in the south and Norðurá and Hvítá at Kljáfoss in the west. Moulton et al. (2000): the small catchments at Skorradalur in western Iceland. Stefánsson and Gíslason (2001): the small tributaries to Laxá í Kjós in western Iceland.*

It has only been about 10,000 years since Iceland was deglaciated, setting the chemical weathering clock of the surface rocks to zero. The glaciers scraped away the chemical weathering residue, organic matter and secondary minerals such as clays. The ratio of present mechanical versus present chemical denudation rate in Iceland is 13 (500:37 t km^{-2}yr^{-1}, Tómasson, 1986; Gíslason *et al.*, 1996, and this study), that is, higher than the overall ratio for the Earth, 8.8 (230:26 t km^{-2}yr^{-1} Milliman and Syvitski, 1992; Berner and Berner, 1996). This ratio varies considerably in Iceland as it does worldwide. Where the ratio is low, one would expect to see *in situ* the actual results of chemical weathering.

Carbonate and evaporate minerals weather usually congruently, that is they dissolve completely. Conversely most silicates dissolve incongruently during weathering, that is some elements do not leave the weathering site; they are tied up in secondary minerals such as clays and hydroxides or their amorphous precursors. The elements that do not leave the weathering site are referred to as immobile elements. The relative tendency for the elements to leave the weathering site, in solution, can be studied by the so-called relative mobility of the elements. Note that the immobile elements can leave the weathering site by mechanical denudation of the secondary minerals. The importance of identifying specific minerals that contribute solutes to soils, rivers and the oceans is underscored by the following typical net weathering reactions, assuming the carbonic acid to be the only significant overall proton donor. CO_2 is derived directly or indirectly (decomposition of organic matter [Eq. 1]) from the atmosphere and dissolved ions are transported via rivers to the ocean where some of them, such as Ca^{++}, Mg^{++}, Sr^{++} and HCO_3^-, can precipitate in the form of carbonates:

Primary production (photosynthesis, respiration and decay)

$$106\,CO_2 + 16\,NO_3^- + HPO_4^{2-} + 122\,H_2O + 18\,H^+ = C_{106}H_{263}O_{110}N_{16}P + 138\,O_2 \tag{1}$$

CO_2 consumed by this reaction is returned to the atmosphere during respiration and decay of organic matter. On a geological time scale, thousands to millions of years, this process is important when the organic matter is permanently buried in sediments, sealed off from oxygen.

Carbonates

$$CaCO_3 + CO_2 + H_2O \longrightarrow Ca^{++} + 2\,HCO_3^- \longrightarrow CaCO_3 + CO_2 + H_2O \tag{2}$$

CO_2 consumed by this reaction is returned to the atmosphere upon reprecipitation in the ocean. On a geological time scale this process simply transfers carbonates from land via rivers to the ocean resulting in no net consumption of atmospheric CO_2.

Na-K silicates (e.g. feldspars, white-mica) to clays (e.g kaolinite, halloysite, imogolite, allophane)

$$KAlSi_3O_8 + CO_2 + 5.5H_2O \longrightarrow 0.5Al_2Si_2O_5(OH)_4 + K^+ + HCO_3^- + 2H_4SiO_4 \tag{3}$$

This type of reaction consumes CO_2 but there is no net removal of CO_2 from the surface reservoirs on a geological time scale because Na and K do not form carbonates that can precipitate in the ocean. During the weathering process, as written here, K is mobile, Si is less mobile, some of it stays behind at the weathering site, and Al is immobile, all of it is left behind at the weathering site.

Ca-Mg silicates (e.g. plagioclase, olivine, pyroxene, volcanic glass) to clays (e.g. kaolinite, halloysite, imogolite, allophane)

$$CaAl_2Si_2O_8 + 2\,CO_2 + 3H_2O \longrightarrow Al_2Si_2O_5(OH)_4 + Ca^{++} + 2HCO_3^- \ldots \longrightarrow \ldots CaCO_3 + CO_2 + H_2O \tag{4}$$

This type of reaction causes net removal of CO_2 from the atmosphere due to transport of Mg, Ca and HCO_3^- from the weathering sites via rivers to the oceans, where it eventually precipitates as carbonates. The overall stoichiometry is one mole of CO_2 for each mole of Ca or Mg released by weathering. During the weathering process on land, as written here, Ca is mobile, but Si and Al are immobile, they are left behind at the weathering site.

Of these reactions, only reaction (4) and sometimes (1), result in net removal of CO_2 from the atmosphere on a geological time scale. However, reactions (2) and (3) may still have important effects, because they can lead to transient removal of CO_2 from the atmosphere over the time scale needed for the system to stabilise; years to thousands of years.

12.2 CHEMICAL WEATHERING IN ICELAND

The relative mobility of elements during chemical weathering of basalt has been studied using the chemistry of rivers and rocks in the catchment area of the rivers (e.g. Gíslason *et al.*, 1996) and by comparison of the composition of 'fresh' and weathered rocks (e.g. Craig and Loughnan, 1964). The former approach (river approach) yields, geologically speaking, a near instantaneous look at the relative mobility, whereas the latter (rock approach) gives integrated mobilities, often over millions of years. Furthermore, the river approach defines mobilities over tens to hundreds of kilometres, whereas the rock approach yields mobilities over nanometers to some meters. The discharge weighted relative mobility for southwest Iceland ($R_{mobility}$) in Fig. 12.2 is calculated by the water/rock concentration ratio, normalised to sodium, as shown below for Ca:

$$R_{mobilityCa} = (Ca_{water}/Na_{water})/(Ca_{rock}/Na_{rock}) \tag{5}$$

where Ca_{water} and Na_{water} are the concentrations of Ca and Na in river water, but Ca_{rock} and Na_{rock} are the average concentrations of Ca and Na in unaltered rocks in the catchment area of the river. All water concentrations are corrected for atmospheric contribution. The sequence of mobility of the elements in the youngest rocks is different from that of the oldest rocks. The relative mobility of Ca, Mg and Sr increases with the

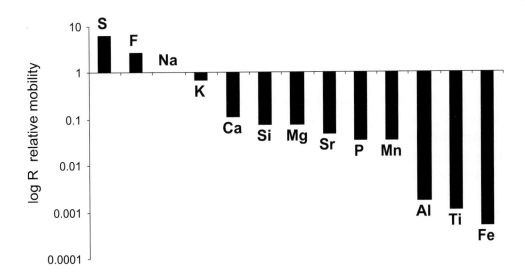

Fig. 12.2 The discharge weighted average relative mobility of some elements during the present weathering of basalt in southwest Iceland (from Gíslason et al., *1996)*

age of the rocks and vegetative cover. The relative mobility of the least mobile elements (Fe, Ti, Al, and Mn) during weathering in Iceland is similar to that observed elsewhere in the world under remarkably variable climatic conditions. These elements are 10^3 to 10^4 times less mobile than Na. Thus, one would expect a similar chemical composition of the end product of chemical weathering in Iceland as, for example, in Hawaii.

Arnalds (1990, 1999), Wada *et al.* (1992), Arnalds *et al.* (1995), and Arnalds and Kimble (2001) identified directly and indirectly three phases as the dominant ones for the clay size fraction of the soils in Iceland (>700 g kg⁻¹ of the clay-size fraction), that is the weathering minerals that are left behind at the weathering site: allophane and/or imogolite (variable Si-Al ratios and variable hydration and crystallinity) and poorly crystalline ferrihydrite. These phases are mostly amorphous to X-rays. Layer silicate minerals and laminar opaline silicas were not identified or occurred in only small quantities. The organic carbon content of the Icelandic soils varies from 0.2–30 wt.% (Arnalds, 1999). According to Douglas (1987) manganese-rich coatings are common on the surface of weathered basalts in Iceland. He also identified poorly crystalline iron oxides and hydroxides, smectites, amorphous Al-Si minerals, and perhaps illite or kaolinite in some cases. Crovisier *et al.* (1992) studied the mineralogy and geochemistry of the weathering product of basaltic glass in Iceland. They identified two types of weathering material: Si-, Ca-, Mg-rich particles with smectite structure and amorphous Fe-, Ti- and Al-rich material. On the basis of equilibrium calculations between clay mineral solid solutions and aqueous solution, Crovisier *et al.* (1992) concluded that these two phases were formed simultaneously at equilibrium in the same aqueous solution. Further, they observed that the youngest sample contained only amorphous secondary phases. With increasing age crystalline weathering phases become more abundant, namely Ca-, Mg-, Fe-smectites and in the most evolved samples zeolites like chabazite and erionite were identified.

12.3 CHEMICAL DENUDATION RATES

The area weighted average total chemical denudation rate (TCDR, sum of major rock derived river fluxes excluding carbon) and the rate for individual elements or major chemical species are shown for southwestern Iceland in Fig. 12.3 (Gíslason *et al.*, 1996). The fluxes for NO_3 and NH_4 are negative since there is a net fixation of these nutrients within the catchments.

Excluding the geothermally affected rivers in Iceland, the average temperature variation between individual rivers in Iceland measured at the gauging stations in the lowlands is small, 3.7 °C, or from 1.6 to 5.3. (Gíslason *et al.*, 1996, 1998, 2002a,b). The total chemical denudation rate for the major catchments (Gíslason *et al.*, 1996; Louvat 1997; Fig. 12.4a and Table 12.1) and small catchments (Stefánsson and Gíslason, 2001) in Iceland increases with increasing runoff similar to that found in studies all over the world (e.g. Meybeck, 1979; Bluth and Kump, 1994). The total chemical denudation rate decreases with increasing age of the rocks (Fig. 12.4b and Table 12.1). A considerable part of the rocks that are younger than 1 m yr are glassy (hyaloclastites), but glassy

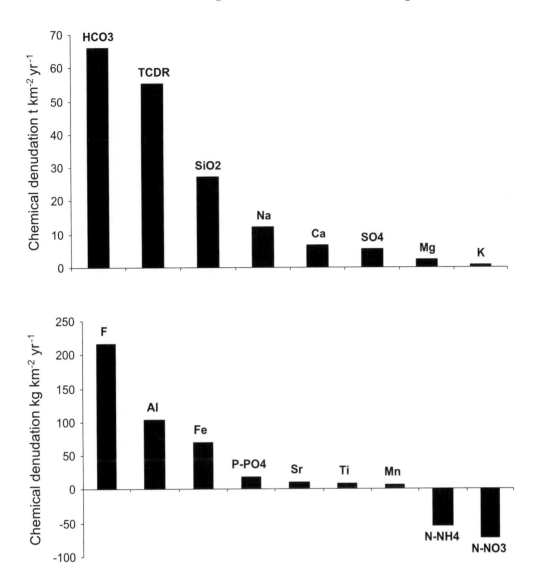

Fig. 12.3 The area weighted average total chemical denudation rate (TCDR; calculated from the sum of individual major rock derived dissolved element fluxes excluding the bicarbonate flux) and the area weighted average for individual elements or major chemical species in south-west Iceland (from Gíslason et al., 1996).

basalt dissolves about 10 times faster than crystalline basalt at pH 9–10 (Gíslason and Eugster, 1987a). In old rocks, glass is less common and if present is often altered as is olivine and Ca-rich plagioclase. Furthermore, the permeability of the young rocks is higher than the old ones (Sigurðsson and Ingimarsson, 1990), resulting in more reactive surface areas in contact with the water. As shown by the data in Fig. 12.4a and Table 12.1, the rivers draining the oldest rock, Fossá in Berufjörður, Geithellnaá, and Þverá at Draghals (Fig. 12.1) reflect low denudation rates at a particular runoff. This suggests that there is different runoff dependence in old rocks as compared to younger ones. There is a conspicuous correlation between chemical denudation rate and atmospheric CO₂ fixation in the catchments (Fig. 12.4c). There is relatively little atmospheric CO₂

Table 12.1 Age, runoff, total chemical denudation rate of rock derived elements (TCDR), and CO_2 fixation of the Icelandic catchments. (The upper part of the table is from Gíslason *et al.* (1996) based on the average of 17–23 samples for each catchment. The lower part of the table is from Louvat (1997) based on a single sample from individual catchments. Tributaries are in italics after the main river.

Main River *Tributary*	Age m yr	Runoff mm yr^{-1}	Chemical denudation t km^{-2} yr^{-1}	CO_2 fixation t km^{-2} yr^{-1}
Þjórsá	0.6	1750	58	51.2
Ölfusá	0.6	2408	62	53.5
Stóra Laxá	1.9	1478	40	31.0
Hvítá, Gullfoss	0.5	2096	58	47.4
Tungufljót	0.3	2074	48	36.4
Brúará, Dynjandi	0.3	3480	89	73.1
Brúará, Efstidalur	0.2	6043	146	68.4
Sog	0.3	3492	80	79.3
Ellidaár	0.2	3157	69	63.3
Laxá, Vogatunga	6.7	1732	29	29.7
Þverá, Dragháls	5.2	3906	56	52.2
Hvítá, Ferjukot	3.7	1680	38	33.8
Grímsá	2.9	2176	47	45.4
Flókadalsá	2.7	1695	40	36.6
Hvítá, Kljáfoss	1.0	1769	38	30.9
Þverá, Borgarfjörður	6.6	1664	39	44.1
Norðurá, Stekkur	7.7	2222	41	38.8
Hvítá, Kljáfoss	1.0	1404	31	19.8
Norðurá, Stekkur	7.7	2208	36	25.1
Blanda	3.0	914	19	15.4
Austari Jökulsá	3.0	1795	31	19.4
Vestari Jökulsá	3.0	803	24	22.9
Fnjóská	7.0	1239	31	21.1
Skjálfandafljót	0.6	1088	29	20.7
Jökuldá á Fjöllum	0.3	910	30	28.6
Jökuldá á Dal	0.2	1829	29	25.1
Jökulsá í Fljótsdal	4.0	1865	33	28.6
Fossá	7.5	2473	20	7.0
Geithellnaá	6.5	2742	39	20.7
Djúpá	1.0	2547	43	24.2
Hverfisfljót	0.6	3839	77	53.3
Skaftá	0.3	1822	57	35.2
Ytri Rangá	0.5	1488	91	93.3
Þjórsá	0.6	1962	48	34.3
Ölfusá	0.6	2655	64	51.1
Hvítá, Gullfoss	0.5	2208	53	35.2

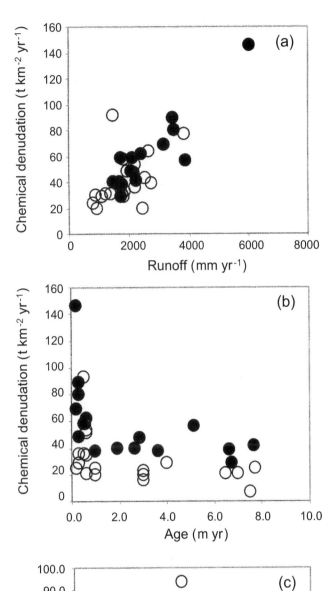

Fig. 12.4 (a) The total chemical denudation rate (TCDR) versus runoff. (b) The total chemical denudation rate (TCDR) versus age of rocks. (c) Net transient CO$_2$ fixation by chemical weathering and denudation versus total chemical denudation rate (TCDR) for the Icelandic catchments. The filled circles represent average rates for each catchment based on 17–23 samples taken over 2 years (Gíslason et al., 1996). The open circles depict rates based on a single sample from individual catchments (Louvat, 1997; Fig. 12.1 and Table 12.1).

fixation at the high denudation rate in Brúará at Efstidalur (Fig. 12.1) because all the runoff from the catchment is in the form of groundwater, with chemical reaction proceeding sealed off from the atmosphere resulting in pH 10 waters with partial pressure of CO_2 as low as 10^{-6} bars. As soon as this water comes in contact with the atmosphere it starts to consume CO_2 (Gíslason 1989; Gíslason *et al.*, 1996). There is no general relationship between the total chemical denudation rate and vegetative cover and glacial cover in Iceland (Gíslason *et al.*, 1996; Louvat, 1997). Thus, vegetation and glacial cover seems to be a secondary variable for TCDR in Iceland. In general there is an increase in the chemical denudation rate for all major elements with increasing runoff in southwestern Iceland, nitrate and ammonium fluxes are increasingly negative, and phosphate fluxes are mostly independent of runoff, but the greater the phosphate flux the more negative are the nitrate and ammonium fluxes (Gíslason *et al.*, 1996). The chemical denudation rates for Si, Na , K, F, Al, P, Fe, Ti, and Mn decrease with increasing age of rocks (Figs. 12.5a,b) as does the total chemical denudation rate shown in Fig. 14.4b. Conversely, the chemical denudation rates for Ca, Mg, and Sr increase and the fluxes of nitrate and ammonium are less negative with increasing age of rocks in catchments that are 'older' than 1 m yr (Figs. 12.5c,d). Where the variation in the primary variables, runoff and age of rocks, is small (older than 1 m yr, runoff 1700 to 2200 mm yr^{-1}), the denudation rate for Ca, Mg and Si, increases with increased vegetative cover (Figs. 12.5e,f).

The discharge weighted average total flux (TDS, sum of individual major elemental fluxes) for southwest Iceland is 121 t km^{-2} yr^{-1} of which 66 t km^{-2} yr^{-1} is bicarbonate (HCO_3^-), mostly derived, directly or indirectly, from the atmosphere. The total rock derived chemical denudation rate (TCDR) for the major Icelandic catchments excluding bicarbonate ranges from 19 to 146 and the discharge weighted average for southwestern Iceland is 55 t $km^{-2}yr^{-1}$ (Gíslason *et al.*, 1996). These rates are lower than that determined for the volcanic island Réunion Island in a tropical climate but similar to those for Sao Miguel, Azores, in a temperate climate (Louvat and Allègre, 1997; 1998). The chemical denudation rate is at a maximum in the mountainous regions of the humid temperate and humid tropical zones, 80 t $km^{-2}yr^{-1}$ and 67 t $km^{-2}yr^{-1}$, respectively (Meybeck, 1979). These are the only morphoclimatic zones that have higher average total chemical denudation rates than the 55 t $km^{-2}yr^{-1}$ for southwest Iceland. The TCDR of the Hvítá catchment at Ferjukot in western Iceland (Fig. 12.1) is 38 (t $km^{-2}yr^{-1}$), probably close to the average for Iceland since the average runoff (1680 mm yr^{-1}) and age (3.7 m yr) of this catchment is similar to the average for Iceland. Alternatively, the chemical denudation rate for all of Iceland can be calculated by a linear regression of the runoff dependence of old and young catchments in Fig. 12.4a and the runoff for Iceland, 1650 mm yr^{-1} (total discharge from Iceland (Rist, 1956) divided by the size of Iceland). This yields 37 TCDR (t $km^{-2}yr^{-1}$). The TCDR for the continents is 26 t $km^{-2}yr^{-1}$ and is dominated by carbonate weathering (Berner and Berner, 1996).

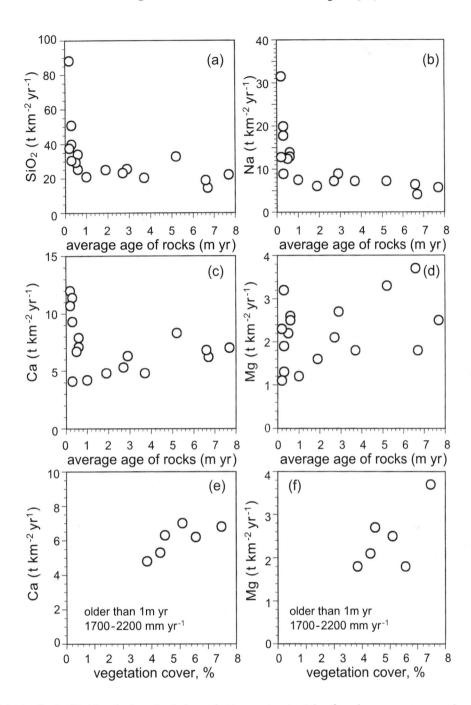

Fig. 12.5 (a to f). Individual chemical denudation rates in t km⁻² yr⁻¹ versus age and vegetative cover (modified from Gíslason et al., 1996).

12.4 THE SATURATION STATE (ΔG_R) OF RIVER AND GROUND WATERS RELATIVE TO PRIMARY AND SECONDARY BASALTIC MINERALS

The importance of the saturation state (ΔG_r; the Gibbs free energy of reaction) is twofold: (1) In general it determines whether a mineral of a given composition tends to precipitate (ΔG_r positive) or dissolve (ΔG_r negative) in water of a specific composition and temperature. In the case of igneous primary minerals it determines whether they dissolve or not. (2) Within a critical range it determines the rate (driving force) of dissolution and precipitation.

The saturation state of olivine (Fig. 12.6) and the primary basaltic mineral's end-members and average compositions is summarised in Fig. 12.7 for river and ground waters in Iceland (Stefánsson *et al.*, 2001). The saturation state of these minerals in soil solutions and natural glasses of variable composition, in all solutions, has not jet been assessed. River water and cold groundwater in Iceland are undersaturated with respect to olivine and presumably even more so with basaltic glass (Fig. 12.6). They are generally undersaturated with respect to pyroxenes and plagioclase (Fig. 12.7). They are more undersaturated with respect to magnesium-rich olivine and pyroxene than iron-rich olivine and pyroxene and the saturation state is dependent on the composition of the plagioclase (Fig. 12.7; Gíslason and Arnórsson, 1990, 1993; Stefánsson *et al.*, 2001; Stefánsson and Gíslason, 2001; Arnórsson *et al.*, 2002). Mg-rich olivine and Ca-rich plagioclase are most undersaturated. The Fe-Ti-oxides of the average composition of the rocks are the most stable primary minerals (Fig. 12.7). These observations are generally in accord with the results of mineralogical studies on the pattern of weathering susceptibility: that is, olivine > pyroxene > plagioclase > sanidine (Craig and Loughnan, 1964); glass > olivine > pyroxene > amphibole > plagioclase > K-feldspar (Colman, 1982); glass ~ olivine > plagioclase > pyroxene > opaque minerals (Eggleton *et al.*, 1987); glass, olivine > laihunite > clinopyroxene > orthopyroxene > plagioclase > K-feldspar > magnetite > apatite > rutile > quartz (Banfield *et al.*, 1991) and Nesbitt and Wilson's (1992) study of Australian basalts, that is, olivine > glass > plagioclase > clinopyroxene > Fe-Ti-oxides. The sequence is dependent on pH, composition of the minerals and the concentration of organic anions in the soil solutions. For example the pyroxene could be stabilized relative to plagioclase in high pH waters.

It is difficult to assess the saturation state of weathering products with respect to soil and river water since the weathering minerals are poorly crystalline, fine grained, and of variable composition. The findings of Gíslason and Arnórsson (1990), Gíslason *et al.* (1996), and Stefánsson and Gíslason (2001) can be summarised as follows. The rivers and groundwaters are supersaturated with several weathering minerals. At the early stage of weathering, basaltic glass, olivine, pyroxene and plagioclase dissolve in Icelandic river and ground waters, releasing elements that are consumed as follows: Ca by smectite and/or Ca zeolite; Si by a fine-grained mixture of quartz and moganite (Gíslason *et al.*, 1997), and in allophane, imogolite, smectites and zeolites; Mg by smectite and/or talc. Phosphorus is probably retained in the biomass and adsorbed on clays and hydroxides. Mn and Fe are retained in poorly crystalline oxides and hydroxides and smectites. Aluminium is retained in allophane, imogolite, smectites and zeolites. And if titanium is mobile, it is probably retained in poorly crystalline Fe and/or Mn oxides and hydroxides.

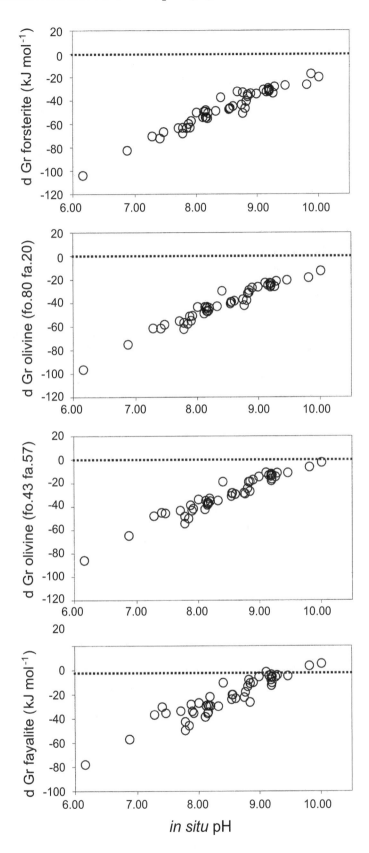

Fig. 12.6 The pH dependence of the saturation state of olivines in selected cold ground waters and river waters in Iceland depicted in terms of the Gibbs free energy of the dissolution reactions. The dashed lines represent equilibrium conditions (modified from Gíslason and Arnórsson, 1993)

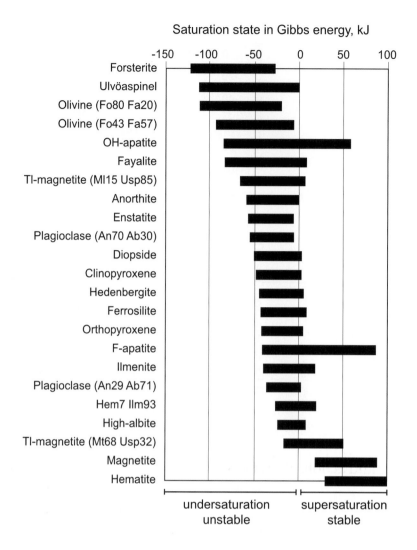

Fig. 12.7 The stability of primary basaltic minerals under weathering conditions as inferred from the composition of surface and cold groundwaters in Iceland (Stefánsson et al., 2001).

12.5 DISSOLUTION RATE OF BASALTIC MINERALS AND GLASS

The solution properties that affect the dissolution and precipitation rates of minerals and glasses, and therefore overall chemical weathering and denudation rates, are: temperature, pH, ionic strength, saturation state ($\triangle G_r$), and activities of individual aqueous species (Lasaga *et al.*, 1994; Oelkers, 2001a). On average the dissolution rate of silicates increases by an order of magnitude, from 0° to 25°C (apparent activation energy of 63 kJ mol[-1]; Lasaga *et al.*, 1994). The average temperature variation between individual rivers in southwest Iceland is small, 2.6°C, excluding the geothermally affected rivers. This variation translates to a factor of 1.3 for the dissolution rate of minerals, or more specifically the chemical denudation rate of the mobile elements in the glass and minerals.

The effect of pH on the dissolution rate of basaltic glass far from equilibrium is similar to the one for feldspars (Guy and Schott, 1989; Oelkers and Schott ,1998; Oelkers, 2001a; Oelkers and Gíslason, 2001; Gíslason and Oelkers, 2003). The dissolution rates decrease

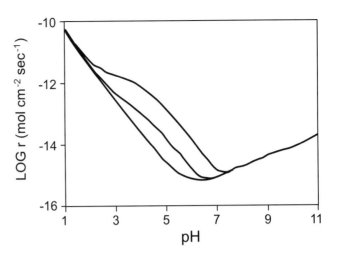

Fig. 12.8 The logarithm of the BET dissolution rate of basaltic glass (MORB) at 25 °C as a function of pH and a total Al concentration of 10^{-6} mole kg^{-1} in the presence of 0, 0.5 and 1 (mmol kg^{-1}) concentration of oxalate ion (modified from Oelkers and Gíslason, 2001).

dramatically with increasing pH at acid conditions, reach minimum at near neutral pH, and increase with increasing pH at basic conditions (Fig. 12.8). The pH at which basaltic glass dissolution reaches minimum decreases with increasing temperature. It is at pH 7 at 0°C but around pH 6 at 25°C (Gíslason and Oelkers, 2003). Conversely, the dissolution rate of olivine and pyroxene decreases rapidly from low to high pH (Pokrovsky and Schott, 2000; Oelkers 2001b; Oelkers and Schott, 2001), and furthermore, at pH above 7 an increase in the activity of the carbonate species in solution decreases the dissolution rate of forsterite (Pokrovsky and Schott, 2000). The presence of organic oxalate anion increases the rates of plagioclase and basaltic glass dissolution at pH 3 to 6 (Fig. 12.8; Oelkers and Schott, 1998; Oelkers, 2001a; Oelkers and Gíslason, 2001). The maximum effect at 25°C is at pH 4.5 where 1 mmol kg^{-1} oxalate ion increases basaltic glass dissolution rate by a factor of 50. The concentration of dissolved organic carbon in Icelandic soil waters is as high as 1 mmol kg^{-1} (Gíslason *et al.*, 1999). The dependence of the dissolution rate of minerals and glasses and precipitation rate of secondary minerals on the saturation state ($\triangle G_r$) is complicated (Lasaga *et al.*, 1994; Daux *et al.*, 1997; Oelkers, 2001a). The dissolution rate is near independent of the saturation state when the quotient $-\triangle G_r/sRT$ is a greater negative number than 2.3 to 6 (Walther and Wood, 1986; Daux *et al.*, 1997; Oelkers, 2001a; Gíslason and Oelkers, 2003), where *s* stands for Temkin's average stoichiometric number, R is the gas constant and T is temperature in Kelvin. According to this, the dissolution rate is independent of the saturation state when $\triangle G_r$ is more negative than –5 to –15 (kJ mol^{-1}) in the weathering environment, when the Temkin's number is equal to one. It is important to note that aqueous species, such as Al^{+++} and Mg^{++} for example, continue to affect the dissolution rates even at much lower undersaturation (Oelkers, 2001a). As shown in Fig. 12.7, several of the primary minerals of basalt are at this critical undersaturation (0 to –15 kJ mol^{-1}) in some of the waters.

In summary, the observed variation in the chemical denudation rate in Iceland (Fig. 12.4a) is not the result of temperature differences. The dissolution rate of the main primary minerals and glass is high in soil solution due to relatively low pH (4–7), and in the case of plagioclase and glass, high concentration of organic ligands (1–0.1 mmol kg^{-1}). The expected dissolution rate enhancement for Ca-poor plagioclase at a high pH (8–10) is cancelled out to some extent by the small undersaturation ($-\triangle G_r$) at high pH (Figs. 12.7,8) but glass dissolution is high at the high pH. The dissolution rate of olivine and pyroxene decrease continuously from low to high pH. The state of undersaturation ($-\triangle G_r$) affects the dissolution rate for Ca-poor plagioclase and pyroxene at pH higher than 8. Furthermore, the observed state of oversaturation ($+\triangle G_r$) with respect to weathering minerals controls the rate of precipitation of these minerals. High precipitation and runoff results in, lowering of pH where the concentration of organic acids are low, decrease in the concentration of total dissolved Al, and Mg, affecting the dissolution rate of plagioclase and basaltic glass, and olivine and pyroxene at all undersaturations. And finally increasing runoff increases undersaturation ($-\triangle G_r$) of primary minerals, decreases oversaturation ($+\triangle G_r$) of weathering minerals and increases the reactive surface area between rock and water. The dominant effect of runoff on the chemical denudation rate in Iceland (Fig. 12.4a) is similar to what has been observed elsewhere (Meybeck, 1979; Bluth and Kump, 1994).

12.6 THE EFFECT OF VEGETATION ON CHEMICAL DENUDATION RATE

Where the variation in the primary variables, runoff and age of rocks, is small (older than 1 m yr, runoff 1700 to 2200 mm yr^{-1}), the chemical denudation rate for Ca, Mg (Figs. 12.5e,f) and Si increases with increased vegetative cover (Gíslason *et al.*, 1996).

Moulton *et al.* (2000) quantified the plant effect on weathering of 5 m yr old crystalline tholeiitic basaltic lava flows in the Skorradalur valley in southwestern Iceland (Fig. 12.1) by comparing the weathering rates in small catchments of five types: 1. Unvegetated with soil. 2. Unvegetated with no soil, 3. Natural birch trees. 4. Planted birch trees. 5. Planted conifer trees. All catchments had similar lithology, elevation, slope and microclimate. Chemical denudation rates, including data on cation uptake by growing trees, indicate that the rate of weathering release of Ca and HCO_3 to streams is about 3 times higher in vegetated areas than unvegetated. The release is 4 times higher for Mg, and 2–3 times for Si (Moulton *et al.*, 2000). The annual flux of dissolved Ca and Mg out of the catchments via streams is in all cases much higher than the rate of tree up-take of these elements. The storage of K and probably S in trees is a significant component of the chemical denudation flux of K and S from the basalt to trees and streams. Birch weathering fluxes normal to biomass are higher than those of the conifers, suggesting that angiosperms are greater weathering agents than gymnosperms, possibly because of less efficient nutrient scavenging (Moulton *et al.*, 2000).

12.7 THE EFFECT OF VOLCANIC GLASS ON THE CHEMICAL DENUDATION RATE

As shown in Fig. 12.4b, the chemical denudation rate in Iceland decreases with increasing age of rocks. This was attributed to the greater abundance of un-weathered glassy rocks and higher runoff in the younger catchments. Gíslason and Eugster (1987a) measured by laboratory experiments the dissolution rate of basalt, both glassy and crystalline, from 25 to 60 °C. The starting solutions were deionised water, saturated initially with the CO_2 of the atmosphere, and then sealed off from the atmosphere. The pH of the solutions increased from the initial 5.6 to 9 in one week and then remained at pH 9 to 10 for the reminder of the experiments that lasted for up to three months. The pH of most spring waters in Iceland draining relatively young rock (< 1 m yr) range from 9 to 10. As shown in Fig. 12.9, the BET surface area normalised dissolution rate of the glass is about 10 times higher than that of crystalline basalt of the same chemical composition. Stefánsson and Gíslason (2001) studied the effect of rock crystallinity on the chemical denudation rate of basalts in the Laxá í Kjós catchment in southwest Iceland (Fig. 12.1). The age of the rocks was similar (0.7–3.1 m yr) but runoff, rock crystallinity and vegetative cover of the bedrock were variable. Total chemical denudation rate (TCDR) normalised to runoff increased with the abundance of glassy rocks in the catchment areas. Basaltic glass was observed to enhance the fluxes of the mobile elements like Na, Ca, F and S by a factor of 2 to 5 at constant runoff of 2000 mm yr⁻¹, constant vegetative cover and 0 to 100% glass content of the rocks. The K and Mg fluxes were found to be independent of rock crystallinity. Runoff alone cannot explain the fluxes of these elements and it seems that vegetative cover and seasonal variations in biomass activities influence the K and Mg fluxes. Chemical denudation rates of solutes, such as Al and Fe that were highly influenced or even controlled by secondary minerals were independent of the glass content of the catchment rocks.

12.8 LONG TERM AND TRANSIENT CO₂ BUDGET FOR ICELAND

Global rates of CO_2 uptake by surficial processes must be essentially balanced, on a million year time scale, by global rates of degassing to avoid environmentally unacceptable fluctuations in atmospheric CO_2 (Holland, 1978; Berner *et al.*, 1983).

The global features of Iceland's geology are the product of the Icelandic Mantle Plume and crustal accretion at the European and North American lithosphere plate boundary. This 'double feature' results in an anomalously high lava production rate in Iceland, 3.4 km³ per century, compared to other segments of the Mid-Atlantic ridge (Jakobsson, 1972). The CO_2 degassing from Iceland has been estimated to be 70 to 39 kg s⁻¹ (Arnórsson and Gíslason, 1994). This translates to 1.2–2.2 m t CO_2 yr⁻¹.

As shown above, runoff, glacier cover and age of the catchment area for the Hvítá River at Ferjukot in western Iceland (Fig. 12.1) is close to the average for Iceland. The

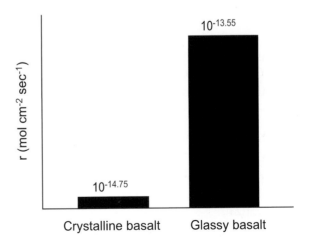

Fig. 12.9 The logarithm of the BET normalised dissolution rate of basaltic glass (qtz saturated) compared with crystalline basalt of the same composition at 25°C and pH 9–10 (modified from Gíslason and Eugster, 1987a).

surface area normalised fluxes for this catchment and the total surface area for Iceland can therefore be used to assess the transient uptake of CO_2 by chemical weathering in Iceland (3.5 m t CO_2 yr^{-1}), Ca and Mg fluxes, and fixation of CO_2 in organic matter in Iceland. Alternatively, the runoff dependence of chemical denudation, transient CO_2 fixation, and Ca and Mg fluxes in old and young catchments (excluding Ytri Rangá and Brúará at Efstidalur in the correlation, Fig. 12.1) can be used together with the total runoff from Iceland yielding similar results. For example, the transient fixation of CO_2 by chemical weathering calculated this way is 3.1 m t CO_2 yr^{-1}. The total calculated chemical denudation uptake is transient since CO_2 uptake resulting in the weathering of Na and K silicates would in the long run, in the absence of weathering of Ca and Mg silicates and reverse weathering, lead to the CO_2 oversaturation of the ocean with respect to the atmosphere (Eq. 3). Furthermore, the stoichiometry of the Ca and Mg silicate weathering alone, on land, is two moles of CO_2 for each mole of Ca or Mg released (first half of Eq. 4). Permanent uptake on a geological time scale is the consumption associated with the weathering of Ca and Mg silicates and their precipitation as carbonates in the ocean (all of Eq. 4). The net stoichiometry of the weathering and carbonate precipitation is one mole of CO_2 for each mole of Ca or Mg released by weathering (all of Eq. 4). The permanent CO_2 uptake in Iceland due to Ca and Mg is 0.9 m t CO_2 yr^{-1} using the Hvítá at Ferjukot (Fig. 12.1) fluxes but it is slightly higher using the runoff correlation for the Icelandic catchments, 1.0 m t CO_2 yr^{-1} (Table 12.2). The net transient geochemical CO_2 flux for Iceland is negative (Table 12.2). That is, more CO_2 is consumed during chemical weathering and transported by rivers to the ocean than is released to the atmosphere from Icelandic volcanoes and geothermal systems. However, the net long-term CO_2 flux is positive (Table 12.2). In other words, more CO_2 is released from the Icelandic Mantle Plume and the segment of the mid-Atlantic ridge in Iceland to the atmosphere than will eventually precipitate in the ocean as Ca and Mg carbonates, as the result of Ca and Mg released during chemical weathering of Ca-Mg silicates in Iceland.

The transient overall CO_2 budget for Iceland includes CO_2 fixation by chemical weathering and primary producers (Eqs. 1, 3, and the first part of Eq. 4), and the anthropogenic release of CO_2 and the release from volcanoes and geothermal fields. The net CO_2 fixation by primary producers (Eq. 1) can be assessed in two ways.

1) The negative flux of NO_3 and NH_4 in the catchment of Hvítá River at Ferjukot, resulting from net fixation of organic material in the catchment, can be used. Assuming the C/N mole ratio to be equal to that of organic matter in soil, 21 (Likens *et al.*, 1981,

Table 12.2 CO_2 released to the atmosphere from magma and geothermal fields (Arnórsson and Gíslason 1994) and transient and permanent fixation of atmospheric CO_2 by chemical weathering and denudation in Iceland.

From volcanoes tCO$_2$ yr^{-1}	Transient uptake tCO$_2$ yr^{-1}	Permanent uptake Ca + Mg tCO$_2$ yr^{-1}	Net transient tCO$_2$ yr^{-1}	Net permanent tCO$_2$ yr^{-1}
2.21	-3.12	-1.05	-0.92	1.15
1.23	-3.12	-1.05	-1.89	0.18

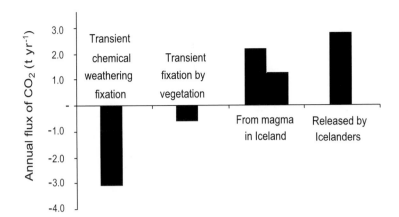

Fig. 12.10 The transient CO₂ budget for Iceland, including CO₂ fixation by chemical weathering and vegetation, and the anthropogenic and magmatic (maximum and minimum estimate) release of CO₂.

considerably higher than in Eq. 1), and furthermore, assuming the importance of import by precipitation and export via rivers to be equal. This is similar to the estimated average for the Earth (Berner and Berner, 1996). According to this, the net CO_2 fixation in organic carbon is of the order of 0.6 million t CO_2 yr^{-1}.

2) The mean annual yield of Icelandic plant communities in the lowlands, 8.4–6.9 t of dry matter km^{-2} yr^{-1} (Þorsteinsson, 1980; Gísladóttir, 1995) together with the extent of vegetated areas in Iceland, 24,000 km^2 (The Statistical Bureau of Iceland, 1984), give 0.2 million t CO_2 yr^{-1} maximum net fixation. No attempt was made to take into account the forested, cultivated and fertilised fields in Iceland. Furthermore, the net annual carbon fixation above ground and in roots is assumed to be of the same order as the gross annual yield above ground, and all vegetation communities are assumed to be progressive. Added to this is the net primary production of lakes (Eq. 1) covering 3000 km^2 of Iceland. As a representative of lakes, Lake Ellidavatn was used, its net production is 35 g C m^{-2} yr^{-1} (equal to t km^{-2} yr^{-1}; Gíslason *et al.* 1998). The net primary production of Icelandic lakes so calculated equals to 0.4 m t CO_2 yr^{-1}. Lake Ellidavatn has the lowest net primary production of the three major lakes measured so far: Lake Þingvallavatn 135 g C m^{-2} yr^{-1} and Lake Mývatn 220 g C m^{-2} yr^{-1} (Ólafsson, 1979; Jónasson, 1992; Jónasson and Hersteinsson, 2002). The net fixation of CO_2 by primary production on land and in lakes is therefore of the order 0.6 million t CO_2 yr^{-1}.

The release of CO_2 due to the activity of man in Iceland calculated according to the Kyoto protocol was 2.8 m t CO_2 yr^{-1} in 1990 (Ministry of the Environment, Iceland, 1992), but the year 1990 was the reference year for the Kyoto protocol. The transient overall CO_2 budget for Iceland is shown in Fig. 12.10. The uncertainty is probably greatest for the CO_2 fixation by vegetation and the volcanic release of CO_2. The net transient budget in Fig. 12.10 is 'positive'. That is there is a net CO_2 release to the atmosphere from Iceland and Icelanders of the order 0.3–1.3 m t CO_2 yr^{-1}.

ACKNOWLEDGEMENTS

Bergur Sigfússon, Eric Oelkers, Ólafur Arnalds and Þórarinn Jóhannsson assisted in many ways during the course of this work. Jórunn Harðardóttir, Stefán Arnórsson and Hrefna Kristmannsdóttir are thanked for reviewing and improving an early version of this paper.

13. Icelandic soils

Ólafur Arnalds

Agricultural Research Institute, Keldnaholt, IS-112 Reykjavík, Iceland

Most Icelandic soils form in volcanic parent materials. The majority of the soils are classified as Andosols or Vitrisols according to an Icelandic soil classification scheme. Drainage and eolian deposition play a major role in soil formation in Iceland in addition to periodic tephra deposition. Wetlands cover about 23,000 km^2, but organic Histosols only about 1,300 km^2. This limited extent of Histosols results from rapid eolian and tephra deposition that reduces the organic content of the soils. The Andosols are dominated by allophane and ferrihydrite clays and Al/Fe–humus complexes, which give the soils typical Andosol properties such as low bulk density, high water retention and hydraulic conductivity and P-retention. The various types of Andosols comprise about 48% of Icelandic soils. The Vitrisols are soils of the deserts. Some of the Vitrisols have considerable allophane contents and much of the Vitrisols are classified as Andisols according to Soil Taxonomy. They cover >30% of Icelandic soil surfaces. Cryoturbation is intense in Iceland because of frequent freeze-thaw cycles, mild winters, a large amount of available water and the nature of the soils. Soil erosion is intense in Iceland and has caused large-scale environmental change over the past 1–2 k yrs. Determination of eolian deposition between tephra layers of known age has been used extensively to theorize about environmental change during the Holocene.   Soil properties, such as organic content and clay minerals, of current and past soil surfaces, can be used to draw conclusions about past environments in Iceland.

13.1 INTRODUCTION

The widespread eolian and tephra mantle that overlies older surfaces in Iceland is sometimes looked upon as sediments that have not undergone much weathering. This view of the mantle is far from being true. The surface mantle makes up a system where pedogenesis has significantly altered the parent materials and added new constituents to create a new system, a soil. Most Icelandic soils are influenced by the volcanic origin of the parent materials. Soils that form in tephra parent materials are most often Andosols according to the FAO World Reference Base (WRB, see FAO, 1998) or Andisols according to the US Soil Taxonomy (Soil Survey Staff, 1998).

Andosols exhibit unique properties, called andic soil properties, that set them apart from other soils of the world. These properties are largely derived from the colloidal constituents that form in the soils as a result of pedogenesis, including allophane, imogolite, and ferrihydrite clay minerals, various organic compounds and organo-metal complexes. The distinctive properties of Andosols include high water retention, low bulk density, rapid hydraulic conductivity and infiltration, lack of cohesion, and high P-retention, as was reviewed by Wada (1985), Shoji *et al.* (1993), and Kimble *et al.* (2000). Andosols are varied and are often divided into soils dominated by allophanic materials, organic constituents, or volcanic glass (see Shoji *et al.*, 1993, 1996).

Icelandic soils are to a large extent Andosols, but they are quite varied: soils with permafrost, soils of deserts and barren rocks, typical Andosols, and organic soils are examples of different soil types in Iceland. Soil erosion is intense in Iceland, occuring both on desert surfaces and affecting Andosols with vegetation cover, and was recently surveyed for all of the country (Arnalds *et al.*, 2001a). Land degradation and soil erosion has caused a pronounced environmental change over the past 1000–2000 years.

The purpose of the paper is to give a general overview of Icelandic soils. The discussion is partly based on a recent classification scheme for Icelandic soils and a discussion provided by Arnalds (2003). A short review of cryoturbation and soil erosion in Iceland will also be given.

13.2 SOILS

13.2.1 The Icelandic soil environment

The location of Iceland on the Mid-Atlantic Ridge results in frequent volcanic eruptions. A discussion on volcanism and tephra in Iceland can be found elsewhere (e.g. Larsen *et al.*, 1999; Larsen, 2000). The repeated, but small additions of tephra are one of the characteristics of the Icelandic soil environment. The basaltic composition of the tephra further enhances weathering of the parent materials (Dahlgren *et al.*, 1993), and weathering rates have been shown to be rapid in the Icelandic soil environment (see e.g., Gíslason *et al.*, 1996; Stefánsson and Gíslason, 2001; Gíslason, this volume).

An influential factor for soil formation in Iceland is the steady flux of eolian materials, which are constantly being added to the top of the soils at a rate of $<0.001->1$ mm yr^{-1} (e.g. Þórarinsson, 1961; Arnalds, 2000). The eolian materials are mainly made of tephra, such as glass and pumice fragments but also some basaltic crystalline materials. The sources of the eolian materials are mostly unstable desert surfaces associated with glacial margins, floodplains of glacial rivers, and thick tephra deposits. The eolian materials are also andic soil materials re-distributed by wind erosion (see Arnalds, 2000; Arnalds *et al.*, 2001b). The rate of eolian sedimentation is mostly dependent on the distance from the eolian source.

The presence of vegetation cover has a dominating influence on the classification of Icelandic soils. The barren areas, or 'deserts', have limited vegetation cover and have contrasting characteristics to soils that have vegetation on the surface. The reason for the lack of vegetation cover varies between areas as will be discussed later. Landscape location and drainage control the wetness of the soils that have vegetation cover. The bedrock of the Tertiary basalt-stack has slower permeability than the volcanic rocks and tephra of the volcanic belt. Wetlands are therefore commonly found on Tertiary bedrock, but they also dominate large plains such as in southern Iceland.

13.2.2 Soil classification and major soil types

The main soil types in Iceland were recently redefined by the Agricultural Research Institute (Arnalds, 2003) and these types and their major characteristics are presented in Table 13.1. A soil map of Iceland can be viewed on the Agricultural Research Institute web site (www.rala.is/desert; www.rala.is/ymir).

Histosols are soils with >20% C in surface horizons. They are found at poorly drained sites far from eolian sources. Histic Andosols have 12–20% C while the Gleyic Andosols have <12% C but evidence of reduced conditions or oxidation–reduction processes such as mottling or greyish colours. Histosols, Histic Andosols and Gleyic Andosols are together the soils of Icelandic wetlands. Brown Andosols are typical Andosols of freely drained sites, often with high allophane content (Wada *et al.*, 1992; Arnalds *et al.*, 1995; see below).

Eolian deposition rates and wetness of the soils are factors that determine to a large extent the classification of the soils with vegetation cover (Andosols and Histosols). Organic soils, which meet the criterion for Histosols (>20% C) are only found in wet locations far away from eolian sources with little eolian input (Fig. 13.1). As the eolian input increases, there is a sequence from Histosols, through Histic Andosols and Gleyic

Table 13.1 Major soil types in Iceland; main criteria for division, extent and relation to the WRB soil classification system.

Soil type (symbol)	Icelandic term	Criteria	Extent km²	WRB[1]
Histosols (H)	Mójörð	>20% C	1,090	(Andic) Cryic Histosol
Histic Andosols (HA)	Svartjörð	12-20% C	4,920	Gleyic/Histic Andosol
Hydric Andosols (WA)	Blautjörð	wet, andic	2,390	Gleyic Andosol
Brown Andosols (BA)	Brúnjörð	dry, andic	13,360	Haplic Andosol
Vitrisols (V)	Glerjörð	<1% C[2]	28,190	Vitric Andosol, Arenosol, Leptosol
Leptosols	Bergjörð	shallow/ bedrock	7,310	Leptosol
Cryosols	Frerajörð	permafrost		Cryosol
Complexes				
C—WA			140	
BA—WA			28,080	
V—L			4,790	

1: FAO, 1998. 2: Other criteria can apply.

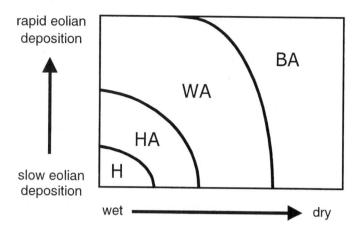

Fig. 13.1 The influence of eolian deposition and drainage on the separation of Icelandic soil types. From Arnalds, 2003. H: Histosols; HA: Histic Andosols; WA: Gleyic Andosols; BA: Brown Andosols. See also Table 13.1.

Andosols to Brown Andosols. The same sequence is found along the drainage gradient from wet to dry. The combined extent of Icelandic wetlands is about 25% of Icelandic soils. The relatively small extent of the Histosols (1.2%) is surprisingly low considering the high latitude (63°–66° N) and the extent of wetlands.

Soils of barren gravel and sandy areas with limited vegetation cover in Iceland are defined as Vitrisols, soils dominated by tephra, according to the Agricultural Research Institute classification scheme (Arnalds, 2003). These soils classify as Andisols according to Soil Taxonomy (Arnalds and Kimble, 2001), but Vitric Andosols, Arenosols and Leptosols according to the WRB (FAO, 1998). The Vitrisols are low in available nitrogen and can become water deficient during periods of drought. Some of the Vitrisols (Cambic Vitrisols) have considerable allophane contents (up to 15%). The surface of the Vitrisols is unstable and they are susceptible to erosion by wind and water, cryoturbation, and needle-ice formation. Arnalds and Kimble (2001) provided a recent overview about soils of Icelandic deserts.

Cryosols, soils with permafrost, occur in Iceland as is evident by palsas (mounds with permanently frozen ice-core) in central Iceland (e.g. Þórhallsdóttir, 1997). The extent of permafrost has not been surveyed, but efforts to map Icelandic Cryosols have been initiated in relation to the newly made soil map of Iceland (Arnalds and Grétarsson, 2001).

13.2.3 Some general properties

The clay content of Andosols is commonly estimated using chemical dissolution of the soils with ammonium oxalate (see Blakemore *et al.*, 1987). The amount of Al, Fe and Si extracted (Al_{ox}, Fe_{ox} and Si_{ox}) can be used to calculate allophane and ferrihydrite in the soils. Al_{ox} and Fe_{ox} are also associated with organic complexes. The main central concept of Andosols as defined by the WRB and Soil Taxonomy is that of soils that contain more than 2% $(Al+0.5Fe)_{ox}$. In addition, Andosols need to have low bulk density (<0.9 g cm^{-3}) and retain phosphorus to qualify as Andosols. Vitric materials (tephra) are

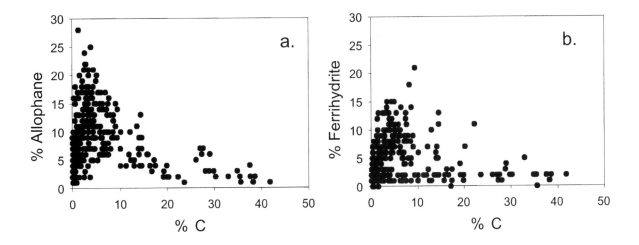

Fig. 13.2 Allophane (a) and ferrihydrite (b) content of the soils as related to carbon content. Both values decrease as organic content increases. The two highest points for ferrihydrite are mottles in poorly drained soils. Few Vitrisols are included in the data set with cluster of points <1% C. From Arnalds, 2003.

also included if $(Al+0.5Fe)_{ox}$ is >0.4% and volcanic glass is >30% (for details see FAO, 1998; Soil Survey Staff, 1998).

The organic content of the world Andosols is generally high even though the soils are well drained and in warm climates (see Shoji *et al.*, 1993). The organic content of Icelandic Andosols is relatively high, and the organic materials extend deep down into the profile with erratic distribution. The organic content of the Histosols is, in contrast to the Andosols, relatively low compared to other Histosols in the world, because of eolian and tephra additions. Soils of wetlands at a similar latitude are usually much higher in organic content in surface horizons.

The allophane content in Iceland is calculated by multiplying Si_{ox} by 6 using a method described by Parfitt (1990). The ferrihydrite content is obtained by multiplying Fe_{ox} by 1.7 according to the method outlined by Parfitt and Childs (1988). Imogolite is also present in Icelandic soils (Wada *et al.*, 1992), but is ignored in the following discussion. Layer silicates are not present or are minor components of the clay fraction of the soils (Wada *et al.*, 1992). Fig. 13.2 shows allophane and ferrihydrite of >200 soil horizons in Icelandic soils. There are two distinct trends that can be observed in both graphs. At low organic contents (Brown Andosols) clay content rises with increasing carbon levels. As the carbon content exceeds about 10%, allophane tends to decrease. This relationship is well known and is explained by the bonding of Al with organic materials and lower pH of the soil (see e.g. Parfitt and Kimble, 1989; Shoji *et al.*, 1993), either of which reduce or hinder allophane formation. Lower ferrihydrite contents in organic horizons are related to losses of reduced iron at poorly drained locations (Arnalds, 2003).

Soil reaction of Icelandic soils is highly related to the organic content and generally ranges between 5 and 7. Organic soils have the lowest pH values, but the reaction of Vitrisols is commonly near or slightly above neutral. The soils show high water content measured at 15 bar water tension ('wilting point'), as is characteristic of Andosols (Fig. 13.3).

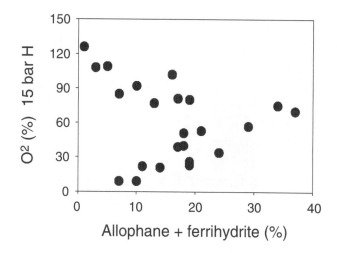

Fig. 13.3 15 bar water content of selected soil horizons. Two trends are observed: increased water retention with increased clay content, and increased 15 bar water content (reaches > 100% in Histic soil horizons) as the organic content rises (not shown) with corresponding decrease in clay content (shown). From Arnalds, 2003.

Water retention increases with allophane content (Si_{ox}) up to a point, but as the soils become more organic in poorly drained positions, with lower allophane and ferrihydrite content, the retention increases still further (>100%). Hydraulic conductivity and infiltration is rapid in Andosols (Orradóttir, 2002). The bulk density of the Andosols and Histosols ranges between <0.2 g cm^{-3} and about 0.9 g cm^{-3} (Guicharnaud, 2002; Arnalds, 2003), but the Vitrisols have higher values, often near 1 g cm^{-3}.

13.3 CRYOTURBATION

The effect of freeze-thaw in soils is evident on most soil surfaces. Several factors enhance cryoturbation processes in Iceland. Freeze-thaw cycles are numerous and the soil temperature stays near 0°C for extensive periods in winter, which results in a relatively stationary freezing front. Winter thaw is common, with more moisture added to the soil during winter months. Hydraulic conductivity of Andosols is rapid, both saturated and unsaturated (Nanzyo *et al.*, 1993), which has also been shown for Icelandic Andosols (Orradóttir, 2002). Rapid water conductivity intensifies the freezing effects in soils by enhancing water transfer to the freezing front. As a result of all these factors, much of the winter precipitation tends to be incorporated into the Andosols during winter, allowing more water to freeze near the surface.

The capacity of some of the soils to store great quantities of water is important because it can be transferred to the freezing front (Histosols and Andosols). The lack of cohesion when saturated (thixotropy) allows for easy deformation of the soil by ice, and this may be an additional factor that needs to be considered in relation to cryoturbation in Iceland. The effect of cryoturbation is seen on the surface as well as in soil profiles. Slopes commonly have well-developed solifluction features, both terraces and lobes (see Arnalds *et al.*, 2001a). More level vegetated surfaces (Brown, Histic and Gleyic

Andosols and Histosols) have well-developed hummocks, termed 'þúfur' in Icelandic (see Schunke and Zoltai, 1988).

Jóhannesson (1960) showed that the largest hummocks normally form over a shallow water table. This complies with the classical model of water pumping from a water table to a freezing front. However, large hummocks occur in Iceland where the water table is at a considerable depth (>10 m) and cannot feed a stationary freezing front. The reasons for the formation of þúfur (hummocks) in the absence of a groundwater table are most likely related to the andic properties of the soils and, in some cases, to the trampling of the surface by grazing animals as noted by Arnalds (1993) and van Vliet-Lanoe *et al.* (1998). Important factors for the formation of þúfur without a water table include: 1) the high water retention of Andosols; 2) difference in insulation between the top of the hummocks and the depressions between them with water transfer to the center of the hummocks; 3) repeated thaws and water additions throughout winter; and 4) energy release in depressions upon freezing of water on the surface. This model is currently being tested at the Agricultural Research Institute in co-operation with Texas A&M University.

Evidence from soil profiles often shows that þúfur have not formed in lowland locations until the middle ages, with relatively level tephra layers in the lower part of soil profiles, while the uppermost part is cryoturbated. The reason is possibly the combination of colder climate and land cover change. Desert soils usually have a gravelly surface horizon created by frost heave of coarse fragments. Patterned ground is also common on desert surfaces.

13.4 SOIL EROSION

Soil erosion in Iceland is intense and has reshaped the surface of country, especially over the past 1200 years, making some areas barren deserts, while other surfaces continue to receive eolian deposition. This erosion still continues. The Andic properties of Icelandic soils have important implications for erosion. They lack phyllosilicates such as smectite, that provide cohesion for other soils. The allophane tends to form stable silt-sized aggregates that are vulnerable to erosion both by wind and water.

Soil erosion processes in Iceland are extremely varied, and many erosion processes can occur at the same site. The Agricultural Research Institute and the Soil Conservation Service conducted a national survey of erosion and desertification at the scale of 1:100,000 over the period 1991–1996. The results were published in Icelandic in 1997, but the publication has been translated into English (Arnalds *et al.*, 2001a; see also www.rala.is/desert). The following discussion is mostly based on the assessment, and related publications by Arnalds (2000) and Arnalds *et al.* (2001a,b).

The assessment of soil erosion in Iceland is based on a classification of erosion forms that can be identified in the landscape. The Icelandic erosion classification system is presented in Table 13.2. Erosion severity is estimated for each of the erosion forms on a scale from zero to five, five being considered extremely severe erosion. The underlying database is made of about 18,000 polygons with information about erosion forms and erosion severity. Each polygon carries one or more erosion forms.

Table 13.2. The Icelandic erosion classification system (erosion forms).

Erosion forms associated with Andosols/Histosols	Desert erosion forms erosion of (Vitrisols, Leptosols)
Rofabards	Melar (lag gravel, till surfaces
Advancing erosion fronts (sand encroachment)	Lavafield surfaces
Erosion/Isolated spots Sandy lava fields	Sandur (bare sand, sand sources)
Isolated spots and solifluction features on slopes	Sandy melar (sandy lag gravel)
Water channels	Scree slopes
Landslides	Andosol remnants

One of the most distinctive erosions forms in Iceland is an erosion escarpment, termed 'rofabard' in Icelandic. Rofabards are formed in relatively thick non-cohesive Andosols (mostly Gleyic and Brown Andosols), which overlie more cohesive materials such as glacial till or lava. The relatively loose Andosols beneath the root mat are undermined, creating escarpments, or rofabards. The rofabards retreat as a unit, with fully vegetated ecosystems on top, but leaving barren deserts in their place. Rofabards are common within an area of about 20,000 km^2, and the erosion database suggests that 15,000–30,000 km^2 of land that was previously fully vegetated and had fertile Andosols, have now become deserts as a result of the erosion processes associated with rofabards. Icelandic rofabards were reviewed by Arnalds (2000).

Erosion spots are small bare patches in otherwise vegetated land. They are most often associated with hummocks, and are a clear sign of overgrazing when they occur in lowland areas. Erosion spots are extremely common in Iceland, and are sometimes severe (Gísladóttir, 1998; 2001; Arnalds et al., 2001a). Erosion spots that occur on slopes with solifluction features are a special erosion form (often shortened to solifluction). Solifluction is also associated with landslides, as the soil movement by solifluction moves soil to points of weakness, which later can fail, resulting in landslides.

Advancing sand fronts ('áfoksgeirar' in Icelandic) are an entity of the Icelandic erosion classification system. They are active tongue-shaped sandy surfaces extending into vegetated areas. These fronts start as sedimentary features (encroaching sand) that abrade and bury the vegetation with sand and destroy it. Sand fronts continue to move as a continuous flux of sand abrades the Andosol mantle and finally the new surface may be 1–2 m lower than the original surface. The advancing fronts are a major problem in Iceland that threaten fully vegetated systems, and can advance over 300 m in a year (see Arnalds et al., 2001b). Encroaching sand has desertified large areas in south and northeast Iceland, especially during the last part of the 19th century.

Deserts are divided into seven separate erosion forms based on geomorphology and the instability of the surface. Lag gravel ('melur'), lavas and sandy deserts are the most common forms. The presence of large sandy desert areas is somewhat unique, considering the moist climate. The sandy areas are most likely the world largest volcaniclastic sand-fields (Arnalds et al., 2001b), which further highlights their uniqueness.

13.5 CONCLUSIONS

Tephra layers have been used extensively to calculate eolian deposition rates, which reflect environmental conditions in the past (e.g. Þórarinsson, 1944, 1961; Sigbjarnarson, 1969; Guðbergsson, 1975; Dugmore *et al.*, 2000). This research shows that there is a 4–10 fold increase in eolian deposition around the Settlement of Iceland about 1200 years ago. The dramatic environmental change that has occurred since then has been attributed to use of the land by man. Other factors than land use should, however, also be stressed. These factors include a cooling trend that began about 2,500 years ago and also growing sources for eolian sand associated with the advance of glaciers. With larger glaciers, the number of meltwater floods increased, both from subglacial thermal areas and from volcanoes. This has resulted in larger unstable surfaces that cause influx of eolian materials that make the Andosol mantle continually thicker and therefore more unstable than it was before (Arnalds, 2000; Arnalds *et al.*, 2001b). The combination of these factors and the use of the land by man cause a 'snowball effect' that greatly accelerated erosion rates after the Settlement.

The biologically active layer of present soils in Iceland is influenced by many factors, some of which have been identified above, such as drainage conditions and eolian additions. The surface horizons of Histic and Gleyic Andosols often have more organic carbon in subsurface horizons than near the surface (Fig. 13.4). This variability in carbon content reflects changing soil environments throughout the Holocene. It is well known that in some wetland soils, distinct layers with well preserved wood remains can be found which are believed to reflect a climate drier and warmer than today (e.g. Einarsson,

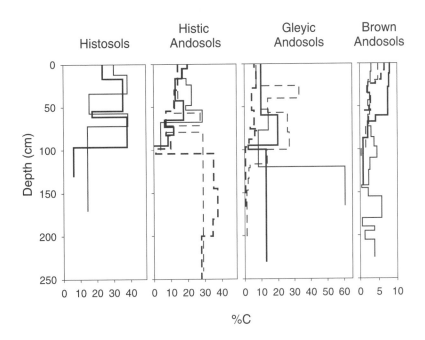

Fig. 13.4 Depth distribution of carbon for 2–4 pedons for Histosols and each of the Andosol soil types. Each line represents one profile. The distribution is quite erratic, often with higher carbon contents at greater depths. Note different scale for Brown Andosols. From Arnalds, 2003.

1978). The carbon content and the amount of clay minerals is influenced both by drainage conditions and additions of fresh parent materials such as eolian and tephra deposition. It is likely that further inference of past environments could be reconstructed by combining data for organic carbon and clay constituents. The Agricultural Research Institute is gradually building a database for Icelandic soils. This database now (Feb. 2003) consists of >70 soil pedons with >500 soil horizons, in addition to data from the analysis of soil cores. This database shows that there is a substantial decrease in carbon content of soil horizons that date after the Settlement (Óskarsson et al., 2003), for Brown and Gleyic Andosols. This reflects a pronounced change in the soil environment after the Settlement, with increased eolian activity and perhaps and removal of biomass by grazing, but this relationship is subjected further investigations.

14. The Holocene vegetation history of Iceland, state-of-the-art and future research

Margrét Hallsdóttir[1] and Chris J. Caseldine[2]

[1]Icelandic Institute of Natural History, Reykjavík, Iceland
[2]Department of Geography, University of Exeter, Exeter, UK.

In the Late Preboreal and Early Boreal Chronozones dwarf-shrub heath and shrub heath, followed by juniper and mountain birch copses, replaced snow beds and fell-field vegetation characteristic of the Lateglacial/Early Preboreal newly deglaciated landscape of Iceland. During the Late Boreal and Early Atlantic Chronozones birch woodland established itself in the more favourable places, especially fjord lowlands and inland valleys. The development of birch woodland suffered a setback due to a transient climatic oscillation some 7500 ^{14}C years ago, but recovered again relatively quickly and more than 6000 ^{14}C years ago birch woodland covered the lowland areas both in northern and southern Iceland. At that time it reached its highest altitude, at least in northern Iceland. During the Late Atlantic and Subboreal Chronozones the birch woodland showed a retrogressive succession towards a more open landscape with expanding mires and heaths. There is some conflict between the evidence from pollen percentages, which indicate that the woodland regenerated several times during the latter half of the Holocene, and pollen influx values which reflect no such regeneration of the woodland. New habitats were created for birch after a period of cool climate and instability during the Early Subatlantic Chronozone as fresh screes and sandur plains became vegetated, at least partly, by woodland. This development was halted at the beginning of the Norse settlement, which resulted in further opening of the woodland. The birch woodland closest to the farms in the lowland of Iceland was cut and utilized for timber and fuel. Grazing of domestic animals opened the landscape still further and the previous woodland never re-established itself. This happened within only half a century from the arrival of the first settlers. During the ensuing 1100 years of human influence the sub-alpine birch woodland has been so intensively utilized that only in fenced, protected areas and at the most inaccessible and remote places has birch survived. The shrub and dwarf-shrub heaths, widespread mires, fell fields, and hay fields so characteristic for the Icelandic landscape at present thus developed as a relatively recent phenomenon in the form in which they appear today. A number of areas are identified for future research: elucidating the origins of the Icelandic flora, deriving palaeoclimatic data from the vegetation record, correlating terrestrial with marine and ice core records and expanding our understanding of the human impact on vegetation.

14.1 INTRODUCTION

14.1.1 Background and research history

Pollen analysis was introduced for the first time in Iceland during the 1940s, when, in his well known doctoral thesis, famous for introducing the method of tephrochronology, Sigurður Þórarinsson (1944) published two pollen diagrams from abandoned farms in Þjórsárdalur, southern Iceland (see Table 14.1 for a detailed list of Icelandic pollen sites). More than ten years later he published the first diagram that covered almost the whole Holocene (Þórarinsson, 1955). Around that time Einarsson (1956, 1957) published his first pollen diagrams of peat sections from southwestern Iceland. His doctoral thesis on the climate history of Iceland during the Lateglacial and Holocene was published in 1961 (Einarsson, 1961), based to a large extent on pollen analysis of peat sections as well as one lake sediment core from 9 sites in Iceland. Others were also working in Iceland, for instance the Finn Okko (1956) published two pollen diagrams from Iceland in his doctoral thesis (cf. Hoffell and Laugardalur) and in the same year the German Straka (1956) published a paper including a pollen diagram from Héðinsvík at Tjörnes, northern Iceland. Most of these analyses were carried out on peat sections with poor time resolution. In the early days of radiocarbon dating absolute dates were few, but the tephra layers proved of great value, both as tools for dating and for correlation. Based on these data Einarsson (1968) published his first zonation scheme of the vegetation history of Iceland, comprising five periods: namely the birch free period, the lower birch period, the lower mire period, the upper birch period, and the upper mire period, the latter with the partly superimposed settlement and historical time.

In the 1970s the first pollen diagrams from Icelandic lakes were published, from Hafratjörn in northern Iceland and Lómatjörn in southern Iceland (Vasari, 1972, 1973) and in these studies Vasari showed that the pollen of *Juniperus* had been an important part of the pollen dispersal at the time of shrub-heath formation in the Boreal Chronozone (1972, p. 243). Later this work was expanded with the first macrofossil diagrams on Icelandic lake sediments and some new radiocarbon dates, which led to a revision of the timing of the pollen zones (1990), although examination of the chronology reveals problems assumed to be due to erroneous radiocarbon dates.

Over the next few years a number of papers on diverse and in some cases specific aspects of the vegetation history were published, covering sites at altitudes from sea level up to the highlands and most often dealing with a part of the Holocene only (Bartley, 1968; Friðriksdóttir, 1973; Skaftadóttir, 1974; Sigþórsdóttir, 1976; Schwar, 1978; Hallsdóttir, 1991; Hansom and Briggs, 1991; Caseldine and Hatton, 1991; Hallsdóttir, 1995, 1996; Wastl *et al.*, 2001), almost all of these studies relying on peat sections. The most recent palynological work of Rundgren (1995, 1998, 1999), with detailed studies on Skagi in northern Iceland which was deglaciated relatively early, looking in particular at the pioneer stages in the vegetational development in Iceland, is however solely based on lake sediment cores using both pollen and macrofossil analysis.

There have also been several studies where the main topic has been human influence on vegetation, thus concentrating on the last 1100 years. These started with Þórarinsson (1944), continued with Einarsson (1963) and Påhlsson (1981), and were significantly enhanced by Hallsdóttir (1982, 1984, 1987, 1992, 1993) and more recently by Zutter (1997).

14.1.2 The situation at the end of the last glacial

As with other isolated islands there has been a lively discussion on the origin of the Icelandic flora. This was originally stimulated by the appearance of Steindór Steindórsson's (1937, 1954, 1962) hypothesis on nunatak refugia, areas where he believed about half of the flora should have survived during the Ice Age. This he supported by geographical and geological research on the extent of the main ice sheet. Contrary to this hypothesis is the *tabula rasa* hypothesis, which is supported in a renewed form by e.g. Buckland *et al.* (1991), where it is argued that little or nothing of the flora could have survived the Last Glacial with colonisation occurring predominantly by transportation as ice-rafted debris during a short episode at the Weichselian/Holocene transition around 10,000 ^{14}C yr BP. Midway between these extreme views is the idea that arctic/alpine floral elements may have survived on nunataks, but that other plants, especially trees, probably immigrated during the Lateglacial and Early Holocene by aerial transport, by ocean currents and with the help of migrating birds (Johansen and Hytteborn, 2001). In recent years it has become more apparent that the Weichselian glacial period was not a continuously cold period, but rather a series of cold spells with intermittent and short lived warm periods (cf. Allen *et al.*, 1999). In this way it may be argued that there were ice-free areas, not only on nunataks but also on lower lying ground adjacent to the sea or submerged at present. The plants had therefore both more time and more space for transport by luck than previously perceived. The latest palaeobotanical research north of Skagi (Rundgren and Ingólfsson, 1999) supports the view that a part of the flora that flourished there during the Alleröd interstadial survived the Younger Dryas stadial. As the climate during that stadial was very harsh, this adds further weight to the argument that plants could have survived conditions prior to the Alleröd thus originating from well before the interstadial.

14.1.3 The current situation - prevailing climatic conditions and altitudinal and latitudinal tree lines.

Birch woodland is the natural climax vegetation in Iceland and lowland areas up to about 300 m asl. lie within the sub-alpine vegetation zone. Above this limit and in the outermost coastal districts in the northwest, north and northeast, arctic-alpine vegetation dominates. This is also the case in the westernmost part of the Snæfellsnes and Reykjanes peninsulas. On the sandur fields in southern Iceland, where the strength of the wind and shifting nature of the substratum prevents the growth of birch trees, plant cover is discontinuous and *Leymus arenarius* dominates.

At present, Iceland is almost devoid of woodlands, mainly due to intensive land use over the last 1100 years. Natural woodlands and plantations cover only 1.2 % (1250 km²) of the country (Sigurðsson, 1977) and the maximum altitude for growing birch trees is to be found in the district of Skagafjörður, northern Iceland, at 620 m asl. (Guðbergsson, 1992). At present, 46% of the total land area is vegetated, in the sense that the vegetation cover is higher than 50% (of which mires cover about 8%, grass heath and dwarf-shrub heath 25%, moss heath 10%, woodland 1%, and cultivated land 1%). Land with less than 50% vegetation cover may be interpreted as fellfield and covers 41% of the country at present, and is thus the largest single landscape type. It mainly occurs at altitudes greater than 500 m, but is also to be seen in the lowlands.

Table 14.1
Pollen sites from Iceland

Location of pollen site	Number of samples	Length of sequence, cm	Reference number	Estimated age/ length of period	Research topic	Average sample density, cm	Average sampling interval, yr
Naustamýri AKUREYRI	16	550	1	c. 9000	Vegetation history	34.4	563
Hoffell NESJUM	30	400	2	c. 9000	Vegetation history	13.3	300
Laugardalur REYKJAVÍK	44	450	2	c. 9000	Vegetation history	10.2	205
Héðinsvík HÚSAVÍK	5	65	3	>3000	Vegetation history	13.0	700
Sogamýri REYKJAVÍK	27	250	4	c. 9000	Vegetation history	9.3	333
Undir Ingólfsfjalli ÖLFUS	32	230	4	c. 9000	Vegetation history	7.2	281
Seltjörn REYKJAVÍK	16	200	5	c. 9000	Vegetation history	12.5	563
Borgarmýri REYKJAVÍK	24	210	5	c. 9000	Vegetation history	8.8	375
Villingavatn GRAFNINGUR	17	520	5	c. 9000	Vegetation history	30.6	529
Ölkelda STADARSVEIT	17	160	5	c. 9000	Vegetation history	9.4	529
Torfalaekur BLÖNDUÓS	14	175	5	c. 9000	Vegetation history	12.5	643
Varmahlíð SKAGAFJÖRÐUR	17	175	5	c. 9000	Vegetation history	10.3	529
Sólheimagerði SKAGAFJÖRÐUR	17	200	5	c. 9000	Vegetation history	11.8	529
Moldhaugar EYJAFJÖRÐUR	28	270	5	c. 9000	Vegetation history	9.6	321
Moldhaugar EYJAFJÖRÐUR	27	275	5	c. 9000	Vegetation history	10.2	333
Moldhaugar EYJAFJÖRÐUR	7	60	5	>8000	Vegetation history	8.6	—
Hallormsstaður HÉRAÐ	33	580	5	c. 9000	Vegetation history	17.6	273
Lómatjörn BISKUPSTUNGUR	49	420	6	c. 9000	Vegetation history	8.6	184
Hafratjörn BLÖNDUÓS	53	520	6	c. 9000	Vegetation history	9.8	170
Ytri Bægisá HÖRGÁRDALUR	25	420	7	8850	Vegetation history	16.8	354
Tjarnarver HÁLENDIÐ	16	150	8	3800	Vegetation history	9.4	238
Sóleyjarhöfði HÁLENDIÐ	17	260	8	6100	Vegetation history	15.3	359
Garðskagi REYKJANES	8	80	9	3300?-7460	Vegetation history	10.0	500
Miðnesheiði REYKJANES	12	120	9	—	Vegetation history	10.0	—
Búrfellshraun HAFNARFJÖRÐUR	16	90	10	7240-8740	Vegetation history	5.6	94
Laugarvatn LAUGARDALUR	31	155	11	—	Vegetation history	5.0	—
Svínavatn GRÍMSNES	46	342	12	c.8000	Vegetation history	7.4	174
Krosshólsmýri FLATEYJARDALUR	23	110	13	2700	Vegetation history	4.8	117
Hvítahlíð BITRUFJÖRÐUR	10	450	14	6900-8830	Sea-level changes	45.0	200
Vatnskotsvatn SKAGAFJÖRÐUR	56	294	15.16	9000	Vegetation history	5.3	161
Torfadalsvatn SKAGI	9 (13)	165	17	2000 (3000)	Vegetation history	12.7	231

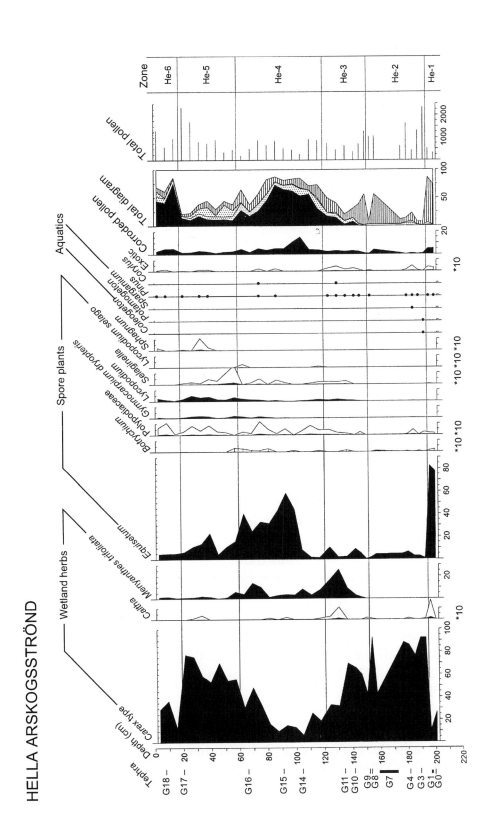

Fig. 14.1 Pollen diagram from Hella Árskógsströnd, Eyjafjörður, Northern Iceland covering the period from ca 10,000 to ca 6200 ^{14}C yr BP. The location of tephra layers is given on the side of the profile (marked G [gjóska=tephra]). Not all layers have been provenanced but G7 is Saksunarvatn and Hekla 5 is either G17 or G18.

that time in the inland valleys of central north Iceland. At Faxaflói birch was expanding in the Late Boreal, as indicated by a rising *Betula* pollen curve in the peat stratigraphy well above the Preboreal/Boreal tephra marker horizon (cf. Kv/Gr in Sigurgeirsson and Leósson, 1993). Birch woodland was rather sparse in western north Iceland (Húnavatnssýsla district) until Atlantic times, not expanding at Efstadalsvatn on southern Ísafjörður until 6750 ^{14}C yr BP (Caseldine *et al.*, 2003), and as previously described, birch was spreading in southern Iceland at the Boreal/Atlantic boundary with dense woodland first appearing well into the Atlantic Chronozone, at ca 7300 ^{14}C yr BP (Vasari and Vasari, 1990). In Flateyjardalur the shrubs and scattered birch trees were replaced by birch woodland close to the boundary of the Early and Mid-Atlantic (6900 ^{14}C BP). A recent study on the upper limit of birch scrub in a valley on Tröllaskagi, northern Iceland, indicates that this was reached between 6700 and 6000 ^{14}C yr BP at an altitude of 450-500 m asl. (Wastl *et al.*, 2001). From pollen concentration and pollen influx values it is evident that the birch woodland in Iceland was at its most extensive before 6000 ^{14}C yr BP (Hallsdóttir, 1996). Since then the peatland has been expanding out from the lowland basins where the birch trees have had few possibilities to recover. In many places peat began to accumulate from this time onwards.

14.2.4 Expanding heaths and mires in the Late Atlantic, Subboreal and Early Subatlantic 6000-1200 ^{14}C yr BP

During the latter half of the Holocene there was a retrogressive succession towards more open birch woodland with widespread mires and heathland. On the drier hills, on the well-drained mountain slopes and along valleys and fjords, birch could thrive. On the other hand, mires were expanding across the lowlands and there woodland was overcome and buried in peat. The landscape became more open and the birch woodland more patchy in favour of shade-intolerant herbs and ferns, as indicated by increasing relative values of the pollen and spores of these two groups. The first evidence of *Sorbus cf. aucuparia* pollen (rowan or mountain ash) is seen during this phase of vegetational development and appears earlier in northern than in southern Iceland. In the pollen diagram from Hegranes in Skagafjörður the pollen curve is almost continuous after 5500 ^{14}C yr BP, while in the south it is first seen just below the tephra layer Hekla-4 ca 4000 ^{14}C yr BP (Hallsdóttir, 1995) (Fig.14.2).

The expansion of mires was not continuous though, as evidenced by wood remains in peat sequences. This is most clearly seen in sections near the edges of the mires, with alternating layers of wood remains and relatively homogeneous *Carex* peat. Recent research, supported by radiocarbon dates, indicates that oscillations in the mire vegetation, which are reflected in macroscopic remains in the peat, have not necessarily been contemporaneous in all parts of the country (Zutter, 1997). However, in some cases there may indeed have been local fluctuations in the groundwater table, which can blur the general picture.

From southern Iceland we can see the lowest birch pollen values, indicating the most drastic opening up of the woodland, in pre-historic times occurring at the same level as the thickest Katla tephra in the Holocene (Kn), which has been radiocarbon dated to 3300 ^{14}C yr BP (Róbertsdóttir, 1992). The *Betula* minimum is more pronounced in the peat stratigraphy than in lakes, sometimes it is seen below the tephra layer and sometimes just above it, indicating that factors other than the volcanic eruption itself must have operated to diminish the woodland cover, probably a deterioration in climate.

Two, and sometimes three, late Holocene woodland regenerations can be seen in southern Iceland younger than the tephra Ke (ca 2850 ^{14}C yr BP, Róbertsdóttir, 1992). The latest is dated to ca 1600 ^{14}C yr BP (Haraldsson, 1981) and appears in most of the so-called landnám profiles as a transient *Betula* pollen peak below the Landnam tephra (Vö, now dated to AD 871±2 by means of ice core chronology, Grönvold *et al.*, 1995).

14.2.5 Post-settlement history of vegetation

Changes in vegetation composition were rapid in the wake of the settlement, and using tephrochronology it can be shown that birch woodland vanished in the vicinity of the farms during only one generation. This happened in the period between the tephra horizons Vö-871 and K-920 in southern Iceland, where the influence of settlement has been most intensely studied (Hallsdóttir, 1987). The woodland could not regenerate itself due to almost unlimited sheep grazing; grass heath, dwarf-shrub heath and mires expanded at the expense of the woodland, as Poaceae, Ericales, *Galium*, *Thalictrum alpinum* and *Carex* pollen types become an important part of the pollen spectra. In the pollen diagrams there are also indications of grain cultivation up to the 16th century AD, probably both barley and oats (Þórarinsson, 1944; Hallsdóttir, 1987, 1993), as well as flax (*Linum usitatissimum*) during the first centuries of settlement (Einarsson, 1962). New pollen types appear and give proxy evidence about cultivation such as the occurrence of weed plants like *Polygonum aviculare*, *Urtica* sp. and *Spergula arvensis*, and some native weed species (apophytes) flourish.

Prior to the 15th to 16th century AD iron was extracted from bog-iron in Iceland and this industry required much birch-charcoal. Making this charcoal was devastating for the birch woods (Þórarinsson, 1974), as is shown by numerous remains of ancient charcoal pits in areas of severe soil erosion, such as Haukadalsheiði in southern Iceland and Bárðardalur in the north. Cooler climate increased the requirements for fuel, which also caused pressure on the woods, although turf and peat cutting for fuel had been undertaken simultaneously for a long time. Grazing all round the year, as it was practised up to the 19th or even 20th century further prevented the regeneration of the woodland. All these adverse influences put together resulted in the woods becoming more and more sparse, and finally the landscape approached the character seen at present, almost treeless and with vegetation covering less than half of that at the time of settlement (Arnalds, 1987).

It appears that this change was slower in the highlands than in the more densely populated lowlands, in valleys or by the coast. This has, however, not been thoroughly investigated as yet. In the vicinity of Tjarnarver by the Þjórsá river, at about 575 m asl., some birch scrub and more abundant dwarf birch survived up to at least 1300 AD, when it was overtaken by willow, which later becomes dominant in the pollen spectrum in the 14th century AD (Friðriksdóttir, 1973). Intensive land use began quite early in the lowlands, but in spite of a vegetation more sensitive to interference the interior highlands changed later, albeit responding more quickly after the initiation of intense land use, resulting in many places in desertification (cf. Dugmore and Buckland, 1991). This was accomplished by the unified action of cooler climate, intense utilization of the scrub

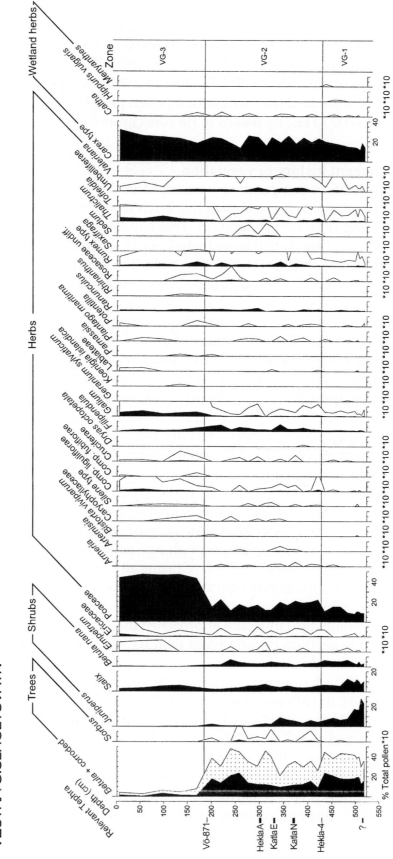

VESTRA GISLHOLTSVATN

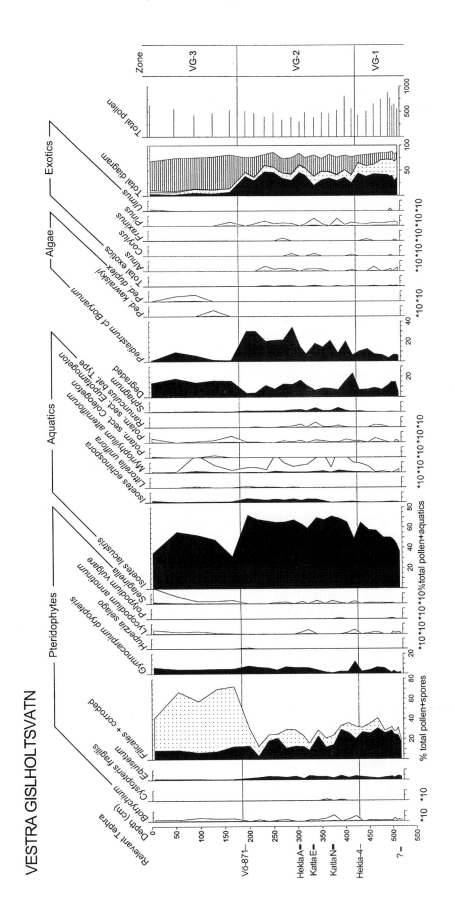

Fig. 14.2 Pollen diagram from Vestra Gíslholtsvatn, South Iceland covering the period from ca 5600 ^{14}C yr BP to the present day with the main tephra horizons marked G, but as yet not fully provenanced.

vegetation, and overgrazing, as the sheep had unlimited access to the rangeland in most districts. The rangeland vegetation, dwarf-shrub heath, willow heath, grasslands and mires, was really a better pastureland than the dense birch scrub, which had an adverse effect on wool production by tearing the fleece of the sheep, but wool products were the main merchandise of the farmers in early times, or up to the 14th and 15th centuries AD when they were overtaken by stockfish (Þorláksson, 1991).

14.3. FUTURE DIRECTIONS IN RESEARCH

From the above review it is clear that considerable progress has been made in developing an understanding of the Holocene vegetation of Iceland but there are still several areas for which further research needs to be undertaken. Underlying current understanding is also a lack of chronological control due to the relative paucity of well dated sequences, although this is perhaps less true of later Holocene profiles which can use well established marker tephra horizons such as the Hekla tephras and the Landnám series. In concluding the review an attempt is made to identify some central research questions for the future suggesting areas which not only require more work but which would potentially repay the research effort.

14.3.1 The origins of the Holocene flora

The work by Rundgren and Ingólfsson (1999) has highlighted the problem of determining the origins and nature of colonisation of the Icelandic Holocene flora. As yet the sites on Skagi remain the only published sites from which Lateglacial pollen and macrofossil records have been published, but as the work of Norðdahl and Pétursson (this volume) has shown there must be several locations around the coastal fringe of Iceland at which Lateglacial sites have survived providing potential data on the geographical distribution of taxa at this important period for ecosystem development. The realisation that almost all of the Icelandic flora present before the Younger Dryas/ Preboreal Oscillations quickly recolonised, implying survival through that period of climatic deterioration, adds strength to the view that refugia may have played an important part in vegetation development but until more well-dated sites are examined any discussion will suffer from a lack of primary data to inform the debate. Improvement of our knowledge of the location of the Icelandic ice sheet(s) through the Late Weichselian as a whole will provide a good basis for testing hypotheses of vegetation development.

What the data from the Holocene show is that the flora is fundamentally northwest European in character and any debate has to concentrate not on from where the species originated but how they arrived in Iceland. A similar origin can be postulated for the fauna such as Coleoptera (Buckland, 1988) and Chironomidae (Caseldine *et al.*, 2003), and to a lesser extent for microfauna such as testate amoebae (Caseldine *et al.*, in press). The question of hybridisation and the special character of Icelandic species has only attracted limited attention although genetic characterisation of species such as birch has recently provided useful new data (Anamthawat-Jónsson, 1994), and Caseldine (2001) has argued on palynological grounds for Icelandic tree birches representing possible hybridisation between tree species originating in northern Fennoscandia and dwarf birch that was already present in Iceland through the Lateglacial. The taxonomy of tree birch in Iceland and changes in tree birches through the Holocene remains a matter of contention and more detailed macrofossil studies would help to inform the debate.

14.3.2 Palaeoclimatic reconstruction from vegetation records in the Holocene

Because of the nature of the Icelandic flora there are only limited opportunities for using the record to derive detailed and precise palaeoclimatic data, although change and directions of change i.e. deterioration or amelioration of temperatures can be inferred from some of the pollen and macrofossil data. Because the altitudinal limits of tree birches are closely linked to summer temperature the tree birch record has been used to estimate periods of treeline expansion and regression (Wastl *et al.*, 2001). The lack of analogue data from current treelines in Iceland, due to the heavy influence of grazing, means that any palaeoclimatic inferences rely on extrapolation from Scandinavia where several recent studies have isolated varying temperature parameters (summarised in Caseldine, 2001), but overall uncertainty over the birch data inevitably leads to significant errors in any temperature reconstructions. Nevertheless long-term potential vegetation-climate links have been modelled using birch for the whole of Iceland by Ólafsdóttir *et al.* (2001) examining the relative importance of climate and human activity for recent historical landscape change. The importance of *Juniperus* in the early Holocene and its interrelationship with birch may have a climatic significance, possibly related to snow cover/precipitation rather than temperature but this remains a speculative observation until more detailed comparison with independent climatic information is forthcoming, as from ice core and/or marine data, or perhaps from lake level data, a source of possible palaeoclimatic information as yet virtually untapped in Iceland.

Providing a rigorous terrestrial palaeoclimatic record is a major challenge to future palaeoecological research and will require a broad multi-proxy record to be effective, both in establishing the reliability of individual proxies and in understanding the importance of non-climatic factors in determining vegetation change. The first work comparing pollen, lithostratigraphic and chironomid data from a site in NW Iceland has provided an encouraging basis for the future (Caseldine *et al.*, 2003), but also underlines the importance of establishing Icelandic training sets and transfer functions as a foundation for temperature reconstruction, rather than relying on extrapolation from apparently similar environments such as northern Fennoscandia.

14.3.3 Wider correlations – marine and ice core records

The close proximity of high resolution palaeoclimatic records from marine sediments, especially to the north and west of Iceland (Eiríksson *et al.*, 2000a; Jiang *et al.*, 2002) and from the Greenland ice sheet (Johnsen *et al.*, 2001) offers significant opportunities for establishing links between terrestrial, marine and atmospheric systems in and around Iceland. Despite the potential limitations of the floral record it still has an important role in combination with other proxies to understand how systems operated in the past, and how they may develop in the future. Accurate correlation of the records requires a rigorous chronological framework which in Iceland is facilitated by the detailed tephrochronology available for both onshore and offshore records (Eiríksson *et al.*, 2000a).

Reconstruction of offshore and nearshore water temperatures and the location of currents through time (Eiríksson *et al.*, 2000a; Andrews *et al.*, 2001a, Jiang *et al.*, 2002) now provides a valuable backdrop against which vegetation change can be evaluated leading into questions of the importance of lags and thresholds in the terrestrial ecosystem. Here again the objective needs to be to focused on more specific issues using the marine and ice core records to generate questions for examination in Iceland e.g. was there a terrestrial response to the 8.2 cal. yr BP event (Alley *et al.*, 1997)? Pollen evidence from a lake core in northern Iceland hints at such an evidence, but needs to be followed up by further research at other sites (Hallsdóttir, 1996 p. 211).

14.3.4 Human record – fleshing out the picture

Despite the relatively low resolution of traditional peat and lake records for determining the detailed impact of human settlement on the biota a broad picture has emerged as reviewed above, assisted especially by reconstruction of patterns of soil erosion anchored firmly by the tephrochronological framework available for southern Iceland (Dugmore *et al.*, 2000). In recent years these records have been supplemented by results derived from archaeological sites at which palaeobotanical analyses have been carried out (Zutter, 1997, 2000). Integration of well-dated excavations with local palaeoecological records offers an important opportunity for a much more informative understanding of how communities interacted with their local environments, developing ideas first examined by Hallsdóttir (1987).

14.4 CONCLUSION

Almost 60 years of palaeoecological study in Iceland may not have produced a history of vegetation change comparable to that derived for the more intensively studied areas of North West Europe. Nevertheless it has laid the foundations for future work which will be able to focus on more specific questions, the answers to which will form a key part of our understanding regarding the evolution and future of North Atlantic terrestrial, marine and atmospheric systems. The challenge for researchers in this field is to ensure that the terrestrial record is not overlooked as attention is paid increasingly to the perhaps more attractive high resolution marine and ice core records. Understanding terrestrial responses to climatic change and hence how such systems will evolve in the future cannot be achieved without detail from the terrestrial record itself.

REFERENCES

(note that the ordering does not follow alphabetical arrangement as in Icelandic, although in the main Icelandic characters, including "Þ" (th) and "ð" (d), are used)

Aagaard, K. and Carmack, E.C. 1989. The role of sea ice and other fresh water in the Arctic circulation. *Journal of Geophysical Research*, 94 (C10), 14485-14498.

Aagaard, K. and Malmberg, S.-A. 1978. Low-frequency characteristics of Denmark Strait overflow. *ICES CM* C:47, 1-22.

Alho, P. 2003. Landcover characteristics in NE Iceland with special references to jökulhlaup geomorphology. *Geografiska Annaler*, 85A, 213-227.

Allen, C.C. 1980. Icelandic subglacial volcanism: Thermal and physical studies. *Journal of Geology*, 88, 108-117.

Allen, J.R.L. 1984. *Sedimentary Structures, Their Character and Physical Basis, Vols. 1 and 2*. Amsterdam, Elsevier.

Allen, J.R.M., Brandt, U., Brauer, A., Hubberten, H-W., Huntley, B., Keller, J., Kraml, M., Mackensen, A., Mingram, J., Negendank, J.F.W., Nowaczyk, N.R., Oberhänsli, H., Watts, W.A., Wulf, S. and Zolitschka, B. 1999. Rapid environmental changes in southern Europe during the Last Glacial period. *Nature*, 400, 740-743.

Alley, R. B., Ágústsdóttir, A. M., and Fawcett, P. J. 1999a. Ice-core evidence of late Holocene reduction in North Atlantic Ocean Heat transport. In, Clark, P.U., Webb, R.S. and Keigwin, L. D. (eds.), *Mechanisms of Global Climate Change at Millennial Time Scales*. American Geophysical Union, Washington, DC, 301-312.

Alley, R.B., Lawson, D.E., Evenson, E.B., Strasser, J.C. and Larson G.J. 1998. Glaciohydraulic supercooling: a freeze-on mechanism to create stratified, debris-rich basal ice. II. Theory. *Journal of Glaciology*, 44, 563-569.

Alley, R. B., Mayewski, P. A., Sowers, T., Stuiver, M., Taylor, K. C. and Clark, P. U. 1997. Holocene climatic instability: a prominent, widespread event 8200 yr ago. *Geology*, 25, 483-486.

Alley, R. B., Meese, D. A., Shuman, C. A., Gow, A. J., Taylor, K. C., Grootes, P. M., White, J. W. C., Ram, M., Waddington, E. D., Mayewski, P. A. and Zielinski, G. A., 1993. Abrupt increase in Greenland snow accumulation at the end of the Younger Dryas event. *Nature*, 362, 527-529.

Alley, R.B., Strasser, J.C., Lawson, D.E., Evenson, E.B. and Larson G.J. 1999b. Glaciological and geological implications of basal-ice accretion in overdeepenings. In, Mickelson D.M. and Attig J.W. (eds.), *Glacial Processes: Past and Present*. Geological Society of America, Special Paper 337, 1-9.

Anamthawat-Jónsson, K. 1994. Genetic variation in Icelandic birch. *Norwegian Journal of Agricultural Sciences*, Supplement 18, 9-14.

Andrésdóttir, A. 1987. Glacial Geomorphology and Raised Shorelines in the Skarðsströnd – Saurbær Area West Iceland. *Examensarbeten i geologi vid Lunds Universitet, Kvartärgeologi 18*, 25 pp.

Andrews, J. T. 2000. Rates of marine sediment accumulation at the margin of the Laurentide Ice Sheet ≤ 14 ka. *Journal Sedimentary Research*, 70, 782-787.

Andrews, J. T. and Helgadóttir, G. 1999. Late Quaternary glacial and marine environments off North Iceland: Where was the LGM? *Abstract volume, Geological Society of America*, 31(7), A314.

Andrews, J. T. and Syvitski, J. P. M. 1994. Sediment fluxes along high latitude continental margins (NE Canada and E. Greenland). In, Hay, W. (ed.), *Material Fluxes on the Surface of the Earth*. Washington, D.C., National Academy Press, 99-115.

Andrews, J. T., Barber, D. C. and Jennings, A. E. 1999. Errors in generating time-series and dating events at late Quaternary (radiocarbon) time-scales: examples from Baffin Bay, Labrador Sea, and East Greenland. In, Clark, P.U., Webb, R.S. and Keigwin, L. D. (eds.), *Mechanisms of Global Climate Change at Millennial Time Scales*. American Geophysical Union, Washington, D.C.

Andrews, J. T., Caseldine, C., Weiner, N. J. and Hatton, J. 2001a. Late Quaternary (~4 ka) marine and terrestrial environmental change in Reykjarfjörður, N. Iceland: Climate and/or settlement? *Journal of Quaternary Science*, 16, 133-144.

Andrews, J. T., Cooper, T. A., Jennings, A. E., Stein, A. B. and Erlenkeuser, H. 1998a. Late Quaternary iceberg-rafted detritus events on the Denmark Strait-Southeast Greenland continental slope (65N): related to North Atlantic Heinrich events? *Marine Geology*, 149, 211-228.

Andrews, J.T. Geirsdóttir, Á., Harðardóttir, J., Principato, S., Kristjánsdóttir, G.B., Helgadóttir, G., Grönvold, K., Drexler, J. and Sveinbjörnsdóttir, Á. 2002a. Distribution, age, sediment magnetism, and geochemistry of the Saksunarvatn (10.18 ± cal ka) tephra in marine, lake, and terrestrial sediment, NW Iceland. *Journal of Quaternary Science*, 17, 731-745.

Andrews, J.T., Harðardóttir, J., Geirsdóttir, Á. and Helgadóttir, G. 2002b. Late Quaternary ice extent and glacial history from the Djúpáll trough off Vestfirðir peninsula, north-west Iceland: a stacked 36 cal. ky environmental record. *Polar Research*, 21, 211-226.

Andrews, J. T., Harðardóttir, J., Helgadóttir, G., Jennings, A. E., Geirsdóttir, Á., Sveinbjörnsdóttir, Á. E., Schoolfield, S., Kristjánsdóttir, G. B., Smith, L. M., Thors, K. and Syvitski, J. P. M. 2000. The N and W Iceland Shelf: Insights into Last Glacial Maximum ice extent and deglaciation based on acoustic stratigraphy and basal radiocarbon AMS dates. *Quaternary Science Reviews*, 19, 619-631.

Andrews, J. T., Helgadóttir, G., Geirsdóttir, Á. and Jennings, A. E. 2001b. Multicentury-scale records of carbonate (hydrographic?) variability on the N. Iceland margin over the last 5000 yrs. *Quaternary Research*, 56, 199-206.

Andrews, J. T., Jennings, A. E., Cooper, T., Williams, K. M. and Mienert, J. 1996. Late Quaternary sedimentation along a fjord to shelf (trough) transect, East Greenland (ca. 68°N). In, Andrews, J. T., Austin, W., Bergsten, H. and Jennings, A. E. (eds.), *Late Quaternary Paleoceanography of North Atlantic Margins*. London, Geological Society of London, 153-166.

Andrews, J. T., Kihl, R., Jennings, A. E., Kristjánsdóttir, G. B., Helgadóttir, G. and Smith, L. M. 2002c. Contrast in Holocene sediment properties on the shelves bordering Denmark Strait (64-68°N), North Atlantic. *Sedimentology*, 49, 5-24.

Andrews, J. T., Kirby, M. E., Jennings, A. E. and Barber, D. C. 1998b. Late Quaternary stratigraphy, chronology, and depositional processes on the SE Baffin Island slope. Detrital carbonate and Heinrich events: implications for onshore glacial history. *Geographie physique et Quaternaire*, 52, 91-105.

Andrews, J. T., Kristjánsdóttir, G., Geirsdóttir, Á., Harðardóttir, J., Helgadóttir, G., Sveinbjörnsdóttir, Á., Jennings, A.E. and Smith, L. M. 2001c. Late Holocene (~5 cal ka) trends and century-scale variability of N. Iceland marine records: measures of surface hydrography, productivity, and land/ocean interactions. In, Seivdov, D., Maslin, M. and Haupt, B. (eds.), *Oceans and Rapid Past and Future Climate Change: North-South Connections*. Washington D.C., American Geophysical Union.

Andrews, J. T., Tedesco, K., Briggs, W. M. and Evans, L. W. 1994. Sediments, sedimentation rates, and environments, SE Baffin Shelf and NW Labrador Sea 8 to 26 KA. *Canadian Journal Earth Sciences*, 31, 90-103.

Angell, J. K. and Korshover, J. 1985. Surface temperature changes following the six major volcanic episodes between 1780 and 1980. *Journal of Climate and Applied Meteorology*, 24, 937-951.

Annálar 1400-1800, I-. 1922-: Reykjavík, Hið íslenzka bókmenntafélag.

Anonymous 1992. *ERS-1 User Handbook*.

Anonymous 1995. Environmental conditions in Icelandic waters 1995. *Hafrannsóknastofnunin Fjölrit* 44 (Marine Research Institute), 36 pp (In Icelandic with english summary).

Arnalds, A. 1987. Ecosystem disturbance in Iceland. *Arctic and Alpine Research*, 19, 508-513.

Arnalds, Ó. 1990. *Characterization and erosion of Andisols in Iceland*. Unpublished Ph.D. Thesis, Texas A&M University, College Station, 174 pp.

Arnalds, Ó. 1993. Notes on clay minerals in Icelandic soils. *Náttúrufræðingurinn*, 63, 73–85 (in Icelandic, English summary).

Arnalds, Ó. 1999. Soils and soil erosion in Iceland. In, Ármannsson, H. (ed.), *Geochemistry of the Earth's Surface*, Rotterdam, Balkema, 135–138.

Arnalds, Ó. 2000. The Icelandic 'rofabard' soil erosion features. *Earth Surface Processes and Landforms*, 25, 17–28.

Arnalds, Ó. 2003. Volcanic soils of Iceland. In, Arnalds, Ó. and Stahr, K. (eds.), *Volcanic Soil Resources, Occurrence, Development and Properties*, Catena, Special Issue, Elsevier, in press.

Arnalds, Ó. and Grétarsson, E. 2001. *Soil Map of Iceland*. 2nd edition. Agricultural Research Institute, Reykjavík, Iceland. Available in digital format, ww.rala.is/desert.

Arnalds, Ó. and Kimble, J. 2001. Andisols of deserts in Iceland. *Soil Science Society of America Journal*, 65, 1778–1786.

Arnalds, Ó., Gísladóttir, F.Ó. and Sigurjónsson, H. 2001b. Sandy deserts of Iceland: an overview. *Journal of Arid Environments*, 47, 359-371.

Arnalds, Ó., Hallmark, C.T. and Wilding, L.P. 1995. Andisols from four different regions of Iceland. *Soil Science Society of America Journal*, 58, 61-169.

Arnalds, Ó., Þórarinsdóttir, E.F., Metúsalemsson, S., Jónsson, A., Grétarsson, E. and Árnason, A. 2001a. *Soil Erosion in Iceland*. Soil Conservation Service and Agricultural Research Institute, Reykjavík, Iceland. Translated from original book published in Icelandic in 1997.

Arnórsson, S. and Gíslason, S.R. 1994. CO_2 from magmatic sources in Iceland. *Mineralogical Magazine*, 58A, 27–28.

Arnórsson, S., Gunnarsson, I., Stefánsson, A., Andrésdóttir, A. and Sveinbjörnsdóttir Á. 2002. Major element chemistry of surface- and ground water in basaltic terrain, N-Iceland. I. Primary mineral saturation. *Geochimica et Cosmochimica Acta*, 66, 4015–4046.

Ásbjörnsdóttir, L. and Norðdahl, H. 1995. Götungar í sjávarsetlögum við Mela á Skarðsströnd. In, Hróarsson, B., Jónsson, D. and Jónsson, S. S. (eds.), *Eyjar í Eldhafi*, Reykjavík, Gott mál, 179-188.

Ashley, G.M. and Warren W.P. 1997. The ice-contact environment. *Quaternary Science Reviews*, 16, 629-634.

Ashwell, I. Y. 1967. Radiocarbon ages of shells in the glaciomarine deposits of Western Iceland. *Geographical Journal*, 133, 48-50.

Ashwell, I. Y. 1975. Glacial and Late Glacial processes in Western Iceland. *Geografiska Annaler*, 57, 225-245.

Ashwell, I. Y. 1996. The geomorphology of Fnjóskadalur, N. Iceland: Re-interpretation in terms of sub-glacial hydrologic processes, with an introduction ' The Deglaciation of Iceland'. *Personal publication*.

Áskelsson, J. 1936. On the last eruptions in Vatnajökull. *Societas Scientia Islandica*, 18, Reykjavík. 68 pp.

Axelsdóttir, H. 2002. *Búðaröðin. Landmótunarfræðileg kortlagning jökulgarða milli Þjórsár og Eystri-Rangár*. Unpublished B.S. Thesis, University of Iceland, Reykjavík, 41 pp.

Axelson, S. 1960. De Danske Annalerna och Hakonar-sagan som källor för de äldre Islandske Annalernas Danska Uppgifter 1203-1264. *Historisk Tidsskrift*, I, 217-27.

Banfield, J.F., Jones, B.F. and Veblen, D.R. 1991. An AEM-TEM study of weathering and diagenesis, Abert Lake, Oregon: I. Weathering reactions in the volcanics. *Geochimica et Cosmochimica Acta*, 55, 2781–2793.

Banks, Sir J. 1950. Dagbókarbrot úr Íslandsferð, 1772. Trans. J. Benediktsson. *Skírnir*, 124, 210-22.

Barber, D. C., Dyke, A., Hillaire-Marcel, C., Jennings, A. E., Andrews, J. T., Kerwin, M. W., Bilodeau, G., McNeely, R., Southon, J., Moorehead, M. D. and Gagnon, J.-M. 1999. Forcing of the cold event of 8200 years ago by catastrophic drainage of Laurentide lakes. *Nature*, 400, 344-348.

Bard, E., Arnold, M., Mangerud, J., Paterne, M., Labeyrie, L., Duprat, J., Melieres, M.-A., Sonstegaard, E. and Duplessy, J.-C. 1994. The North Atlantic atmosphere-sea surface ^{14}C gradient during the Younger Dryas climatic event. *Earth and Planetary Science Letters*, 126, 275-287.

Bárðarson, G. G. 1921. Fossile skalaflejringer ved Breiðafjörður i Vest-Island. *Geologiska Föreningens i Stockholm Förhandlingar*, 22, 323-380.

Bárðarson, G. G. 1923. Fornar sjávarminjar við Borgarfjörð og Hvalfjörð. *Rit Vísindafélags Íslendinga*, 1, 1-116.

Bárðarson, G. G. 1934. Islands Gletscher. Beiträge zur Kenntnis der Gletscherbewegungen und Schwankungen auf Grund alter Quellenschriften und neuester Forschung. *Vísindafélag Íslendinga*, XVI, 60 pp.

Barlow, L.K., Sadler, J.P., Ogilvie, A.E.J., Buckland, P.C., Amorosi, T., Ingimundarson, J.H., Skidmore, P., Dugmore, A.J. and McGovern, T.H. 1997. Interdisciplinary investigations of the end of the Norse Western Settlement in Greenland. *The Holocene*, 7, 489-499.

Bartley, D.D. 1973. The stratigraphy and pollen analysis of peat deposits at Ytri-Bægisá near Akureyri, Iceland. *Geologiska Föreningens i Stockholm Förhandlingar*, 95, 410-414.

Beckman, N. 1912. Annalstudier. *Studier i Nordisk Filologi* III 4, 1-12.

Behrendt, J.C., Blankenship, D.D., Damaske, D. and Cooper, A.K. 1995. Removal of late Cenozoic subglacially emplaced volcanic edifices by the West Antarctic Ice Sheet. *Geology*, 23, 527-530.

Bekker-Nielsen, H., Olsen, T.D. and Widding, O. 1965. *Norrøn Fortællekunst: Kapitler af den Norsk-Islandske Middelalder-litteraturs Historie*. Copenhagen, Akademisk Forlag.

Belkin, I. M., Levitus, S., Antonov, J. and Malmberg, S.-A. 1998. 'Great Salinity Anomalies' in the North Atlantic. *Progress in Oceanography*, 41, 1-68.

Bell, W.T. and Ogilvie, A.E.J. 1978. Weather compilations as a source of data for the reconstruction of European climate during the medieval period. *Climatic Change*, 1, 331-348.

Bemmelen, R.W. and Rutten, M.G. 1955. *Tablemountains of Northern Iceland*. Leiden, E.J. Brill.

Benediktsson, B. 1881/1884-1930/1932. *Sýslumannsævi*. 5 Vols. Reykjavík, Hið íslenzka bókmenntafélag.

Benediktsson, J. 1948. Ole Worm's Correspondence with Icelanders. In, Helgason, J. (ed.) *Bibliotheca Arnamagnæana*, Vol. VII. Copenhagen, Ejnar Munksgaard.

Benediktsson, J. (ed.) 1950. Arngrimi Jonæ Opera Latine Conscripta, Vol. I. In, Helgason, J. (ed.), *Bibliotheca Arnamagnæana*, Vol. IX. Copenhagen, Ejnar Munksgaard.

Benediktsson, J. 1956. Hver samdi Qualiscunque descriptio Islandiæ? *Nordæla Afmæliskveðja Til Sigurðar Nordals*. Reykjavík, Helgafell.

Benediktsson, J. (ed.) 1957. Arngrimi Jonæ opera latine conscripta. Vol. IV Introduction and notes. In, Helgason, J. (ed.), *Bibliotheca Arnamagneana*, Vol. XII. Copenhagen, Ejnar Munksgaard.

Benediktsson, J. 1968. Íslendingabók. Landnámabók. *Íslenzk Fornrit* I. Reykjavík, Hið íslenzka fornritafélag.

Benediktsson, K. 1972. 3. Byggðasaga Mýrahrepps. *Byggðasaga Austur-Skaftafellssýslu.* II, Bindi. Reykjavík, Bókaútgáfa Guðjóns Ó. Guðjónssonar, 9-67.

Benn, D.I. 1989. Debris transport by Loch Lomond Readvance glaciers in northern Scotland: basin form and the within-valley asymmetry of lateral moraines. *Journal of Quaternary Science*, 4, 243-254.

Benn, D.I. 1994. Fluted moraine formation and till genesis below a temperate glacier: Slettmarkbreen, Jotunheimen, Norway. *Sedimentology*, 41, 279-292.

Benn, D.I. 1995. Fabric signature of subglacial till deformation, Breiðamerkurjökull, Iceland. *Sedimentology*, 42, 735-747.

Benn, D.I. and Evans D.J.A. 1996. The interpretation and classification of subglacially deformed materials. *Quaternary Science Reviews*, 15, 23-52.

Benn, D.I. and Evans D.J.A. 1998. *Glaciers and Glaciation*. London, Arnold.

Bennett, M.R. and Bullard, J.E. 1991. Iceberg tool marks: an example from Heinabergsjökull, southeast Iceland. *Journal of Glaciology*, 37, 181-183.

Bennett, M.R., Huddart, D. and McCormick, T. 2000. An integrated approach to the study of glaciolacustrine landforms and sediments: a case study from Hagavatn, Iceland. *Quaternary Science Reviews*, 19, 633-665.

Berger, G., Claparols, C., Guy, C. and Daux, V. 1994. Dissolution rate of basalt glass in silica rich solutions. Implications for long-term alteration. *Geochimica et Cosmochimica Acta*, 58, 4875–4886.

Bergh, S.G. and Sigvaldason, G.E. 1991. Pleistocene mass-flow deposits of basaltic hyaloclastites on a shallow submarine shelf, South Iceland. *Bulletin of Volcanology*, 53, 597-611.

Bergsson, J. 1997 (written in 1839). Einholtsprestakall. (Description of the parish of Einholt.) In, Jónsson, J.A. and Sigmundsson, S. (Eds.), *Skaftafellssýsla. Sýslu- og sóknalýsingar Hins íslenska bókmenntafélags.* Reykjavík, Sögufélag, 105-122.

Bergþórsson, P. 1967. Kuldaskeið um 1300? (A cold period around 1300?). *Veðrið*, 12, 55-58.

Bergþórsson, P. 1969a. An estimate of drift ice and temperature in Iceland in 1000 years. *Jökull*, 19, 95-101.

Bergþórsson, P. 1969b. Hafís og hitastig á liðnum öldum. In, Einarsson, M.Á. (ed.), *Hafísinn.* Reykjavík, Almenna bókafélagið, 333-345.

Berner, R.A. 1992. Weathering, plants, and the long-term carbon cycle. *Geochimica et Cosmochimica Acta*, 56, 3225–3231.

Berner, E.K. and Berner, R.A. 1996. *Global Environment. Water, Air and Geochemical Cycles.* New Jersey, Prentice-Hall.

Berner, R.A., Lasaga, A.C. and Garrels, R.M. 1983. The carbonate-silicate geochemical cycle and its effect on atmospheric carbon dioxide over the past 100 million years. *American Journal of Science*, 283, 641–683.

Bjarnar, V. 1965. The Laki eruption and the famine of the mist. In, Bayerschmidt, C.A.F. and Friis, E.J. (eds.), *Scandinavian Studies*. The American-Scandinavian Foundation, Washington, University of Washington Press, 410-21.

Bjarnason, B. 1941. *Brandstaðannáll*. Reykjavík, Sögufélagið Húnvetningur og Húnvetningafélagið í Reykjavík.

Björck, S., Ingólfsson, Ó., Hafliðason, H., Hallsdóttir, M. and Anderson, N. J. 1992. Lake Torfadalsvatn: a high resolution record of the North Atlantic ash zone I and the last glacial-interglacial environmental changes in Iceland. *Boreas*, 21, 15-22.

Björck, S., Rundgren, M., Ingólfsson, Ó. and Funder, S. 1997. The Preboreal Oscillation around the Nordic Seas: terrestrial and lacustrine responses. *Journal of Quaternary Science*, 12, 455-465.

Björnsson, F. 1956. Kvíárjökull. *Jökull*, 6, 20-22.

Björnsson, F. 1998. Samtíningur um jökla milli Fells og Staðarfjalls. (Glacier variations between Fell and Staðarfjall, southeastern Iceland.) *Jökull*, 46, 49-61.

Björnsson, H. 1969. Sea ice conditions and the atmospheric circulation north of Iceland. *Jökull*, 19, 11-28.

Björnsson, H. 1974. Explanation of jökulhlaups from Grímsvötn, Vatnajökull, Iceland. *Jökull*, 24, 1-26.

Björnsson, H. 1975. Subglacial water reservoirs, jökulhlaups and volcanic eruptions. *Jökull*, 25, 1–14.

Björnsson, H. 1976. Marginal and supraglacial lakes in Iceland. *Jökull*, 26, 40-51.

Björnsson, H. 1977. The cause of jökulhlaups in the Skaftá river, Vatnajökull. *Jökull*, 27, 71-78.

Björnsson, H. 1979. Glaciers in Iceland. *Jökull*, 29, 74-80.

Björnsson, H. 1988. Hydrology of ice caps in volcanic regions. *Societas Scientia Islandica*, 45, 139 pp.

Björnsson, H. 1992. Jökulhlaups in Iceland: prediction, characteristics and simulation. *Annals of Glaciology*, 16, 95-106.

Björnsson, H. 1996. Scales and rates of glacial sediment removal: a 20 km long, 300 m deep trench created beneath Breiðamerkurjökull during the Little Ice Age. *Annals of Glaciology*, 22, 141-146.

Björnsson, H. 1997. Grímsvatnahlaup fyrr og nú. In, Haraldsson, H. (ed.) *Vatnajökull: Gos og hlaup*. Vegagerðin, 61-77.

Björnsson, H. 1998. Hydrological characteristics of the drainage system beneath a surging glacier. *Nature*, 395, 771-774.

Björnsson, H. 2002. Subglacial lakes and jökulhlaups in Iceland. *Global and Planetary Change*, 35, 255-271.

Björnsson, H. and Einarsson, P. 1990. Volcanoes beneath Vatnajökull, Iceland: Evidence from radio-echo sounding, earthquakes and jökulhlaups. *Jökull*, 40, 147-168.

Björnsson, H. and Guðmundsson, M.T. 1993a. Variations in the thermal output of the subglacial Grímsvötn Caldera, Iceland. *Geophysical Research Letters*, 20, 2127–2130.

Björnsson, H. and Guðmundsson, M.T. 1993b. Interaction of a surge and jökulhlaup: the surge of Skeiðarárjökull and the jökulhlaup from Grímsvötn in 1991. In, Collins, D. (ed.), *Abstracts of International Workshop on Glacier Hydrology*. Jesus College, Cambridge.

Björnsson, H., Pálsson, F. and Guðmundsson, M.T. 1992. *Vatnajökull, Northwestern Part, 1:100,000, Subglacial Surface*. Reykjavík, National Power Company and Science Institute.

Björnsson, H., Pálsson, F. and Guðmundsson, M.T. 2000. Surface and bedrock topography of the Mýrdalsjökull ice cap, Iceland: the Katla caldera, eruption sites and routes of jökulhlaups. *Jökull*, 49, 29-46.

Björnsson, H., Pálsson, F. and Magnússon, E. 1999. *Skeiðarárjökull: 'Landslag og rennslisleiðir vatns undir sporði'*, Raunvísindastofnun Háskólans, RH-11-99.

Björnsson, H., Pálsson F., Flowers, G.E. and Guðmundsson M.T. 2001. The extraordinary 1996 jökulhlaup from Grímsvötn, Vatnajökull, Iceland. *Eos*, 82, 47.

Björnsson, H., Pálsson F., Sigurðsson, O. and Flowers, G.E. 2003. Surges of glaciers in Iceland. *Annals of Glaciology*, 36, 82-90.

Björnsson, H., Rott, H., Guðmundsson, S., Fischer, A., Siegel, A. and Guðmundsson, M.T. 2001. Glacier-volcano interactions deduced by SAR interferometry. *Journal of Glaciology*, 47, 58-70.

Björnsson, S. 1991. Símalína lögð um Skaftafellssýslur 1929. (Telephone wire installed in the Skaftafell counties 1929.) *Skaftfellingur*, 7, 76-119.

Black, T.A. 1990. *The Holocene Fluctuations of the Kvíárjökull Glacier, Southeastern Iceland*. Unpublished M.A. Thesis, University of Colorado.

Blakemore, L.C., Searle, P.L. and Daly, B.K. 1987. Methods of chemical analysis of soils. *New Zealand Soil Bureau Science* Report No. 80.

Blöndal, S. and Gunnarsson, S. B. 1991. *Íslands Skógar: Hundrað Ára Saga*. Reykjavík, Mál og Mynd.

Bluth, G.J.S. and Kump, L.R. 1994. Lithologic and climatologic controls of river chemistry. *Geochimica et Cosmochimica Acta*, 58, 2341–2359.

Bodéré, J.-C. 1977. Les kettles du sud-est de L'islande. *Revue de Geographie Physique et de Geologie Dynamique*, 19, 259-270.

Bogacki, M. 1973. Geomorphological and geological analysis of the proglacial of Skeiðarárjökull, central western and eastern sections. *Geographica Polonica*, 26, 57-88.

Bond, G. C. and Lotti, R. 1995. Iceberg discharges into the North Atlantic on millennial time scales during the last glaciation. *Science*, 267, 1005-1009.

Bond, G., Mandeville, C. and Hoffman, S. 2001. Were rhyolitic glasses in the Vedde ash and in the North Atlantic Ash Zone 1 produced by the same volcanic eruption? *Quaternary Science Reviews*, 20, 1189-1199.

Bond, G., Showers, W., Cheseby, M., Lotti, R., Almasi, P., de Menocal, P., Priore, P., Cullen, H., Hajdas, I. and Bonani, G. 1997. A pervasive millennial-scale cycle in North Atlantic Holocene and glacial climate. *Science*, 278, 1257-1266.

Bond, G. C., Showers, W., Elliot, M., Evans, M., Lotti, R., Hajdas, I., Bonani, G. and Johnson, S. 1999. The North Atlantic's 1-2 kyr climate rhythm: relation to Heinrich Events, Dansgaard/Oeschger Cycles and the Little Ice Age. In, Clark, P. U., Webb, R. S. and Keigwin, L. D. (eds.), *Mechanisms of Global Climate Change at Millennial Time Scales*. Washington, D.C., American Geophysical Union, 35-58.

Boothroyd, J. C. and Nummedal, D. 1978. Proglacial braided outwash: a model for humid alluvial fan deposits. In, Miall, A. D. (ed.), *Fluvial Sedimentology*. Canadian Society of Petroleum Geologists Memoir, 5, 641-668.

Boulton, G.S. 1976. The origin of glacially fluted surfaces: observations and theory. *Journal of Glaciology*, 17, 287-309.

Boulton, G.S. 1979. Processes of glacier erosion on different substrata. *Journal of Glaciology*, 23, 15-38.

Boulton, G.S. 1986. Push moraines and glacier contact fans in marine and terrestrial environments. *Sedimentology*, 33, 677-698.

Boulton, G.S. 1987. A theory of drumlin formation by subglacial sediment deformation. In, Menzies J. and Rose J. (eds.), *Drumlin Symposium*. Rotterdam, Balkema, 25-80.

Boulton, G.S. and Dent D.L. 1974. The nature and rates of post-depositional changes in recently deposited till from south-east Iceland. *Geografiska Annaler*, 56A, 121-134.

Boulton, G.S. and Eyles N. 1979. Sedimentation by valley glaciers: a model and genetic classification. In, Schlüchter C. (ed.), *Moraines and Varves*. Rotterdam, Balkema, 11-23.

Boulton, G.S. and Hindmarsh R.C.A. 1987. Sediment deformation beneath glaciers: rheology and sedimentological consequences. *Journal of Geophysical Research*, 92, 9059-9082.

Boulton, G.S and Paul M.A. 1976. The influence of genetic processes on some geotechnical properties of tills. *Journal of Engineering Geology*, 9, 159-194.

Boulton, G. S., Jarvis, J. and Thors, K. 1988. Dispersal of glacially derived sediment over part of the continental shelf of south Iceland and the geometry of the resultant sediment bodies. *Marine Geology*, 83, 193-223.

Bourgeois, O., Dauteuil, O. and van Vleit-Lanoe, B. 1998. Pleistocene subglacial volcanism in Iceland: tectonic implications. *Earth and Planetary Science Letters*, 164, 165-178.

Bourgeois, O., Dauteil, O. and van Vliet-Lanoe, B. 2000. Geothermal control on ice stream formation: flow patterns of the Icelandic Ice Sheet at the Last Glacial Maximum. *Earth Surface Processes and Landforms*, 25, 59-76.

Bradley, R. S. and Jones, P.D. (eds.) 1992. *Climate since A.D. 1500*. London, New York, Routledge.

Brady, P.V. and Gíslason, S.R. 1997. Seafloor weathering controls on atmospheric CO_2 and global climate. *Geochimica et Cosmochimica Acta*, 61, 965–997.

Branney, M. J. and Gilbert, J.S. 1995. Ice-melt collapse pits and associated features in the 1991 lahar of Volcán Hudson, Chile: criteria to distinguish eruption-induced glacier melt. *Bulletin of Volcanology*, 57, 293-302.

Brayshaw, A.C. 1985. Bed microtopography and entrainment thresholds in gravel bed rivers. *Bulletin of the Geological Society of America*, 96, 218-223.

Briffa, K. R., Jones, P. D., Pilcher, J. R. and Hughes, M. K. 1988. Reconstructing summer temperatures in northern Fennoscandinavia back to AD 1700 using tree-ring data from Scots pine. *Arctic and Alpine Research*, 20, 385-394.

Brown, U. (ed.) 1952. *Þorgils Saga Og Hafliða*. London, Oxford University Press.

Bruun, D. 1925. Fjældvej gennem Islands indre Højland ved Daniel Bruun. (A mountain track through central Iceland with Daniel Bruun.) *Turistruter paa Island IV*. Copenhagen.

Buckland, P.C. 1988. North Atlantic faunal connections - introduction or endemics? *Entomologica Scandinavica*, Supplement 32, 7-29.

Buckland, P. and Dugmore, A. 1991. 'If this is a Refugium, why are my feet so bloody cold?' The origins of the Icelandic biota in the light of recent research. In, Maizels, J.K. and Caseldine, C. (eds.), *Environmental Change in Iceland: Past and Present*. Dordrecht, Kluwer Academic Publishers, 107-125.

Burg, F. (ed.) 1928. *Qualiscunque descriptio Islandiæ*. Hamburg, Veröffentlichungen aus der Hamburger Staats und Universitäts Bibliothek, Band I.

Carrivick, J. L., Russell, A.J. and Tweed, F.S. (subm.) Geomorphological evidence for jökulhlaups from Kverkfjöll volcano, Iceland. *Geomorphology*.

Carrivick, J. L., Russell, A.J., Tweed, F.S. and Knudsen, Ó. 2002. Determining the routeways and flow characteristics of jökulhlaups from Kverkfjöll, Iceland. In, Benito, G., Thorndycraft, V., Llasat, M-C. and Barriendos, M. (eds.), *Paleofloods, Historical data and climatic variability: applications in flood risk assessment*. International workshop, Barcelona, Spain, October 16-19, 2002, 67.

Cartee-Schoolfield, S. 2000. *Late Pleistocene Sedimentation in the Denmark Strait Region*. Unpublished M.Sc. Thesis, University of Colorado.

Carus-Wilson, E. M. 1933. The Iceland trade. In, Power, E. and Postan, M.M. (eds.), *Studies in English Trade in the Fifteenth Century*. London, Routledge and Kegan Paul, 155-82.

Caseldine, C.J. 2001. *Betula* as a palaeoclimatic indicator in Iceland and/or evidence for Holocene hybridisation? Evidence from northern Iceland. *Review of Palaeobotany and Palynology*, 117, 139-152.

Caseldine, C. and Hatton, J. 1994. Interpretation of Holocene climatic change for the Eyjafjörður area of Northern Iceland from pollen-analytical data: comments and preliminary results. In, Stötter, J. and Wilhelm, F. (eds.), Environmental Change in Iceland. *Münchener Geographische Abhandlungen*, Reihe B 12, 41-62.

Caseldine, C. and Stötter, J. 1993. 'Little Ice Age' glaciation of Tröllaskagi peninsula, northern Iceland: climatic implications for reconstructed equilibrium line altitudes (ELAs). *The Holocene*, 3, 357-366.

Caseldine, C.J., Geirsdóttir, Á., and Langdon, P.G. 2003. Efstadalsvatn - a multi-proxy study of a Holocene lacustrine sequence from NW Iceland. *Journal of Paleolimnology*, 30, 55-73.

Caseldine, C.J., Dinnin, M.H., Hendon, D. and Langdon, P. 2003. The Holocene development of the Icelandic biota and its palaeoclimatic significance. *British Archaeological Reports*, in press.

Cassidy, N.J., Russell, A.J., Marren, P.M., Fay, H., Rushmer, E.L., Van Dijk, T.A.G.P. and Knudsen Ó. 2003. GPR-derived architecture of November 1996 jökulhlaup deposits, Skeiðarársandur, Iceland. In, Bristow, C.S. and Jol, H.M. (eds.), *Ground Penetrating Radar in Sedimentation*. Special Publication of the Geological Society, 211, 153-166.

Castaneda, I. S. 2001. *Holocene Paleoceanographic and Climatic Variations of the Inner North Iceland Continental Shelf, Reykjarfjörður Area*. Unpublished M.Sc. Thesis, University of Colorado.

Cawley, J.L., Burruss, R.C. and Holland H.D. 1969. Chemical weathering in central Iceland: an analog of Pre-Silurian weathering. *Science*, 165, 164–165.

Chapman, M.G., Allen, C.C., Guðmundsson, M.T., Gulick, V.C., Jakobsson, S.P, Lucchitta, B.K., Skilling, I.P. and Waitt, R.B. 2000. Volcanism and ice interactions on Earth and Mars. In, Zimbelman, J.R. and Gregg, T.K.P. (eds.), *Environmental Effects of Volcanic Eruptions*. New York, Kluwer/Plenum, 39-73.

Churski, Z. 1973. Hydrographic features of the proglacial area of Skeiðarárjökull. *Geographica Polonia*, 26, 209-254.

Clark, C.D. 1999. Glaciodynamic context of subglacial bedform generation and preservation. *Annals of Glaciology*, 28, 23-32.

Clark, P.U. and Walder, J.S. 1994. Subglacial drainage. Eskers, and deforming beds beneath the Laurentide and Eurasian ice sheets. *GSA Bulletin*, 106, 304-314.

Clarke, G.K.C. 1982. Glacier outburst floods from 'Hazard Lake', Yukon Territory, and the problem of flood magnitude prediction. *Journal of Glaciology*, 28, 3-21.

Clarke, G.K.C. 1987. Subglacial till: a physical framework for its properties and processes. *Journal of Geophysical Research*, 92, 9023-9036.

Clarke, G.K.C., Collins S.G. and Thompson D.E. 1984. Flow, thermal structure and subglacial conditions of a surge-type glacier. *Canadian Journal of Earth Sciences*, 21, 232-240.

Clarke, R.P.K. 1969. Kettle holes. *Journal of Glaciology*, 8, 485-486.

Clausen, H. B. and Hammer, C. U. 1988. The Laki and Tambora eruptions as revealed in the Greenland ice cores from 11 locations. *Annals of Glaciology*, 10, 16-22.

Clayton, L. and Moran S.L. 1974. A glacial process-form model. In, Coates D.R. (ed.), *Glacial Geomorphology*. Binghamton, State University of New York, 89-119.

Collinson, J.D. 1971. Some effects of ice on a river bed. *Journal of Sedimentary Petrology*, 41, 557-564.

Colman, S. M. 1982. Chemical weathering of basalts and andesites. *U.S. Geological Survey Professional Paper*, 1246, 51 pp.

Cooper, L.H.N. 1955. Deep water movements in the North Atlantic as a link between climatic changes around Iceland and biological productivity of the English Channel and Celtic Sea. *Journal of Marine Research*, 14, 347-362.

Courtillot, V., Jaupart, C., Manighetti, I., Tapponnier, P. and Besse J. 1999. On causal links between flood basalts and continental breakup. *Earth and Planetary Science Letters*, 166, 177- 195.

Craig, D.C. and Loughnan, F.C. 1964. Chemical and mineralogical transformation accompanying the weathering of basic volcanic rocks from New South Wales. *Australian Journal of Soil Research*, 2, 218–234.

Croot, D.G. 1988. Morphological, structural and mechanical analysis of neoglacial ice-pushed ridges in Iceland. In, Croot, D.G. (ed.), *Glaciotectonics: Forms and Processes*. Rotterdam, Balkema, 33-47.

Crovisier, J.L. 1989. *Dissolution des Verres Basaltiques dans L'eau de Mer et dans L'eau Douce. Essai de modélisation*. Unpublished Ph.D. Dissertation, L'Université Louis Pasteur.

Crovisier, J.L., Honnorez, J., Fritz, B. and Petit, J.C. 1992. Dissolution of subglacial volcanic glasses from Iceland: laboratory study and modelling. *Applied Geochemistry, Supplementary Issue*, 1, 55–81.

Cuffey, K. M., and Clow, G. D. 1997. Temperature, accumulation, and ice sheet elevation in central Greenland through the last deglacial transition. *Journal of Geophysical Research*, 102, 26,383-26,396.

Dahlgren, R.A., Shoji, S. and Nanzyo, M. 1993. Mineralogical characteristics of volcanic ash soils. In, Shoji, S., Nanzyo, M. and Dahlgren, R.A. (eds.), *Volcanic Ash Soils. Genesis, properties and utilization*. Developments in Soil Science, Amsterdam, Elsevier, 101–143.

Danish General Staff, 1905. Maps No. 59 NA; 69 NV; 77 SA; 87 NV, NA, SV, SA; 96 SA; 97 NV, NA, SV. Scale 1:50,000. Surveyed in 1903-1904.

Dansgaard, W., White, J. W. C. and Johnsen, S. J. 1989. The abrupt termination of the Younger Dryas climate event. *Nature*, 339, 532-533.

Dansgaard, W., Johnsen, S. J., Reeh, N., Gundestrup, N., Clausen, H. B. and Hammer, C. U. 1975. Climatic changes, Norsemen and modern man. *Nature*, 225, 24-28.

Daux, V., Guy, C., Advocat T., Crovisier, J. and Stille, M. 1997. Kinetic aspects of basaltic glass dissolution at 90° C: Role of silicon and aluminium. *Chemical Geology*, 142, 109–128.

Demarée, G. R. and Ogilvie, A.E.J. 2001. *Bon baisers d'Islande:* climatological, environmental and human dimensions impacts in Europe of the *Lakagígar* eruption (1783-1784) in Iceland. In, Jones, P.D., Ogilvie, A.E.J., Davies, T.D. and Briffa, K.R. (eds.), *History and Climate: Memories of the Future?* New York, Boston, Dordrecht, London, Moscow, Kluwer Academic/Plenum Publishers, 219-246.

Demarée, G. R., Ogilvie, A. E. J. and Zhang, D. 1998. Further evidence of Northern Hemispheric coverage of the Great Dry Fog of 1783. *Climate Change*, 39, 727-730.

Dessert, C., Dupre, B., Francois, L.M., Schott, J., Gaillardet, J., Chakarapani, G. and Bajpai, S. 2001. Erosion of Deccan traps determined by river geochemistry: impact on the global climate and the $^{87}Sr/^{86}Sr$ ratio of seawater. *Earth and Planetary Science Letters*, 188, 459–474.

Dickson, R.R. and Brown, J. 1994. The production of North Atlantic Deep Water: sources, rates and pathways. *Journal of Geophysical Research*, 99 (C6), 12319-12341.

Dickson, R. R., Meincke, J., Malmberg, S. and Lee, A. 1988. The 'Great Salinity Anomaly' in the northern North Atlantic 1968-1982. *Progress in Oceanography*, 20, 103-151.

Dickson, R., Lazier, J., Meincke, J., Rhines, P. and Swift, J. 1996. Long-term co-ordinated changes in the convective activity of the North Atlantic. *Progress in Oceanography*, 38, 241-295.

Dixon, J.E., Filiberto, J.R., Moore, J.G. and Hickson, C.J. 2002. Volatiles in basaltic glasses from a subglacial volcano in northern British Columbia (Canada): implications for ice sheet thickness and mantle volatiles. In, Smellie, J.L. and Chapman, M. (eds.), *Ice-Volcano Interaction on Earth and Mars*. Geological Society of London Special Publication, 202, 319-335.

Douglas, G.R. 1987. Manganese-rich rock coatings from Iceland. *Earth Surface Processes and Landforms*, 12, 301–310.

Dowdeswell, J.A. and Sharp, M.J. 1986. Characterization of pebble fabrics in modern terrestrial glacigenic sediments. *Sedimentology*, 33, 699-710.

Dowdeswell, J. A., Kenyon, N. H., Elverhoi, A., Laberg, J. S., Hollender, F.-J., Mienert, J. and Siegert, M. J. 1996. Large-scale sedimentation on the glacier-influenced Polar North Atlantic margins: long-range side-scan sonar evidence. *Geophysical Research Letters*, 23, 3535-3538.

Dredge, L.A. 1982. Relict ice-scour marks and late phases of Lake Agassiz in northernmost Manitoba. *Canadian Journal of Earth Science*, 19, 1079-1087.

Dugmore, A.J. 1987. *Holocene Glacier Fluctuations around Eyjafjallajökull,south Iceland: a Tephrochronological Study*. Unpublished Ph.D. Thesis, University of Aberdeen.

Dugmore, A.J. 1989. Tephrochronological studies of Holocene glacier fluctuations in south Iceland. In, Oerlemans, J. (ed.), *Glacier fluctuations and climatic change*. Dordrecht, Kluwer Academic Publishers, 37-55.

Dugmore, A.J. and Buckland, P. 1991. Tephrochronology and late Holocene soil erosion in South Iceland. In, Maizels, J.K. and Caseldine, C. (eds.), *Environmental Change in Iceland: Past and Present*. Dordrecht, Kluwer Academic Publishers, 147-160.

Dugmore, A.J. and Kirkbride, M.P. 2000. A 2,000 year long record of prehistoric, medieval and 'Little Ice Age' glacier advances in Iceland. In, Russell, A.J. and Marren, P.M. (eds.), Iceland 2000, Modern processes and past environments, International conference Keele University April 27th - 29th, 2000, Abstracts, *Keele University Department of Geography Occasional Papers Series*, 21, 27.

Dugmore, A. J., and Sugden, D.E. 1991. Do the anomalous fluctuations of Sólheimajökull reflect ice-divide migration? *Boreas*, 20, 105-113.

Dugmore, A.J., Newton, A.J., Larsen, G. and Cook, G.T. 2000. Tephrochronology, environmental change and the Norse settlement of Iceland. *Environmental Archaeology*, 5, 21-34.

Dyke, A.S. 1993. Landscapes of cold-centred Late Wisconsinan ice caps, Arctic Canada. *Progress in Physical Geography*, 17, 223-247.

Eggertsson, S. 1919. Ýmislegt smávegis viðvíkjandi Kötlugosinu 1918. (Notes on the Katla eruption in 1918.) *Eimreiðin*, 1919, 212-222.

Eggleton, R.A., Foudoulis C. and Farkevisser D. 1987. Weathering of basalt: changes in rock chemistry and mineralogy. *Clays and Clay Minerals*, 35, 161–169.

Egil's Saga, The Sagas of Icelanders, vol. I. 1997. Reykjavík, Leif Eiriksson Publishing.

Egilsson, J. 1856. Biskupa-annálar Jóns Egilssonar með formála, athugasemdum og fylgiskjölum eftir Jón Sigurðsson. *Safn til Íslands og íslenzkra bókmennta að fornu og nýju* I. Copenhagen, Hið íslenzka bókmenntafélag, 15-36.

Egloff, J. and Johnson, G. L. 1978. Erosional and depositional structures of the Southwest Iceland insular margin: thirteen geophysical profiles. In, Watkins, J. S., Montadert, L. and Dickerson, P. W. (eds.), *Geological and Geophysical Investigations of Continental Margins*. Tulsa, Oklahoma, AAPG, 43-63.

Einarsdóttir, Ó. 1964. *Studier i Kronologisk Metode i Tidlig Islandsk Historieskrivning*. Stockholm, Natur og Kultur, Bibliotheca Historica Lundensis.

Einarsson, Í. 1918. Skrá frá 1712 um eyddar jarðir í Öræfum. (An account from 1712 of deserted farms in the district Öræfi). *Blanda*, I. Reykjavík.

Einarsson, O. 1971. *Íslandslýsing Qualiscunque descriptio Islandiæ*. Reykjavík, Bókaútgáfa Menningarsjóðs.

Einarsson, P. 1991a. Umbrotin í Kröflu 1975-89. (The events at Krafla 1975-89.) In, Garðarsson, A, and Einarsson, Á. (eds.), Náttúra Mývatns. *Hið íslenska náttúrufræðifélag*, 96-139.

Einarsson, P. 1991b. Earthquakes and present-day tectonism in Iceland. *Tectonophysics*, 189, 261-279.

Einarsson, P., Brandsdóttir, P., Guðmundsson, M.T., Björnsson, H., Sigmundsson, F. and Grönvold, K. 1997. Center of the Iceland hotspot experiences volcanic unrest. *Eos*, 78, 374-375.

Einarsson, Þ. 1956. Frjógreining fjörumós úr Seltjörn. *Náttúrufræðingurinn*, 26, 194-198.

Einarsson, Þ. 1957. Tvö frjólínurit úr íslenzkum mýrum. *Ársrit Skógræktarfélags Íslands*, 89-97.

Einarsson, Þ. 1961. Pollenanalytische Untersuchungen zur spät- und postglazialen Klimageschichte Islands. *Sonderveröffentlichungen des Geologischen Institutes der Universität Köln*, 6, 5-52.

Einarsson, Þ. 1963. Vitnisburður frjógreiningar um gróður, veðurfar og landnám á Íslandi. *Saga*, 1962, 442-469.

Einarsson, Þ. 1964. Aldursákvarðanir á fornskeljum. (English Summary: Radiocarbon dating of subfossil shells.) In, Kjartansson, G., Þórarinsson, S. and Einarsson, Þ. (eds.), C14-aldursákvarðanir á sýnishornum varðandi íslenzka kvarterjarðfræði (English Summary: C14 datings of Quaternary deposits in Iceland). *Náttúrufræðingurinn*, 34, 127-134.

Einarsson, Þ. 1966. Physical aspects of subglacial eruptions. *Jökull*, 16, 167-174.

Einarsson, Þ. 1967. Zu der Ausdehnung der weichselzeitlichen Vereisung Nordislands. *Sonderveröffentlichungen des Geologischen Institutes der Universität Köln*, 13, 167-173.

Einarsson, Þ. 1968. *Jarðfræði. Saga bergs og lands*. Mál og menning, Reykjavík.

Einarsson, Þ. 1978. *Jarðfræði*. Mál og menning, Reykjavík.

Einarsson, Þ. 1994. *Myndun og mótun lands, jarðfræði*. Mál og menning, Reykjavík.

Einarsson, Þ. and Albertsson, K. J. 1988. The glacial history of Iceland during the past three million years. *Philosophical Transactions of the Royal Society of London*, 318, 637-644.

Eiríksson, J. and Símonarson, L. A. 1997. Fluctuations of the Weichselian Ice Sheet in SW Iceland: a glaciomarine sequence from Suðurnes, Seltjarnarnes. *Quaternary Science Reviews*, 16, 221-240.

Eiríksson, J., Geirsdóttir, Á. and Símonarson, L. A. 1991. A review of the Pleistocene stratigraphy of Reykjavík, Iceland. *Quaternary International*, 10-12, 143-150.

Eiríksson, J., Knudsen, K. L., Hafliðason, H. and Heinemeier, J. 2000b. Chronology of the late Holocene climatic events in the northern North Atlantic based on AMS ^{14}C dates and tephra markers from the volcano, Hekla, Iceland. *Journal of Quaternary Science*, 15, 573-580.

Eiríksson, J., Knudsen, K. L., Hafliðason, H. and Henriksen, P. 2000a. Late-glacial and Holocene paleoceanography of the North Iceland Shelf. *Journal of Quaternary Science*, 15, 23-42.

Eiríksson, J., Símonarson, L. A., Knudsen, K. L. and Kristensen, P. 1997. Fluctuations of the Weichselian ice sheet in SW Iceland: a glaciomarine sequence from Sudurnes, Seltjarnarnes. *Quaternary Science Reviews*, 16, 221-240.

Elíasson, S. 1977. Molar um Jökulsárhlaup og Ásbyrgi. *Náttúrufræðingurinn*, 47, 160-179.

Elliott, M., Labeyrie, L., Bond, G., Cortijo, E., Turon, J.-L., Tiseray, N. and Duplessy, J.-C. 1998. Millennial-scale iceberg discharges in the Irminger Basin during the last glacial period: relationship with the Heinrich events and environmental settings. *Paleoceanography*, 13, 433-446.

Erlendsson, Þ. 1997 (written 1840). Bjarnaness- og Hoffellssóknir 1840. In, Jónson, J. and Sigmundsson, S. (eds.) *Skaftafellsýsla. Sýslu- og sóknalýsingar Hins íslenska bókmenntafélags*. Reykjavík, Sögufélag, 41-69.

Eskilsson, C., Árnason, J.I. and Rosbjerg, D. 2002. Simulation of the jökulhlaup on Skeiðarársandur, Iceland, in November 1996 using MIKE-21 HD. In, Snorasson, Á., Finnsdóttir, H.P. and Moss, M. (eds.), *The Extremes of the Extremes: Extraordinary Floods*. IAHS Publication, 271, 37-43.

Evans, D.J.A. 1990. The effect of glacier morphology on surficial geology and glacial stratigraphy in a high arctic mountainous terrain. *Zeitschrift für Geomorphologie*, 34, 481-503.

Evans, D.J.A. 1999. Glacial debris transport and moraine deposition: a case study of the Jardalen cirque complex, Sogn-og-Fjordane, western Norway. *Zeitschrift für Geomorphologie*, 43, 203-234.

Evans, D.J.A. 2000. A gravel outwash/deformation till continuum, Skálafellsjökull, Iceland. *Geografiska Annaler*, 82A, 499-512.

Evans, D.J.A. 2003. Ice-marginal terrestrial landsystems: active temperate glacier margins. In, Evans, D.J.A. (ed.), *Glacial Landsystems*. London, Arnold, 12-43.

Evans, D.J.A. and Rea B.R. 1999. Geomorphology and sedimentology of surging glaciers: a land-systems approach. *Annals of Glaciology*, 28, 75-82.

Evans, D.J.A. and Rea, B.R. 2003. Surging glacier landsystem. In, Evans D.J.A. (ed.), *Glacial Landsystems.* London, Arnold, 258-288.

Evans, D.J.A. and Twigg D.R. 2000. *Breiðamerkurjökull 1998.* 1:30,000 Scale Map. University of Glasgow and Loughborough University.

Evans, D.J.A. and Twigg, D.R. 2002. The active temperate glacial landsystem: a model based on Breiðamerkurjökull and Fjallsjökull, Iceland. *Quaternary Science Reviews,* 21, 2143-2177.

Evans, D.J.A., Archer S. and Wilson D.J.H. 1999b. A comparison of the lichenometric and Schmidt hammer dating techniques based on data from the proglacial areas of some Icelandic glaciers. *Quaternary Science Reviews,* 18, 13-41.

Evans, D.J.A., Lemmen D.S. and Rea B.R. 1999a. Glacial landsystems of the southwest Laurentide Ice Sheet: modern Icelandic analogues. *Journal of Quaternary Science,* 14, 673-691.

Evans, D.J.A., Owen L.A. and Roberts D. 1995. Stratigraphy and sedimentology of Devensian (Dimlington Stadial) glacial deposits, east Yorkshire, England. *Journal of Quaternary Science,* 10, 241-265.

Evans, D.J.A., Rea B.R. and Benn D.I. 1998. Subglacial deformation and bedrock plucking in areas of hard bedrock. *Glacial Geology and Geomorphology* rp04/1998 - http:// ggg.qub.ac.uk/ggg/papers/full/1998/rp041998/rp04.html

Evans, D.J.A., Salt K.E. and Allen C.S. 1999c. Glacitectonized lake sediments, Barrier Lake, Kananaskis Country, Canadian Rocky Mountains. *Canadian Journal of Earth Sciences,* 36, 395-407.

Evenson, E.B. and Clinch J.M. 1987. Debris transport mechanisms at active alpine glacier margins: Alaska case studies. In, Kujansuu, R. and Saarnisto, M. (eds.), *INQUA Till Symposium, Finland 1985.* Geological Society of Finland, Special Paper 3, Espoo, 111-136.

Evenson, E.B., Lawson, D.E., Strasser, J.C., Larson, G.J., Alley, R.B., Ensminger S.L. and Stevenson W.E. 1999. Field evidence for the recognition of glaciohydrologic supercooling. In, Mickelson, D.M. and Attig, J.W. (eds.), *Glacial Processes: Past and Present.* Geological Society of America, Special Paper 337, 23-35.

Evenson, B., Lawson, D.E., Larson, G., Roberts, M., Knudsen, Ó., Russell, A.J., Alley, R.B. and Burkhart, P. 2001. *Glaciohydraulic supercooling and basal ice in temperate glaciers of Iceland.* GSA Annual Meeting and Exposition Abstracts with Programs, The Geological Society of America, 33, A-440.

Everest, J. and Bradwell, T. 2003. Buried glacier ice in southern Iceland and its wider significance. *Geomorphology,* 52, 347-358.

Eyles, N. 1979. Facies of supraglacial sedimentation on Icelandic and alpine temperate glaciers. *Canadian Journal of Earth Sciences,* 16, 1341-1361.

Eyles, N. 1983a. The glaciated valley landsystem. In, Eyles N. (ed.), *Glacial Geology.* Oxford, Pergamon, 91-110.

Eyles, N. 1983b. Modern Icelandic glaciers as depositional models for 'hummocky moraine' in the Scottish Highlands. In, Evenson E.B., Schlüchter C. and Rabassa J. (eds.), *Tills and Related Deposits.* Rotterdam, Balkema, 47-59.

Eyles, N., Sladen H.A. and Gilroy S. 1982. A depositional model for stratigraphic complexes and facies superimposition in lodgement till. *Boreas*, 11, 317-333.

Eyþórsson, J. 1931. On the present position of the glaciers in Iceland. Some preliminary studies and investigations in the summer 1930. *Vísindafélag Íslendinga*, X, 35 pp.

Eyþórsson, J. 1935. On the variations of glaciers in Iceland. Some studies made in 1931. *Geografiska Annaler* 1935, 1-2, 121-137.

Eyþórsson, J. 1950. Þykkt Vatnajökuls. (The thickness of Vatnajökull.) *Jökull*, 1, 1- 6.

Eyþórsson, J. 1956. Frá Norðurlandsjöklum. (On some Alpine Glaciers in the Northland.) *Jökull*, 6, 23-29.

Eyþórsson, J. 1957. Frá Norðurlandsjöklum. (On some Alpine Glaciers in the Northland.) *Jökull*, 7, 52-58.

Eyþórsson, J. 1963. Variation of Iceland glaciers 1931-1960. *Jökull*, 13, 31-33.

FAO, 1998. World reference base for soil resources. *World Soil Resources Reports*, 84, Rome, FAO.

Fahnestock, R. K. and Bradley, W. C. 1973. Knik and Matanuska Rivers, Alaska: a contrast in braiding. In, Morisawa, M. (ed.), *Fluvial Geomorphology*. Binghamton Symposia in Geomorphology, 4, 220-250.

Fairbanks, R. G. 1989. A 17,000-year glacio-sea level record: influence of glacial melting rates on the Younger Dryas event and deep-ocean circulation. *Nature*, 342, 637-642.

Fay, H. 2001. *The Role of Ice Blocks in the Creation of Distinctive Proglacial Landscapes During and Following Glacier Outburst Floods (Jökulhlaups)*. Unpublished Ph.D. Thesis, Keele University.

Fay, H. 2002a. Formation of ice block obstacle marks during the November 1996 glacier-outburst flood (jökulhlaup), Skeiðarársandur, southern Iceland. In, Martini I.P., Baker V.R. and Garzon G. (eds.), *Flood and Megaflood Deposits: Recent and Ancient*. Special Publication of the International Association of Sedimentologists, 85-98.

Fay, H. 2002b. Formation of kettle holes following a glacial outburst flood (jökulhlaup), Skeiðarársandur, southern Iceland. In, Snorrason, Á., Finnsdóttir, H.P. and Moss, M. (eds.), *The Extremes of the Extremes: Extraordinary Floods*. IAHS Publication Number 271, 205-210.

Fiacco, J.J., Þórðarson, Þ., Germani, M.S., Self, S., Palai, J.M., Withlow, S. and Grootes, P.M. 1994. Atmospheric aerosol loading and transport due to the 1783-84 Laki eruption in Iceland, interpreted from ash particles and acidity in the GISP2 ice core. *Quaternary Research*, 42, 231-240.

Finnsson, H. 1831. *Om Folkemangdens Formindskelse ved Uaar i Island. Oversat ved Haldor Einarsen*. Copenhagen, i Commission i den Schubotheske Boghandling.

Finnsson, H. 1970. *Mannfækkun af Hallærum*. Eyþórsson, J. and Nordal, J. (eds.). Reykjavík, Almenna Bókafélagið.

Foote, P. G. 1950-51. Sturlu Saga and its background. *Saga Book of the Viking Society* XIII, 207-37.

Foote, P. G. and Simpson, J. (eds.) 1961-64. *Snorri Sturluson, Heimskringla*. Translated by Samuel Laing, Revised edition with Introduction and Notes. 3 vols. London, Everyman's Library, 717, 847.

352

References

Fowler A.C. 1987. A theory of glacier surges. *Journal of Geophysical Research*, 92, 9111-9120.

Fowler, A. C. 1999. Breaking the seal at Grímsvötn. *Journal of Glaciology*, 45, 506-516.

Fowler, A.C. and Ng, F.S.L. 1996. The role of sediment transport in the mechanics of jökulhlaups. *Annals of Glaciology*, 22, 255–259.

Freysteinsson, S. 1969. Tungnárjökull. *Jökull*, 18, 371-388.

Friðriksdóttir, S.P. 1973. *Frjógreining á Jarðvegi úr Tjarnarveri og Sóleyjarhöfða.* Unpublished B.Sc. Thesis, Verkfræði- og raunvísindadeild H.Í.

Frisius, G. 1582. *Opera collecta*, Frankfurt 1582. 8vo.

Frogner, P., Gíslason, S.R. and Óskarsson N. 2001. Fertilizing potential of volcanic ash in ocean surface water. *Geology*, 29, 487–490.

Fuller, M. L. 1914. Geology of Long Island, New York. *US Geological Survey Professional Paper*, 82, 1-231.

Furnes, H. 1975. Experimental palagonitization of basaltic glass of varied composition. *Contribution to Mineralogy and Petrology*, 50, 105.

Furnes, H., Friðleifsson, I.B. and Atkins, F.B. 1980. Subglacial volcanics: on the formation of acid hyaloclastites. *Journal of Volcanology and Geothermal Research*, 8, 95-110.

Galon, R. 1973a. A synthetic description of deposits and landforms observed on the proglacial area of Skeiðarárjökull. Conclusions with regard to the age of the deposits and the way in which deglaciation is proceeding. *Geographica Polonica*, 26, 141-150.

Galon, R. 1973b. Geomorphological and geological analysis of the proglacial area of Skeiðarárjökull. *Geographica Polonica*, 26, 15-56.

Garðarsson, H. 1999. *Saga Veðurstofu Íslands*. Reykjavík, Mál og Mynd.

Geirsdóttir, Á. and Eiríksson, J. 1994. Growth of an intermittent ice-sheet in Iceland during the late Pliocene and early Pleistocene. *Quaternary Research*, 42, 115-130.

Geirsdóttir, Á., Harðardóttir, J. and Eiríksson, J. 1997. The depositional history of the Younger Dryas-Preboreal Búði moraines in South-Central Iceland. *Arctic and Alpine Research*, 29, 13-23.

Geirsdóttir, Á., Harðardóttir, J. and Sveinbjörnsdóttir, Á. 2000. Glacial extent and catastrophic meltwater events during the deglaciation of Southern Iceland. *Quaternary Science Reviews*, 19, 1749-1761.

Geitner, C., Grießer, M., Moran, A., Stötter, J., Waltle, E. and Wastl, M. 2003. Lichenometrische Erhebungen an aufgelassenen Höfen sowie erste Anwendung der Ergebnisse auf gletschergeschichtliche Fragestellungen in Nordisland. *Norden*, 15, 151-162.

Gellatly, A.F., Whalley, W.B. and Gordon J.E. 1986. Topographic control over recent glacier changes in southern Lyngen Peninsula, north Norway. *Norsk Geografisk Tidsskrift*, 41, 211-218.

Gellatly, A.F., Whalley, W.B., Gordon, J.E. and Hansom J.D. 1988. Thermal regime and geomorphology of plateau ice caps of northern Norway: observations and implications. *Geology*, 16, 983-986.

Germanoski, D. and Schumm, S. A. 1993. Changes in braided river morphology resulting from aggradation and degradation. *Journal of Geology*, 101, 451-466.

Gísladóttir, G. 1995. Ecological effects of grazing on subalpine common in south-western Iceland. *Latin America in the World: Environment, Society and Development.* Regional Conference of Latin America and Caribbean countries, International Geographical Union, Havana, 156.

Gísladóttir, G. 1998. Environmental characterisation and change in South-western Iceland. *Department of Physical Geography, Stockholm University, Dissertation Series,* 10, Stockholm.

Gísladóttir, G. 2001. Ecological disturbance and soil erosion on grazing land in Southwest Iceland. In, Conacher A. (ed.), *Land Degradation.* Dordrecht, Kluwer Academic Publishers, 109-126.

Gíslason, S.R. 1989. Kinetics of water-air interactions in rivers: a field study in Iceland. In, Miles, D.L., (ed.), *Water-Rock Interactions.* Rotterdam, Balkema, 263–266.

Gíslason, S.R. and Arnórsson, S. 1990. Saturation state of natural waters in Iceland relative to primary and secondary minerals in basalts. In, Spencer, R.J. and Chou, I.-M., (eds.), Fluid Mineral Interactions: A Tribute to H.P. Eugster. *Geochemical Society, Special Publication* No. 2. Texas: The Geochemical Society, 373–393.

Gíslason, S.R. and Arnórsson, S. 1993. Dissolution of primary basaltic minerals in natural waters: saturation state and kinetics. *Chemical Geology,* 105, 117–135.

Gíslason, S.R. and Eugster H.P. 1987a. Meteoric water-basalt interactions. I: A laboratory study. *Geochimica et Cosmochimica Acta,* 51, 2841–2855.

Gíslason, S.R. and Eugster H.P. 1987b. Meteoric water-basalt interactions. II: A field study in N.E. Iceland. *Geochimica et Cosmochimica Acta,* 51, 2841–2855.

Gíslason, S.R. and Oelkers, H.E. 2003. The mechanism, rates and consequences of basaltic glass dissolution: II. An experimental study of the dissolution rates of basaltic glass as a function of pH, and temperature. *Geochimica et Cosmochimica Acta,* 67, 3817-3832.

Gíslason, S.R., Arnórsson, S. and Ármannsson, H. 1996. Chemical weathering of basalt in SW Iceland: Effects of runoff, age of rocks and vegetative/glacial cover. *American Journal of Science,* 296, 837-907.

Gíslason, S.R., Eiríksdóttir, E.S., Stefánsdóttir, M.B. and Stefánsson A. 1999. Chemistry of soil and river waters in the vicinity of the industrial centre at Grundartangi, western Iceland. Final report, July 15 1999 (in Icelandic).

Gíslason, S.R, Guðmundsson, B.T. and Eiríksdóttir, E.S. 1998. *Chemistry of River Elliðaár 1997-1998.* Reykjavík: Science Institute, University of Iceland, RH–19–98.

Gíslason, S.R., Veblen, D.R. and Livi, K.J.T. 1993. Experimental meteoric water-basalt interactions: Characterization and interpretation of alteration products. *Geochimica et Cosmochimica Acta,* 57, 1459–1471.

Gíslason, S.R., Heaney, P.J., Oelkers, E.H. and Schott, J. 1997. Kinetic and thermodynamic properties of moganite, a novel silica polymorph. *Geochimica et Cosmochimica Acta,* 61, 1193–1204.

Gíslason, S.R, Snorrason, Á., Eiríksdóttir, E.S., Sigfússon, B. Elefsen, S.Ó., Harðardóttir, J., Gunnarsson, Á., Hreinsson, E.Ö., Torsander, P., Kardjilov M.I. and Óskarsson, N.Ö. 2002a. *Chemistry, Discharge and Suspended Matter in Rivers in Eastern Iceland, II.* Gagnagrunnur Raunvísindastofnunar og Orkustofnunar. Reykjavík: Science Institute, University of Iceland, RH-11-2002.

Gíslason, S.R, Snorrason, Á., Eiríksdóttir, E.S., Sigfússon, B., Elefsen, S.Ó., Harðardóttir, J. Gunnarsson, Á. and Torsander, P. 2002b. *Chemistry, Discharge and Suspended Matter in Rivers in Southern Iceland, V.* Gagnagrunnur Raunvísindastofnunar og Orkustofnunar. Reykjavík, Science Institute, University of Iceland, RH-12-2002.

Gomez, B., Russell, A. J., Smith, L. C. and Knudsen, Ó. 2002. Erosion and deposition in the proglacial zone: the 1996 jökulhlaup on Skeiðarársandur, southeast Iceland. In, Snorasson, Á., Finnsdóttir, H. P. and Moss, M. (eds.), *The Extremes of the Extremes: Extraordinary Floods.* IAHS Special Publication 271, 217-221.

Gomez, B., Smith, L. C., Magilligan, F. J., Mertes, F. A. K. and Smith, N. D. 2000. Glacier outburst floods and outwash plain development: Skeiðarársandur, Iceland. *Terra Nova*, 12, 126-131.

Gould, S.J. 1985. *The Flamingo's Smile. Reflections in Natural History.* Harmondsworth, Penguin Books.

Graf, H.-F. 1992. Arctic radiation deficit and climate variability. *Climate Dynamics*, 7, 19-28.

Grattan, J. P. 1995. The distal impact of volcanic gases and aerosols in Europe: a review of the phenomenon and assessment of vulnerability in the late 20th century. In, Penn, S. and Culshaw, M. G. (eds.), *Geohazards and Engineering Geology.* London, The Geological Society, 123-133.

Grattan, J. P. and Brayshay, M. B. 1995. An amazing and portentous summer: Environmental and social responses in Britain to the 1783 eruption of an Iceland volcano. *Geographical Journal*, 161, 125-134.

Grattan, J. P. and Pyatt, F. B. 1994. Acid damage in Europe caused by the Laki Fissure eruption - an historical review. *Science of the Total Environment*, 151, 241-247.

Grönvold, K. and Jóhannesson, H. 1984. Eruption in Grímsvötn: course of events and chemical studies of the tephra. *Jökull*, 34, 1-11.

Grönvold, K., Óskarsson, N., Johnsen, S.J., Clausen, H.B., Hammer, C.U., Bond, G. and Bard, E. 1995. Ash layers from Iceland in the Greenland GRIP ice core correlated with oceanic and land sediments. *Earth and Planetary Science Letters*, 135, 149-155.

Grove, J. M. 1988. *The Little Ice Age.* London and New York, Routledge.

Groveman, B. S. and Landsberg, H. E. 1979. *Reconstruction of Northern Hemisphere temperature: 1579-1880.* Institute for Fluid Dynamics and Applied Mathematics, University of Maryland.

Guðbergsson, G. 1975. Soil formation in Skagafjörður, northern Iceland. *Journal of Agricultural Research in Iceland*, 7, 20-45.

Guðbergsson, G. 1992. Skógar í Skagafirði. *Skógræktarritið*, 1992.

Guðmundsson, Á. 1981. The Vogar fissure swarm, Reykjanes peninsula, SW-Iceland. *Jökull*, 30, 43-64.

Guðmundsson, H. J. 1997. A review of the Holocene environmental history of Iceland. *Quaternary Science Reviews*, 16, 81-92.

Guðmundsson, H.J. 1998. Holocene glacier fluctuations of the Eiríksjökull ice cap, west central Iceland. *Jökull*, 46, 17-28.

Guðmundsson, M.T. 1989. The Grímsvötn caldera, Vatnajökull: Subglacial topography and structure of caldera infill. *Jökull*, 39, 1-19.

Guðmundsson, M.T. 2003. Melting of ice by magma-ice-water interactions during subglacial eruptions as an indicator of heat transfer in subaqueous eruptions. In, White, J.D.L., Smellie, J.L and Clague, D. (eds.), *Explosive Subaqueous Volcanism*. AGU monograph, in press.

Guðmundsson, M.T. and Björnsson, H. 1991. Eruptions in Grímsvötn, Vatnajökull, Iceland, 1934-1991. *Jökull*, 41, 21-45.

Guðmundsson, M.T., Björnsson, H. and Pálsson, F. 1995. Changes in jökulhlaup sizes in Grímsvötn, Vatnajökull, Iceland, 1934-1991, deduced from in situ measurements of subglacial lake volume. *Journal of Glaciology*, 41, 263-272.

Guðmundsson, M. T., Bonnel, A. and Gunnarsson, K. 2003. Seismic soundings of sediment thickness on Skeiðarársandur, SE-Iceland. *Jökull*, 51, 53-64.

Guðmundsson, M.T., Erlingsson, S. and Björnsson, H. 1996. Effects of variable surface loading on volcanic and intrusive activity at the Grímsvötn volcano, Iceland. In, Þorkelsson, B. (ed.), *European Seismological Commission, XXV General Assembly, Abstracts*, 101.

Guðmundsson, M.T., Sigmundsson, F. and Björnsson, H. 1997. Ice-volcano interaction of the 1996 Gjálp subglacial eruption, Vatnajökull, Iceland. *Nature*, 389, 954-957.

Guðmundsson, M.T., Larsen, G., Sigmundsson, F. and Björnsson, H. 2000. Comment: subglacial eruptions and synthetic aperture radar images. *Eos*, 81, 134-135.

Guðmundsson, M.T., Pálsson, F., Björnsson, H. and Högnadóttir, Þ. 2002. The hyaloclastite ridge formed in the subglacial 1996 eruption in Gjálp, Vatnajökull, Iceland: present day shape and future preservation. In, Smellie, J.L., and Chapman, M. (eds.), *Ice-volcano interaction on Earth and Mars*. Geological Society of London Special Publication, 202, 319-335.

Guðmundsson, M.T., Sigmundsson, F., Björnsson, H. and Högnadóttir, Þ. 2003. The 1996 eruption at Gjálp, Vatnajökull ice cap, Iceland: efficiency of heat transfer, ice deformation and subglacial water pressure. *Bulletin of Volcanology*, DOI: 10.1007/s00445-003-0295-9.

Guðnason, B. (ed.) 1957. Sýslulýsingar 1744-1740. *Sögurit* 28. Reykjavík, Sögufélag.

Guicharnaud, R. 2002. *Bulk Density of Icelandic Soils*. Unpublished B.Sc. Thesis, University of Iceland, Reykjavík.

Gunnarson, G. 1980. *A Study of Causal Relations in Climate and History. With an emphasis on the Icelandic experience*. Lund, Meddelande Från Ekonomisk-Historiska Institutionen Lunds Universitet, 17.

Gustavson, T.C. 1974. Sedimentation on gravel outwash fans, Malaspina foreland, Alaska. *Journal of Sedimentary Petrology*, 44, 374-389.

Gustavson, T.C. and Boothroyd J.C. 1987. A depositional model for outwash, sediment sources, and hydrologic characteristics, Malaspina Glacier, Alaska: a modern analog of the southeastern margin of the Laurentide Ice Sheet. *Geological Society of America Bulletin*, 99, 187-200.

Guy, C. and Schott, J. 1989. Multisite surface reaction versus transport control during the hydrolysis of a complex oxide. *Chemical Geology*, 78, 181–204.

Häberle, T. 1991a. *Spät- und postglaziale Gletschergeschichte des Hörgárdalur-Gebietes, Tröllaskagi, Nordisland*. Ph.D thesis, University of Zürich.

Häberle, T. 1991b. Holocene glacial history of the Hörgárdalur area, Tröllaskagi, northern Iceland. In, Maizels, J.K. and Caseldine, C. (eds.), *Environmental Change in Iceland: Past and Present*. Dordrecht, Kluwer, 193-202.

Häberle, T. 1994. Glacial, Late Glacial and Holocene history of the Hörgárdalur area, Tröllaskagi, Northern Iceland. In, Stötter, J. and Wilhelm, F. (eds.), *Environmental Change in Iceland*. Münchener Geographische Abhandlungen B12, 133-145.

Haflidason, H. 1983. The Marine Geology of Eyjafjörður, North Iceland: Sedimentology, Petrographical and Stratigraphical Studies. Unpublished M.Phil. Thesis, University of Edinbugh.

Haflidason, H., Eiríksson, J. and van Kreveld, S. 2000. The tephrochronology of Iceland and the North Atlantic region during the Middle and Late Quaternary: a review. *Journal of Quaternary Science*, 15, 3-22.

Haflidason, H., King, E. L. and Sejrup, H. P. 1998. Late Weichselian and Holocene sediment fluxes of the northern North Sea Margin. *Marine Geology*, 152, 189-215.

Hagen, S. 1995. *Watermass Characteristics and Climate in the Nordic Seas During the Last 10,200 Years*. Candidatus Scientiarum Thesis, University of Tromsø, Tromsø, Norway.

Hagen, S. 1999. *North Atlantic Paleoceanography and Climate History During the Last ~ 70 cal. ka Years*. Ph.D. Thesis, University of Tromsø.

Håkansson, S. 1983. A reservoir age for the coastal waters of Iceland. *Geologiska Föreningens i Stockholm Förhandlingar*, 105, 65-68.

Hakluyt's Voyages 1928. Vol. 1 of the Foreign Voyages and Vol. 9 of the Principal Navigations, London-Toronto-New York. (First published in 1598.)

Hald, M., and Hagen, S. 1998. Early Preboreal cooling in the Nordic seas region triggered by meltwater. *Geology*, 26, 615-618.

Hallgrímsson, H. 1972. Hlaupið Í Teigadalsjökli Í Svafaðardalur. (The burst of Teigadalsjökull, Svarfaðardalur, N.Iceland.) *Jökull*, 22, 79-82.

Hallsdóttir, M. 1982. Frjógreining tveggja jarðvegssniða úr Hrafnkelsdal. In, Þórarinsdóttir, H., Óskarsson, Ó.H., Steinþórsson. S. and Einarsson, Þ. (eds.), *Eldur er í norðri*. Reykjavík, Sögufélag.

Hallsdóttir, M. 1984. Frjógreining tveggja jarðvegssniða á Heimaey. *Árbók hins íslenzka fornleifafélags* 1983, 48-68.

Hallsdóttir, M. 1987. Pollen analytical studies of human influence on vegetation in relation to the Landnám tephra layer in southwest Iceland. *LUNDQUA thesis* 18, Lund University, Department of Quaternary Geology, 46 pp.

Hallsdóttir, M. 1991. Studies in the vegetational history of north Iceland, a radiocarbon-dated pollen diagram from Flateyjardalur. *Jökull*, 40, 67-81.

Hallsdóttir, M. 1992. Saga lands og gróðurs. In, Nielsen, Ó. K. (ed.), *Tjörnin, saga og lífríki*. Reykjavík, Reykjavíkurborg.

Hallsdóttir, M. 1993. Frjórannsókn á mósniðum úr Viðey. RH-08-93. Reykjavík, *Raunvísindastofnun og Árbæjarsafn.*

Hallsdóttir, M. 1995. On the pre-settlement history of Icelandic vegetation. *Icelandic Agricultural Sciences,* 9, 17-29.

Hallsdóttir, M. 1996. Synthesis of the Holocene history of vegetation in northern Iceland. In, Vorren, K.-D., Alm, T. and Birks, H. (eds.), Holocene Treeline Oscillations, Dendrochronology and Palaeoclimate, *Paläoklimaforschung,* 20, 203-214.

Hallsdóttir, M., Larsen, G., Róbertsdóttir, B., Eiríksson, J., Geirsdóttir, Á. and Norðdahl, H. 1998. Middle and Late Holocene vegetation history of the Holt district, Southern Iceland. In, Wilson, J.R. (ed.), *Nordic Geological Winter Meeting, Abstracts,* 13-16.

Hamilton S.J. and Whalley W.B. 1995. Preliminary results from the lichenometric study of the Nautardalur rock glacier, Tröllaskagi, northern Iceland. *Geomorphology,* 12, 123-132.

Hammer, C. U. 1977. Past volcanism revealed by Greenland ice sheet impurities. *Nature,* 270, 482-486.

Hammer, C. U. 1984. Traces of Icelandic eruptions in the Greenland ice sheet. *Jökull,* 34, 51-65.

Hammer, C. U., Clausen, H. B. and Dansgaard, W. 1980. Greenland ice sheet evidence of post-glacial volcanism and its climatic impact. *Nature,* 288, 230-235.

Hansen, B., Jónsson S., Turrell W.R. and Østerhus, S. 2000. Seasonal variations in the Atlantic water inflow to the Nordic Seas. *ICES CM L:03,* 1-15.

Hansom, J.D. and Briggs, D.J. 1991. Sea-level change in Vestfirðir, northwest Iceland. In, Maizels, J.K. and Caseldine, C. (eds.), *Environmental Change in Iceland: Past and Present.* Dordrecht, Kluwer Academic Publishers, 79-91.

Haraldsson, H. 1981. The Markarfljót sandur area, southern Iceland: sedimentological, petrographical and stratigraphical studies. *Striæ,* 15, 1-65.

Harðardóttir, J., Geirsdóttir, Á. and Sveinbjörnsdóttir, Á. E. 2001. Seismostratigraphy and sediment studies of Lake Hestvatn, southern Iceland: Implications for the deglacial history of the region. *Journal of Quaternary Science,* 16, 167-179.

Harington, C.A.R. (ed.) 1992. *The Year Without a Summer? World Climate in 1816.* Ottawa, Canadian Museum of Nature.

Harris, C. and Bothamley K. 1984. Englacial deltaic sediments as evidence for basal freezing and marginal shearing, Leirbreen, Norway. *Journal of Glaciology,* 30, 30-34.

Harris, R.L. 1978. William Morris, Eiríkur Magnússon, and the Icelandic famine relief efforts of 1882. *Saga-Book of the Viking Society,* 31-41.

Hart, J. K. 1994. Proglacial glaciotectonic deformation at Melarbakkar – Ásbakkar, west Iceland. *Boreas,* 23, 112-121.

Hart, J.K. 1995. Recent drumlins, flutes and lineations at Vestari-Hagafellsjökull, Iceland. *Journal of Glaciology,* 41, 596-606.

Hauksson, Þ. (ed.) 1972. *Árna Saga Biskups,* Reykjavík, Stofnun Árna Magnússonar.

Hávarðsson, S. and Jónsson, T. 1997. *Veðurhandrit. Könnun og skráning veðurfræðilegra gagna Handritadeild Landsbókasafns og Þjóðskjalasafns Ísland.* (Weather observations in Iceland before 1872. A short summary of early Icelandic instrumental observations.) Reykjavík, Veðurstofa Íslands, V-G97039-ÚR30.

Helgadóttir, G. 1984. Senkvartaere Foraminifer og Sedimenter i Faxaflói-Jökuldjúpomradet Vest for Island. Ph.D. Thesis, University of Oslo.

Helgadóttir, G. 1997. *Paleoclimate (0 to >14 ka) of W. and NW Iceland: An Iceland/ USA Contribution to P.A.L.E., Cruise Report B9-97*. Marine Research Institute of Iceland. Hafrannsóknastofnun Fjölrit Nr. 62.

Helgadóttir, G. and Andrews, J. T. 1999. Late Quaternary shifts in sediment north of Iceland: Foraminiferal data and stable oxygen isotope analyses. *Geological Society of America Abstracts*, 31, A314.

Helgadóttir, G. and Thors, K. 1998. Setlög í Ísafjarðardjúpi, Jökulfjörðum og Djúpál. *Geoscience Society of Iceland, Spring Meeting 1998*, 12.

Helgason, A. 2001. *The Ancestry and Genetic History of the Icelanders: an Analysis of MTDNA Sequences, Y Chromosome Haplotypes and Genealogy*. Unpublished M.S. Thesis, Institute of Biological Anthropology, University of Oxford.

Helgason, A., Sigurðardóttir, S., Gulcher, J.R., Ward, R. and Stefánsson, K. 2000. MTDNA and the origin of the Icelanders: deciphering signals of recent population history. *American Journal of Human Genetics*, 66, 999-1016.

Helgason, J. 1978. *Biskupa Sögur*, Copenhagen, Editiones Arnamagnæanæ Series A, vol. 13, 2.

Hellund, A. 1882-83. *Islændingen Sveinn Pálssons beskrivelser af islandske vulkaner og bræer*, 2 pts, Kristiania, Separataftryk af 'Turistforeningens Årbog' for 1882.

Hermannsson, H. 1926. Two cartographers Guðbrandur Þorláksson and Þórður Þorláksson. *Islandica* XVII, Ithaca, New York, Cornell University Library.

Hertzfeld, U.C. and Mayer, H. 1997. Surge of Bering Glacier and Bagley Ice Field, Alaska: an up-date to August 1995 and an interpretation of brittle-deformation patterns. *Journal of Glaciology*, 43, 427-734.

Hesse, R. 1995. Continental slope and basin sedimentation adjacent to an ice-margin: a continous sleeve-gun profile across the Labrador Slope, Rise and Basin. In, Pickering, K. T. *et. al.* (eds.), *Atlas of Deep Water Environments Architectural Style in Turbidite Systems*, 14-17.

Hesse, R. and Khodabakhsh, S. 1998. Depositional facies of late Pleistocene Heinrich events in the Labrador Sea. *Geology*, 26, 103-106.

Hicock, S.R. 1992. Lobal interactions and rheologic superposition in subglacial till near Bradtville, Ontario, Canada. *Boreas*, 21, 73-88.

Hicock, S.R. and Dreimanis, A. 1989. Sunnybrook drift indicates a grounded early Wisconsin glacier in the Lake Ontario basin. *Geology*, 17, 169-172.

Hicock, S.R. and Dreimanis, A. 1992. Sunnybrook drift in the Toronto area, Canada: re-investigation and re-interpretation. In, Clark P.U. and Lea, D. (eds.), *The Last Interglacial-Glacial Transition in North America*. Geological Society of America, Special Paper 270, 139-161.

Hicock, S.R. and Fuller, E.A. 1995. Lobal interactions, rheologic superposition, and implications for a Pleistocene ice stream on the continental shelf of British Columbia. *Geomorphology*, 14, 167-184.

Hjálmarsson, A.H., Jóhannesson, H., Björnsson, H., Kristinsson, H. and Kristinsson, M. 1991. *Ferðafélag Íslands Árbók 1991. Fjalllendi Eyjafjarðar að vestanverðu II.* Reykjavík, Ferðafélag Íslands.

Hjartarson, Á. 1988. Þjórsárhraunið mikla - stærsta nútímahraun jarðar. (The Great Þjórsá lava - largest Holocene lava on Earth.) *Náttúrufræðingurinn*, 58, 1-16.

Hjartarson, Á. 1989. The ages of the Fossvogur layers and the Álftanes end-moraine, SW-Iceland. *Jökull*, 39, 21-31.

Hjartarson, Á. 1993. Ísaldarlok í Reykjavík. (English summary: The deglaciation of Reykjavík.). *Náttúrufræðingurinn*, 62, 209-219.

Hjartarson, Á. 1999. Jökulrákir í Reykjavík. (English summary: Glacial striae in Reykjavík.) *Náttúrufræðingurinn*, 68, 155-160.

Hjartarson, Á. 2000. Arnarfellsmúlar. *Náttúrufræðingurinn*, 70, 57-64.

Hjartarson, Á. and Ingólfsson, Ó. 1988. Preboreal glaciation of Southern Iceland. *Jökull*, 38, 1-16.

Hjartarson, Á., Sigurðsson, F. and Hafstað, Þ. 1981. *Vatnabúskapur Austurlands III. Lokaskýrsla.* OS81006/VOD04, Reykjavík, Orkustofnun.

Hjort, C., Ingólfsson, Ó. and Norðdahl, H. 1985. Late Quaternary geology and glacial history of Hornstrandir, Northwest Iceland: a reconnaissance study. *Jökull*, 35, 9-29.

Hjulström, F. 1954. The geomorphology of the alluvial outwash plains (sandur) of Iceland and the mechanics of braided rivers. *Proceedings 25th Congress, IGU, Washington*, 337-342.

Hoffmann, D. J. 1987. Perturbations to the global atmosphere associated with the El Chichón volcanic eruption of 1982. *Review of Geophysics*, 25, 743-759.

Holland, H. D. 1978. *The Chemistry of the Atmosphere and Oceans.* New York, John Wiley & Sons.

Hon, K., Kauahikaua, J., Denlinger, R. and Mackay, K. 1994. Emplacement and inflation of pahoehoe sheet flows: observations and measurements of active lava flows on Kilauea volcano, Hawaii. *Bulletin of the Geological Society of America*, 106, 351-370.

Hoppe, G. 1968. Grímsey and the maximum extent of the last glaciation of Iceland. *Geografiska Annaler*, 50, 16-24.

Hoppe, G. 1982. The extent of the last inland ice sheet of Iceland. *Jökull*, 32, 3-11.

Hoppe, H. 1940. Untersuchungen an Palagonittuffen und ihre Bildungsbedingungen. *Chemie der Erde*, 13, 484–514.

Horrebow, N. 1752. *Tilforladelige Efterretninger om Island, med et nyt Landkort og 2 Aars meteorologiske Observationer.* Copenhagen.

Horrebow, N. 1758. *The natural history of Iceland ... Interspersed with an account of the island, by Mr. Anderson. To which is added, a meteorological table, with remarks. Translated from the Danish original. And illustrated with a new general map of the island.* London, A. Linde.

Höskuldsson, Á. 2000. Gjóskuflóð úr Eyjafjallajökli? (Pyroclastic flows from Eyjafjallajökull?) *Jarðfræðafélag Íslands*, Febrúarráðstefna 2000, abstracts, 1.

Höskuldsson, Á. and Sparks, R.S.J. 1997. Thermodynamics and fluid dynamics of effusive subglacial eruptions. *Bulletin of Volcanology,* 59, 219-230.

Howarth, P.J. 1968. *Geomorphological and Glaciological Studies, Eastern Breiðamerkurjökull, Iceland.* Unpublished Ph.D. Thesis, University of Glasgow.

Howarth, P.J. 1971. Investigations of two eskers at eastern Breiðamerkurjökull, Iceland. *Arctic and Alpine Research,* 3, 305-318.

Howarth, P.J. and Price R.J. 1969. The proglacial lakes of Breiðamerkurjökull and Fjallsjökull, Iceland. *Geographical Journal,* 135, 573-581.

Howarth, P.J. and Welch R. 1969a. *Breiðamerkurjökull, South-east Iceland, August 1945.* 1:30,000 scale map. University of Glasgow.

Howarth, P.J. and Welch R. 1969b. *Breiðamerkurjökull, South-east Iceland, August 1965.* 1:30,000 scale map. University of Glasgow.

Hróarsson, B. 1992. *Ferðafélag Íslands Árbók 1992. Norðan byggða milli Eyjafjarðar og Skjálfanda.* Reykjavík, Ferðafélag Íslands.

Hu, F.S., Slawinski, D., Wright Jr., H. E., Ito, E., Johnson, R. G., Kelts, K. R., McEwan, R. F. and Boedigheimer, A. 1999. Abrupt changes in North American climate during early Holocene times. *Nature*, 400, 437-440.

Hubbard, A., Sugden, D., Dugmore, A., Norðdahl, H. and Pétursson, H. G. in press: A modelling insight into the Icelandic Late Glacial Maximum ice sheet. *Quaternary Science Reviews.*

Hubbard, B. and Sharp M.J. 1989. Basal ice formation and deformation: a review. *Progress in Physical Geography,* 13, 529-558.

Humlum, O. 1985. Genesis of an imbricate push moraine, Höfðabrekkujökull, Iceland. *Journal of Geology,* 93, 185-195.

Humphrey, N.F. and Raymond, C.F. 1994. Hydrology, erosion and sediment production in a surging glacier: Variegated Glacier, Alaska, 1982-83. *Journal of Glaciology,* 40, 539-552.

Humphrey, N.F., Raymond, C.F. and Harrison, W.D. 1986. Discharges of turbid water during mini-surges of Variegated Glacier, Alaska, U.S.A. *Journal of Glaciology,* 32, 195-207.

Hurrell, J.W. 1995. Decadal trends in the North Atlantic Oscillation and relationships to regional temperature and precipitation. *Science,* 269, 676-679.

Indriðason, H. D. 1997. *Fjörukambar og áflæði á Ásgarðsgrundum í Hvammsfirði,* Unpublished B.S. Thesis, University of Iceland, Reykjavík.

Ingólfsson, Ó. 1984. A review of Late Weichselian studies in the lower part of the Borgarfjörður region, western Iceland. *Jökull,* 34, 117-130.

Ingólfsson, Ó. 1985. Late Weichselian glacial geology of the lower Borgarfjörður region, Western Iceland: A preliminary report. *Arctic,* 38, 210-213.

Ingólfsson, Ó. 1987. The Late Weichselian glacial geology of the Melabakkar – Ásbakkar coastal cliffs, Borgarfjörður, W-Iceland. *Jökull,* 37, 57-80.

Ingólfsson, Ó. 1988. Glacial history of the lower Borgarfjörður area, western Iceland. *Geologiska Föreningens i Stockholm Förhandlingar,* 110, 293-309.

Ingólfsson, Ó. 1991a. A review of the Late Weichselian and Early Holocene glacial and environmental history of Iceland. In, Maizels, J. K. and Caseldine, C. (eds.), *Environmental Changes in Iceland: Past and Present*. Dordrecht, Kluwer Academic Publishers, 13-29.

Ingólfsson, Ó. 1991b. 'On glaciers in general and particular'...The life and works of an Icelandic pioneer in glacial research. *Boreas*, 20, 79-84.

Ingólfsson, Ó. and Norðdahl, H. 1994. A review of the environmental history of Iceland, 13,000-9000 yrs BP. *Journal of Quaternary Science*, 9, 147-150.

Ingólfsson, Ó. and Norðdahl, H. 2001. High relative sea level during the Bölling Interstadial in western Iceland: a reflection of ice-sheet collapse and extremely rapid glacial unloading. *Arctic, Antarctic, and Alpine Research*, 33, 231-243.

Ingólfsson, Ó., Norðdahl, H. and Hafliðason, H. 1995. A rapid isostatic rebound in South-western Iceland at the end of the last glaciation. *Boreas*, 24, 245-259.

Ingólfsson, Ó., Björck, S., Hafliðason, H. and Rundgren, M. 1997. Glacial and climatic events in Iceland reflecting regional North Atlantic climatic shifts during the Pleistocene-Holocene transition. *Quaternary Science Reviews*, 16, 1135-1144.

Ísaksson, S. P. 1985. Stórhlaup í Jökulsá á Fjöllum á fyrri hluta 18. aldar. *Náttúrufræðingurinn*, 54, 165-191.

Ísleifsson, S. 1996. *Ísland, framandi land*. Reykjavík, Mál og Menning.

Ísólfsson, I. 1998. Úr myndasafni Ingólfs Ísólfssonar. (From the photo collection of Ingólfur Ísólfsson.) *Jökull*, 46, 62.

Ives J.D., Andrews, J.T. and Barry, R.G. 1975. Growth and decay of the Laurentide Ice Sheet and comparisons with Fenno-Scandinavia. *Naturwissenschaften*, 62, 118-125.

Jacoby, G. C., Workman, K. W. and d'Arrigo, R.D. 1999. Laki eruption of 1783, tree rings, and disaster for northwest Alaska Inuit. *Quaternary Science Reviews*, 18, 1365-1371.

Jakobsson, S.P. 1972. Chemistry and distribution pattern of Recent basaltic rocks in Iceland. *Lithos*, 5, 365–385.

Jakobsson, S.P. 1978. Environmental factors controlling the palagonitization of the Surtsey tephra, Iceland. *Bulletin of the Geological Society of Denmark*, 27, 91-105.

Jakobsson, S. P. 1979a. Petrology of recent basalts of the Eastern Volcanic Zone, Iceland. *Acta Naturalia Islandica*, 26, 1-103.

Jakobsson, S.P. 1979b. Outline of the petrology of Iceland. *Jökull*, 29, 57-73.

Jakobsson, S.P. and Moore, J.G. 1982. The Surtsey drilling project of 1979. *Surtsey Research Programme Report*, 9, 76-93.

James, E. B. 1999. *Sediment Analysis in Reykjafjarðaráll Trough, Northern Iceland*. Unpublished B.A. Honors Thesis, University of Colorado.

Jennings, A. E., and Helgadóttir, G. 1994. Foraminiferal assemblages from the fjords and shelf of Eastern Greenland. *Journal of Foraminiferal Research*, 24, 123-144.

Jennings, A.E. and Weiner, N.J. 1996. Environmental change in eastern Greenland during the last 1300 years: evidence from foraminifera and lithofacies in Nansen Fjord, 68 degrees N. *The Holocene*, 6, 179-191.

Jennings, A.E., Grönvold, K., Smith, L.M., Hald, M. and Hafliðason, H. 2002. High resolution study of Icelandic tephras in the Kangerlussuaq Trough, Southeast Greenland during the last deglaciation. *Journal of Quaternary Science*, 17, 747-757.

Jennings, A. E., Hagen, S., Harðardóttir, J., Stein, R., Ogilvie, A. E. J. and Jónsdóttir, I. 2001. Oceanographic change and terrestrial human impacts in a post 1400 AD record from the southwest Iceland shelf. *Climatic Change*, 48, 83-100.

Jennings, A.E., Smith, L.M., Andrews, J.T., Syvitski, J.P.M. and Hald, M. (subm.) Late Weischelian ice extent and deglaciation of the Kangerlussuaq region of the East Greenland shelf.

Jennings, A. E., Syvitski, J. P. M., Gerson, L., Grönvold, K., Geirsdóttir, Á., Harðardóttir, J., Andrews, J. T. and Hagen, S. 2000. Chronology and paleoenvironments during the late Weichselian deglaciation of the SW Iceland Shelf. *Boreas*, 29, 167-183.

Jennings, A. E., Tedesco, K. A., Andrews, J. T. and Kirby, M. E. 1996. Shelf erosion and glacial ice proximity in the Labrador Sea during and after Heinrich events (H-3 or 4 to H-0) as shown by formanifera. In, Andrews, J. T., Austin, W. E. N., Bergsten, H., and Jennings, A. E. (eds.), *Late Quaternary Palaeoceanography of the North Atlantic Margins*. Geological Society Special Publications, 29-49.

Jewtuchowicz, S. 1971. The present-day marginal zone of Skeiðarárjökull. *Acta Geographica Lodziensia*, 27, 43-52.

Jewtuchowicz, S. 1973. The present-day marginal zone of Skeiðarárjökull. *Geographica Polonica*, 26, 115-137.

Jiang, J., Seidenkrantz, M.-S., Knudsen, K.L. and Eiríksson, J. 2002. Late-Holocene summer sea-surface temperatures based on a diatom record from the north Icelandic shelf. *The Holocene*, 12, 137-148.

Jóhannesson, B. 1960. *The Soils of Iceland*. University Research Institute, Reykjavík, Iceland (Agricultural Research Institute).

Jóhannesson, H. 1984. Grímsvatnagos 1933 og fleira frá því ári. (The Grímsvötn eruption in 1933.) *Jökull*, 34, 151-158.

Jóhannesson, H. 1985a. Þættir úr sögu Skeiðarárjökuls. (English summary: The advances and retreats of the Skeiðarárjökull glacier in southeast Iceland in the last 250 years.) *Náttúrufræðingurinn*, 54, 31-45.

Jóhannesson, H. 1985b. Um endasleppu hraunin undir Eyjafjöllum og jökla síðasta jökulskeiðs. (English summary: On the ages of two recent lava flows in Eyjafjöll and the late glacial terminal moraines in South Iceland.) *Jökull*, 35, 83-95.

Jóhannesson, H. and Sæmundsson, K. 1995. Aldursgreining á skeljum í Njarðvíkurheiði (English summary: Radiocarbon age of subfossil shells at Njarðvíkurheiði, SW-Iceland.) *Náttúrufræðingurinn*, 65, 107-111.

Jóhannesson, H. and Sæmundsson, K. 1998. *Geological Map of Iceland. 1:500,000. Bedrock Geology*. Icelandic Institute of Natural History, Reykjavík, 2nd edition.

Jóhannesson, H., Jakobsson, S. P. and Sæmundsson, K. 1982. *Geological Map of Iceland, Sheet 6, S-Iceland*, second edition. Icelandic Museum of Natural History and Iceland Geodetic Survey, Reykjavík.

Jóhannesson, H., Sæmundsson, K., Sveinbjörnsdóttir, Á. E. and Símonarson, L. A. 1997. Nýjar aldursgreiningar á skeljum á Reykjanesskaganum. *Geoscience Society of Iceland, Spring Meeting 1997*, 29-30.

Jóhannesson, J. 1941. *Gerðir Landnámabókar*. Reykjavík, Hið íslenzka bókmenntafélag.

Jóhannesson, J. 1955-57. Sannfræði og uppruni Landnámu. *Saga, Tímarit Sögufélags* 2, 217-29.

Jóhannesson, J. 1974. *A History of the Old Icelandic Commonwealth. Íslendinga Saga*. Translated from the Icelandic by Haraldur Bessason. University of Manitoba Icelandic Studies Vol. II, Manitoba, University of Manitoba Press.

Jóhannesson, J., Finnbogason, M. and Eldjárn, K. (eds.) 1946. *Sturlunga Saga* I-II. Reykjavík, Sturlunguútgáfan.

Jóhannesson, T. 2002a. The initiation of the 1996 jökulhlaup from Lake Grímsvötn, Vatnajökull, Iceland. In, Snorasson, Á., Finnsdóttir, H.P. and Moss, M. (eds.), *The Extremes of the Extremes: Extraordinary Floods*. IAHS Publiction, 271, 57-64.

Jóhannesson, T., 2002b. Propagation of a subglacial flood wave during the initiation of a jökulhlaup. *Hydrological Sciences Journal*, 47, 417–434.

Jóhannesson, T. and Sigurðsson, O. 1998. Interpretation of glacier variations in Iceland 1930-1995. *Jökull*, 45, 27-33.

Jóhannesson, Þ. 1950. *Saga Íslendinga Tímabilið 1770-1830. Upplýsingaröld* VII. Reykjavík, Menntamálaráð og Þjóðvinafélag.

Jóhannsson, G. 1919. *Kötlugosið 1918*. (The Katla eruption in 1918.) Reykjavík, Bókaverslun Ársæls Kristjánssonar.

Johansen, S. and Hytteborn, H. 2001. A contribution to the discussion of biota dispersal with drift ice and driftwood in the North Atlantic. *Journal of Biogeography*, 28, 105-115.

Johnsen, S., Clausen, H. B., Dansgaard, W., Gundestrup, N. S., Hansson, M., Johnsson, P., Steffensen, P. and Sveinbjörnsdóttir, Á. E. 1992. A "deep" ice core from East Greenland. *Meddelelser om Grønland, Geoscience*, 29, 22 pp.

Johnsen, S.J., Dahl-Jensen, D., Gunderstrup, N., Steffensen, J.P., Clausen, H.B., Miller, H., Masson-Delmotte, V., Sveinbjörnsdóttir, Á. and White, J. 2001. Oxygen isotope and palaeotemperature records from six Greenland ice-core stations: Camp Century, Dye-3, GRIP, GISP2, Renland and NorthGRIP. *Journal of Quaternary Science*, 16, 299-307.

Johnson, W.H. and Hansel A.K. 1999. Wisconsin episode glacial landscape of central Illinois: a product of subglacial deformation processes? In, Mickelson D.M. and Attig J.W. (eds.), *Glacial Processes: Past and Present*. Geological Society of America, Special Paper 337, 121-135.

Jónasson, P.M. 1992. The ecosystem of Þingvallavatn: a synthesis. *Oikos*, 64, 405–434.

Jónasson, P.M. and Hersteinsson, P. 2002. *Þingvallavatn: Undraheimur í mótun*. Reykjavík, Mál og Menning.

Jones, J.G. 1969. Intraglacial volcanoes of the Laugarvatn region, south-west Iceland, I. *Quarterly Journal of the Geological Society of London*, 124, 197-211.

Jones, J.G. 1970. Intraglacial volcanoes of the Laugarvatn region, south-west Iceland, II. *Journal of Geology*, 78, 127-140.

Jones, P.D., Raper, S.C.B, Santer, B.D., Cherry, B.S.G., Goodess, C., Bradley, R.S., Diaz, H.F., Kelly, P.M. and Wigley, T.M.L. 1985. *A Grid Point Surface Air Temperature Data Set for the Northern Hemisphere, 1851-1984.* U.S. Department of Energy Carbon Dioxide Research Division, 251.

Jónsdóttir, I. 1995. *Sea Ice off the Coasts of Iceland in the Early 20th century.* Unpublished M.Phil. Thesis. Cambridge, Scott Polar Research Institute, University of Cambridge.

Jónsdóttir, H. E. and Björnsdóttir, I. E. 1995. *Setlög og Sædýrafánur frá Síðjökultíma og Byrjun Nútíma í Dalasýslu.* Unpublished B.S. Thesis, University of Iceland, Reykjavík.

Jónsson, J. 1982. Notes on the Katla volcanoglacial debris flows. *Jökull,* 32, 61-68.

Jónsson, P., Sigurðsson, O. Snorrason, Á., Víkingsson, S., Kaldal, I. and Árnason, S. 1998. Course of events of the jökulhlaup on Skeiðarársandur, Iceland, in November 1996. In, Cañaveras, J. C., Ángeles García del Gura, M. and Soria, J. (eds.), *15th International Sedimentological Conference,* International Association of Sedimentologists, 456-457.

Jónsson, S. 1992. Sources of fresh water in the Iceland Sea and the mechanisms governing its interannual variability. *ICES Marine Science Symposia,* 195, 62-67.

Jónsson, S. 1999. Temperature time series from Icelandic coastal stations. *Journal of the Marine Research Institute,* 16, 59-68.

Jónsson, T. 1989. *Afturábak frá Stykkishólmi, Veðurathuganir Jóns Þorsteinssonar landlæknis í Nesi og í Reykjavík.* (The observations of Jón Þorsteinsson in Nes and Reykjavík 1820-1854, and their relation to the Stykkishólmur series.) Reykjavík, Icelandic Meteorological Office Report.

Jónsson, T. 1990. Hvert liggja öskugeirar. (Dispersal directions of volcanic plumes from Icelandic eruptions.) *Náttúrufræðingurinn,* 60, 103-105.

Jónsson, T. 1998. Reconstructing the temperature in Iceland from early instrumental observations. Data availability and a status report. *Päleoklimaforschung,* 23, 87-98.

Jónsson, T. and Garðarsson, H. 2001, Early instrumental meteorological observations in Iceland. In, Ogilvie, A.E.J. and Jónsson T. (eds.), *The Iceberg in the Mist: Northern Research in Pursuit of a 'Little Ice Age'.* Dordrecht, Kluwer Acdemic Publishers, 169-187.

Jull, M. and McKenzie, D. 1996. The effect of deglaciation on mantle melting beneath Iceland. *Journal of Geophysical Research,* 101, 21815-21828.

Kaldal, I. and Víkingsson, S. 1990. Early Holocene deglaciation in Central Iceland. *Jökull,* 40, 51-66.

Kaldal, I. and Víkingsson, S. 2000a. *Kárahnjúkavirkjun. Jarðgrunnskort af Umhverfi Hálslóns.* (Kárahnjúkavirkjun. Map of superficial deposits of the surroundings of Hálslón.) Orkustofnun OS-2000/065.

Kaldal, I. and Víkingsson, S. 2000b. *Jarðgrunnskort af Eyjabökkum.* (Map of superficial deposits of the Eyjabakkar.) Orkustofnun OS-2000/068.

Kamb, B. 1987. Glacier surge mechanism based on linked cavity configuration of the basal water conduit system. *Journal of Geophysical Research,* 92, 9083-9100.

Kamb, B., Raymond, C.F., Harrison, W.D., Engelhardt, H., Echelmeyer, K.A., Humphrey, N., Brugman, M.M. and Pfeffer, T. 1985. Glacier surge mechanism: 1982-1983 surge of Variegated Glacier, Alaska. *Science,* 327, 469-479.

Karcz, I. 1968. Fluviatile obstacle marks from the wadis of the Negev (southern Israel). *Journal of Sedimentary Petrology*, 38, 1000-1012.

Käyhkö, J.A., Alho, P., Hendriks, J.P.M. and Rossi, M.J. 2002. Landsat TM based land cover mapping of Ódáðahraun semi-desert, north-eastern Iceland. *Jökull*, 51, 1-16.

Keith, D. B. and Jones, E. W. 1935. Grímsey, North Iceland. *Geographical Journal*, 86, 143-152.

Kellogg, T. B. 1984. Late-glacial - Holocene high-frequency climatic changes in deep-sea cores from the Denmark Strait. In, Mörner, N. A. and Karlén, W. (eds.), *Climatic Changes on a Yearly to Millennial Basis*. Dordrecht, D. Reidel Publishing Company, 123-133.

Kerguélen-Trémarec, Y. J. de. 1808. Relations of a voyage in the North Sea, along the coasts of Iceland, Greenland, Ferro, Shetland, the Orcades, and Norway, made in the years 1767 and 1768. Extr. fr. *A general collection of the best and most interesting voyages and travels...* By John Pinkerton, Vol. I, London.

Keszthelyi, L. and Self, S. 1998. Some physical requirements for the emplacement of long basaltic lava flows. *Journal of Geophysical Research*, 103, 27447- 27464.

Ketilsson, M. (ed.) 1773-76: *Islandske Maaneds Tidender* I-III Aargang. Hrappsøe and Copenhagen.

Kimble, J.M., Ping, C.L., Sumner, M.E. and Wilding, L.P. 2000. Andisols. In, Sumner, M.E. (ed.), *Handbook of Soil Science*. Boca Raton, Florida, CRC Press, E-209–224.

Kington, J. A. 1988. *The Weather of the 1780s Over Europe*. Cambridge, Cambridge University Press.

Kington, J. A. and Kristjánsdóttir, S. 1978. Veðurathuganir Jóns Jónssonar eldra og yngra og gildi þeirra við daglega veðurkortagerð eftir sögulegum gögnum. (The Value of the Jón Jónssons' Icelandic Weather Diaries to Historical Daily Weather Mapping.) *Veðrið*, 21, 42-51.

Kirkbride, M. P. and Dugmore, A. J. 2001. Can lichenometry be used to date the 'Little Ice Age' glacial maximum in Iceland? *Climatic Change*, 48, 51-167.

Kirkbride M.P. and Spedding N. 1996. The influence of englacial drainage on sediment-transport pathways and till texture of temperate valley glaciers. *Annals of Glaciology*, 22, 160-166.

Kjartansson, G. 1943. Árnesingasaga. (The geology of Árnessýsla.) *Árnesingafélagið*, Reykjavík.

Kjartansson, G. 1951. Water floods and mud flows. In, *The eruption of Hekla 1947-1948*. II, 4, *Societas Scientia Islandica*, 51 pp.

Kjartansson, G. 1952. Meira um Rauðhól. *Náttúrufræðingurinn*, 22, 78-89.

Kjartansson, G. 1955. Fróðlegar jökulrákir. (English Summary: Studies of glacial striae in Iceland.) *Náttúrufræðingurinn*, 25, 154-171.

Kjartansson, G. 1958. *Jarðmyndanir í Holtum og nágrenni*. (English summary: The geology of Holt in Rangárvallasýsla, S.W. Iceland.) Department of Agriculture, Reports 11. University Research Institute, Reykjavík, 22 pp.

Kjartansson, G. 1959. On the geology and geomorphology of Iceland II. The Móberg Formation. *Geografiska Annaler*, 41, 139-143.

Kjartansson, G. 1960. *Geological Map of Iceland, Sheet 3, South-West Iceland*. Reykjavík, The Cultural Fund.

Kjartansson, G. 1962a. Jökulminjar á hálsunum milli Berufjarðar og Hamarsfjarðar. (English summary: Evidence of glaciation on the promontory Hálsar between Berufjörður and Hamarsfjörður in eastern Iceland.) *Náttúrufræðingurinn*, 32, 83-92.

Kjartansson, G. 1962b. *Geological Map of Iceland, Sheet 6, South-Central Iceland*. Reykjavík, The Cultural Fund.

Kjartansson, G. 1964. Ísaldarlok og eldfjöll á Kili. (Termination of the last glaciation and the volcanoes at Kjölur.) *Náttúrufræðingurinn*, 34, 101-113.

Kjartansson, G. 1966. Nokkrar nýjar ^{14}C-aldursákvarðanir. (English Summary: Some new ^{14}C datings in Iceland.) *Náttúrufræðingurinn*, 36, 126-141.

Kjartansson, G. 1967. The Steinholtshlaup, central-south Iceland on January 15th, 1967. *Jökull*, 17, 249-262.

Kjartansson, G. 1969. *Geological Map of Iceland, Sheet 1, North-West Iceland*. Reykjavík, The Cultural Fund.

Kjartansson, G. 1970. Úr sögu berggrunns og landslags á Miðsuðurlandi. *Suðri*, 2, 12-100.

Kjartansson, H. and Arnórsson, S. 1972. *Leirmyndanir í Dalasýslu og Þingeyjarsýslum. Lokaskýrsla um Jarðfræðilega Frumrannsókn*. Reykjavík, Orkustofnun.

KL = Kulturhistorisk Leksikon for Nordisk Middelalder fra Vikingetid Reformationstid I- 1956, Danish edition. Copenhagen, Rosenkilde og Bagger.

Klimek, K. 1972. Present day fluvial processes and relief of the Skeiðarársandur plain (Iceland). *Polska Academia Nauk Institut Geografi*, 94, 129-139.

Klimek, K. 1973. Geomorphological and geological analysis of the proglacial area of Skeiðarárjökull. *Geographia Polonica*, 26, 89-113.

Knudsen, Ó. 1995. Concertina eskers, Brúarjökull, Iceland: an indicator of surge-type behaviour. *Quaternary Science Reviews*, 14, 487-493.

Knudsen, Ó. and Russell, A.J. 2002. Jökulhlaup deposits at Ásbyrgi, northern Iceland: sedimentology and implications or flow type. In, Snorasson, Á., Finnsdóttir, H.P. and Moss, M. (eds.), *The Extremes of the Extremes: Extraordinary Floods*. IAHS Publication, 271, 107-112.

Knudsen, Ó., Jóhannesson, H., Russell, A.J. and Haraldsson, H. 2001a. Changes in the Gígjukvísl river channel during the November 1996 jökulhlaup, Skeiðarársandur, Iceland. *Jökull*, 50, 19-32.

Knudsen, Ó., Roberts, M. J., Tweed, F. S., Russell, A. J., Lawson, D. E., Larson, G. J., Evenson, E. B. and Björnsson, H. 2001b. Five 'supercool' Icelandic glaciers, *Eos*, AGU, 82, Fall Meeting Supplement, IP52A-0750.

Krigström, A. 1962. Geomorphological studies of sandur plains and their braided rivers in Iceland. *Geografiska Annaler*, 44, 328-346.

Kristjánsdóttir, G.B. 1999. *Late Quaternary Climatic and Environmental Changes on the North Iceland Shelf*. Unpublished M.Sc. Thesis, University of Iceland, Department of Geosciences.

Kristjánsdóttir, S. 1998. Sources of climatic information in the manuscript collection of the National Library in Iceland. In, *Abstract Volume. The Second Climate and History Conference*, 7-11 September, 1998, Climatic Research Unit, Norwich, UK, 60.

Kristjánsson, J. 1988. *Eddas and Sagas: Iceland's Medieval Literature*. Translated by Peter Foote. Reykjavík, Hið íslenska bókmenntafélag.

Kristmannsdóttir, H., Snorrason, Á., Gíslason, S.R., Haraldsson, H., Gunnarsson, Á, Hauksdóttir, S. and Elefsen, S.Ó. 2002. Geochemical warning for subglacial eruptions – background and history. In, Snorasson, Á., Finnsdóttir, H.P. and Moss, M. (eds.), *The Extremes of the Extremes: Extraordinary Floods*. IAHS Publication, 271, 231-236.

Krüger, J. 1985. Formation of a push moraine at the margin of Höfðabrekkujökull, south Iceland. *Geografiska Annaler*, 67A, 199-212.

Krüger, J. 1987. Traek af et glaciallandskabs udvikling ved nordranden af Mýrdalsjökull, Iceland. *Dansk Geologisk Foreningens*, Arsskrift for 1986, 49-65.

Krüger, J. 1993. Moraine ridge formation along a stationary ice front in Iceland. *Boreas*, 22, 101-109.

Krüger, J. 1994. Glacial processes, sediments, landforms and stratigraphy in the terminus region of Mýrdalsjökull, Iceland. *Folia Geographica Danica*, 21, 1-233.

Krüger, J. 1995. Origin, chronology and climatological significance of annual-moraine ridges at Mýrdalsjökull, Iceland. *The Holocene*, 5, 420-427.

Krüger, J. 1996. Moraine ridges formed from subglacial frozen-on sediment slabs and their differentiation from push moraines. *Boreas*, 25, 57-63.

Krüger, J. 1997. Development of minor outwash fans at Kötlujökull, Iceland. *Quaternary Science Reviews*, 16, 649-659.

Krüger, J. and Aber, J. S. 1999. Formation of supraglacial sediment accumulations on Kötlujökull, Iceland. *Journal of Glaciology*, 45, 400-402.

Krüger, J. and Thomsen, H.H. 1984. Morphology, stratigraphy and genesis of small drumlins in front of the glacier Mýrdalsjökull, south Iceland. *Journal of Glaciology*, 30, 94-105.

Kugelmann, O. 1989. *Gletschergeschichtliche Untersuchungen im Svarfaðardalur und Skíðadalur, Tröllaskagi, Nordisland*. Unpublished Diplom Thesis, Institute of Geography, University of Munich.

Kugelmann, O. 1990. Datierung neuzeitlicher Gletschervorstöße im Svarfaðardalur/ Skíðadalur (Nordisland) mit einer neue erstellten Flechtenwachstumskurve. *Münchener Geographische Abhandlungen*, B8, 36-58.

Kugelmann, O. 1991. Dating recent glacier advances in the Svarfaðardalur-Skíðadalur area of northern Iceland by means of a new lichen curve. In, Maizels, J.K. and Caseldine, C. (eds.), *Environmental Change in Iceland: Past and Present*. Dordrecht, Kluwer, 203-217.

Kvamme, T., Mangerud, J., Furnes, H. and Ruddiman, W. F. 1989. Geochemistry of Pleistocene ash zones in cores from the North Atlantic. *Norsk Geologiske Tidskrift*, 69, 251-272.

Lacasse, C., Carey, S. and Sigurðsson, H. 1998. Volcanogenic sedimentation in the Iceland Basin: influence of subaerial and subglacial eruptions. *Journal of Volcanology and Geothermal Research*, 1793, 1-27.

Lacasse, C., Sigurðsson, H., Carey, S., Paterne, M. and Guichard, F. 1996. North Atlantic deep-sea sedimentation of Late Quaternary tephra from the Iceland hotspot. *Marine Geology*, 129, 207-235.

Lacasse, C., Sigurðsson, H., Jóhannesson, H., Paterne, M. and Carey, S. 1995. Source of Ash Zone I in the North Atlantic. *Bulletin of Volcanology*, 57, 18-32.

Lamb, H. H. 1970. Volcanic dust in the atmosphere; with a chronology and assessment of its meteorological significance. *Philosophical Transactions of the Royal Society*, 266A, 425-533.

Lamb, H. H. 1977. *Climate: Present, Past and Future. Volume 2, Climatic History and the Future*. London, Methuen.

Landsberg, H. E., Yu, C. S. and Huang, L. 1968. *Preliminary Reconstruction of a Long Time Series of Climatic Data for the Eastern United States*. College Park, Institute for Fluid Dynamics and Applied Mathematics, University of Maryland, 30pp.

Langley, K. 2000. *A Morphological Investigation of Volcanic Activity Beneath Vatnajökull, Iceland, Interpreted from Radio Echo Sounding Data*. MS Thesis, University of Iceland.

Larsen, E., Gulliksen, S., Lauritzen, S.-E., Lie, R., Lövlie, R. and Mangerud, J. 1987. Cave stratigraphy in western Norway; multiple Weichselian glaciations and interstadial vertebrate fauna. *Boreas*, 16, 267-292.

Larsen, G. 2000. Holocene eruptions within the Katla volcanic system, south Iceland. *Jökull*, 49, 1-27.

Larsen, G. 2002. A brief overview of eruptions from ice-covered and ice-capped volcanic systems in Iceland during the past 11 centuries: frequency, periodicity and implications. In, Smellie, J.L. and Chapman, M. (eds.), *Ice-Volcano Interaction on Earth and Mars*. Geological Society of London Special Publication, 202, 81-90.

Larsen, G. and Guðmundsson, M.T. 1997. Gos í eldstöðvum undir Vatnajökli eftir 1200 AD. (Eruptions in volcanoes under Vatnajökull since 1200 AD.) In, Haraldsson, H. (ed.), *Vatnajökull. Gos og hlaup* 1996. Vegagerðin, 23-36.

Larsen, G., Dugmore, A. and Newton, A. 1999. Geochemistry of the historical-age silicic tephras in Iceland. *The Holocene*, 9, 463–471.

Larsen, G., Guðmundsson, M.T. and Björnsson, H. 1998. Eight centuries of periodic volcanism at the center of the Iceland Hot Spot revealed by glacier tephrostratigraphy. *Geology*, 26, 943-946.

Lasaga, A.C., Soler, J.M., Ganor, J., Burch, T.E. and Nagy, K.L. 1994. Chemical weathering rate laws and global cycles. *Geochimica et Cosmochimica Acta*, 58, 2361–2386.

Lawson, D.E., Strasser, J.C., Evenson, E.B., Alley R.B., Larson G.J. and Arcone, S.A. 1998. Glaciohydraulic supercooling: a freeze-on mechanism to create stratified, debris-rich basal ice. I. Field evidence. *Journal of Glaciology*, 44, 547-562.

Leet, R.C. 1988. Saturated and subcooled hydrothermal boiling in groundwater-flow channels as a source of harmonic tremor. *Journal of Geophysical Research*, 93, 4835-4849.

Lehman, S. J. and Keigwin, L. D. 1992. Sudden changes in North Atlantic circulation during the last deglaciation. *Nature*, 356, 757-762.

Levi, S., Auðunsson, H., Duncan, R. A., Kristjánsson, L., Gillot, P.-Y. and Jakobsson, S. P. 1990. Late Pleistocene geomagnetic excursion on Icelandic lavas: confirmation of the Laschamp excursion. *Earth and Planetary Science Letters*, 96, 443-457.

Likens, G.E., Bormann, F. H. and Johnsson, N.M. 1981. Interaction between major biogeochemical cycles in terrestrial ecosystems. In, Likens, G. E. (ed.), *Some Perspectives of the Major Biogeochemical Cycles-SCOPE 17*. New York, John Wiley, 93–112.

Longva, O. and Bakkejord, K.J. 1990. Iceberg deformation and erosion in soft sediments, southeast Norway. *Marine Geology*, 92, 87-104.

Longva, O. and Thoresen, M.K. 1991. Iceberg scours, iceberg gravity craters and current erosion marks from a gigantic Preboreal flood in southeastern Norway. *Boreas*, 20, 47-62.

Louchlin, S.C. 2002. Facies analysis of proximal subglacial and proglacial volcaniclastic successions at the Eyjafjallajökull central volcano, southern Iceland. In, Smellie, J.L., and Chapman, M. (eds.), *Ice-Volcano Interaction on Earth and Mars*. Geological Society of London Special Publication, 202, 149-178.

Louvat, P. 1997. *Étude Geochimique de l'Érosion Fluviale Dîles Volcaniques a l'Aide des Bilans d'Éléments Majeurs et Trace*s. Unpublished Ph.D. Thesis, University of Paris, Paris.

Louvat, P. and Allègre, C.J. 1997. Present denudation rates on the island of Réunion determined by river geochemistry: basalt weathering and mass budget between chemical and mechanical erosion. *Geochimica et Cosmochimica Acta*, 61, 3645–3670.

Louvat, P. and Allègre, C.J. 1998. Riverine erosion rates on Sao Miguel volcanic island, Azores archipelago. *Chemical Geology*, 148, 177–200.

Louvat, P., Gíslason, S.R. and Allégre, C.J. 1999. Chemical and mechanical erosion of major Icelandic rivers: geochemical budgets. In, Ármannsson, H. (ed.), *Geochemistry of the Earth's Surface*. Rotterdam, Balkema, 111–114.

Mackenzie, Sir G. 1811. *Travels in the Island of Iceland During the Summer of the Year 1810*. Edinburgh.

Mackintosh, A.N., Dugmore, A.J. and Hubbard, A.L. 2002. Holocene climatic changes in Iceland: evidence from modelling glacier length fluctuations at Sólheimajökull. *Quaternary International*, 91, 39-52.

Maclennan, J., Jull, M., McKenzie, D., Slater, L. and Grönvold, K. 2002. *The Link Between Volcanism and Deglaciation in Iceland*. G3, 3, 11. DOI:10.1029/2001GC000282.

Magilligan, F.J., Gomez, B., Mertes, L.A.K., Smith, L.C., Smith, N.D., Finnegan, D. and Garvin J.B. 2002. Geomorphic effectiveness, sandur development, and the pattern of landscape response during jökulhlaups: Skeiðarársandur, southeastern Iceland. *Geomorphology*, 44, 95-113.

Magnúsdóttir, B. and Norðdahl, H. 2000. Aldur hvalbeins og fornra fjörumarka í Akrafjalli. (English summary: Re-examination of the deglaciation history of the area around Akrafjall in South-western Iceland.) *Náttúrufræðingurinn*, 69, 177-188.

Magnússon, Á. 1955. Chorographica islandica. In, Lárusson, Ó. (ed.), *Safn til sögu Íslands og íslenskra bókmennta*, annar flokkur I, 2. (Written in 1702-1714.) Reykjavík, Hið íslenzka bókmenntafélag.

Magnússon, Á. and Vídalín, P. 1980. *Jarðabók Árna Magnússonar og Páls Vídalín*. (Farm account by Árni Magnússon and Páll Vídalín (information gathered in 1702-1714).). Vol. 1-13, second edition. Reykjavík, Hið íslenzka fræðafélag.

Magnússon, S.G. 1995. Ég er 479 dögum yngri en Nilli. Dagbækur og daglegt líf Halldórs Jónssonar frá Miðdalsgröf. *Skírnir*, 169, 309-347.

Magnússon, S.G 1997. Menntun, ást og sorg. Einsögurannsókn á íslensku sveitasamfélagi 19. og 20. aldar. *Sagnfræðirannsóknir*, 13. Reykjavík, Sagnfræðistofnun and Háskólaútgáfan.

Magnússon, Þ. 1626. Þorsteins Magnússonar um jökulbrunann fyrir austan 1625. *Safn til Sögu Íslands IV, 1907-1915*. Copenhagen and Reykjavík, 201-215.

Maizels, J.K. 1977. Experiments on the origin of kettle holes. *Journal of Glaciology*, 18, 291-303.

Maizels, J.K. 1979. Proglacial aggradation and changes in braided channel patterns during a period of glacier advance: an Alpine example. *Geografiska Annaler*, 61, 87-101.

Maizels, J.K. 1987. Large-scale flood deposits associated with the formation of coarse-grained braided terrace sequences. In, Ethridge, F. G., Flores, R. M. and Harvey, M. D. (eds.), *Recent Developments in Fluvial Sedimentology*. Society of Economic Paleontologists and Mineralogists, Special Publication No. 39, 135-148.

Maizels, J.K. 1989a. Sedimentology, paleoflow dynamics and flood history of jökulhlaup deposits: paleohydrology of Holocene sediment sequences in southern Iceland sandur deposits. *Journal of Sedimentary Petrology*, 59, 204-223.

Maizels, J.K. 1989b. Sedimentology and palaeohydrology of Holocene flood deposits in front of a jökulhlaup glacier, South Iceland. In, Bevan, K. and Carling, P. (eds.), *Floods, Hydrological, Sedimentological and Geomorphological Implications: An Overview*. Wiley, London, 239-253.

Maizels, J. K. 1991. The orgin and evolution of the Holocene sandur deposits in areas of jökulhlaup drainage, Iceland. In, Maizels, J.K. and Caseldine, C. (eds.), *Environmental Change in Iceland: Past and Present*. Dordrecht, Kluwer, 267-302.

Maizels J. 1992. Boulder ring structures produced during jökulhlaup flows: origin and hydraulic significance. *Geografiska Annaler*, 74A, 21-33.

Maizels, J.K. 1993a. Lithofacies variations within sandur deposits: the role of runoff regime, flow dynamics and sediment supply characteristics. *Geology*, 85, 299-325.

Maizels, J.K. 1993b. Quantitative regime modelling of a fluvial depositional sequence: application to Holocene stratigraphy of humid-glacial braid-plains (Icelandic sandurs). In, North, C.P. and Prosser, D.J. (eds.), *Characterisation of Fluvial and Aeolian Reservoirs*. Geological Society Special Publication, 73, 53-78.

Maizels, J.K. 1995. Sediments and landforms of modern proglacial terrestrial environments. In, Menzies, J. (ed.), *Modern Glacial Environments: Processes, Dynamics and Sediments*. Oxford, Butterworth-Heinemann, 365-416.

Maizels, J.K., 1997. Jökulhlaup deposits in proglacial areas. *Quaternary Science Reviews*, 16, 793-819.

Maizels, J.K. and Russell, A.J. 1992. Quaternary perspectives on jökulhlaup prediction. In, Gray, J.M. (ed.), *Applications of Quaternary Research*. Quaternary Proceedings, 2, 133-153.

Major, J.J. and Newhall, C.H. 1989. Snow and ice perturbation during historical volcanic eruptions and the formation of lahars and floods. *Bulletin of Volcanology*, 51, 1-27.

Malin, M.C. and Eppler, D.B. 1981. Catastrophic floods of the Jökulsá á Fjöllum, Iceland. *Reports of the Planetary Geology Programme*, NASA, 272-273.

Malmberg, S.-A. 1969. Hydrographic changes in the waters between Iceland and Jan Mayen in the last decade. *Jökull*, 19, 30-43.

Malmberg, S.-A. 1985. The water masses between Iceland and Greenland. *Journal of the Marine Research Institute*, 9, 127-140.

Malmberg, S.-A. and Jónsson, S. 1997. Timing of deep convection in the Greenland and Iceland Seas. *ICES Journal of Marine Science*, 54, 300-309.

Malmberg, S.-A. and Kristmannsson, S. S. 1992. Hydrographic conditions in Icelandic waters, 1980-1989. *ICES Marine Science Symposia*, 195, 76-92.

Malmberg, S.-A., and Magnússon, G. 1982. Sea surface temperature and salinity in south Icelandic waters in the period 1868-1965. *Rit Fiskideildar*, 3, 1-31.

Malmberg, S.-A, Valdimarsson, H. and Mortensen, J. 1996. Long-time series in Icelandic waters in relation to physical variability in the northern North Atlantic. *NATO Scientific Council Studies*, 24, 69-80.

Malthe-Sørenssen, A., Walmann, T., Jamtveit, B., Feder, J. and Jøssang, T. 1998. Modeling and characterization of fracture patterns in the Vatnajökull glacier. *Geology*, 26, 931-934.

Mangerud, J., Andersen, S. T., Berglund, B. E. and Donner, J. J. 1974. Quaternary stratigraphy Norden, a proposal for terminology and classification. *Boreas*, 3, 109-126.

Maria, A., Carey, S., Sigurðsson, H., Kincaid, C. and Helgadóttir, G. 2000. Source and dispersal of jökulhlaup sediments discharged to the sea following the 1996 Vatnajökull eruption. *Geological Society of America Bulletin*, 112, 1507-1521.

Marren, P. M. 2002a. Glacier margin fluctuations, Skaftafellsjökull, Iceland: implications for sandur evolution. *Boreas*, 31, 75-81.

Marren, P. M. 2002b. Fluvial-lacustrine interaction on Skeiðarársandur, Iceland: implications for sandur evolution. *Sedimentary Geology*, 149, 43-58.

Marren, P. M. 2003. Present-day sandurs are not representative of the geological record. *Sedimentary Geology*, 152, 1-5.

Marren, P. M., Russell, A. J. and Knudsen, Ó. 2002. Discharge magnitude and frequency as a control on proglacial fluvial sedimentary systems. In, Dyer, F., Thoms, M. C. and Olley, J. M. (eds.), *The Structure, Function and Management Implications of Fluvial Sedimentary Systems*. IAHS Publication 276, 297-303.

Marren, P. M., Russell, A. J., Knudsen, Ó. and Rushmer, E. L. in press. Sedimentology and architecture of a sandur formed by multiple jökulhlaups, Kverkfjöll, Iceland. *Sedimentary Geology*, in revision.

Martin, H.E. and Whalley, W.B. 1987. A glacier ice-cored rock glacier in Tröllaskagi, northern Iceland. *Jökull*, 37, 49-55.

Martin, H.E., Whalley, W.B. and Caseldine, C.J. 1991. Rock glacier and glacier fluctuations in Tröllaskagi, northern Iceland, 1946-1986. In, Maizels, J. and Caseldine C.J. (eds.), *Environmental Change in Iceland: Past and Present*. Dordrecht, Kluwer, 255-265.

Martin, H.E., Whalley W.B., Orr, J. and Caseldine, C.J. 1994. Dating and interpretation of rock glacier using lichenometry: south Tröllaskagi, north Iceland. *Münchener Geographische Abhandlungen*, B12, 205-224.

Mathews, W.H. 1947. 'Tuyas', flat-topped volcanoes in northern British Columbia. *American Journal of Science*, 245, 560-570.

Matthews, J.A. and Petch, J.R. 1982. Within-valley asymmetry and related problems of Neoglacial lateral moraine development at certain Jotunheimen glaciers, southern Norway. *Boreas*, 11, 225-247.

Matthews, J.A., McCarroll, D. and Shakesby, R.A. 1995. Contemporary terminal moraine ridge formation at a temperate glacier: Styggedalsbreen, Jotunheimen, southern Norway. *Boreas*, 24, 129-139.

Matthíasdóttir, M., 1997. *Keflavíkurberg, Setmyndun á Utanverðum Reykjanesskaga*. Unpublished B.S. Thesis, University of Iceland, Reykjavík.

Matthíasson, H. 1980. *Árbók Ferðafélags Íslands 1980*. Langjökulsleiðir. (Yearbook of the Touring Club of Iceland 1980. Routes by Langjökull.) Reykjavík, Ferðafélag Íslands.

Melville, B.W. 1997. Pier and abutment scour: integrated approach. *Journal of Hydraulic Engineering*, 123, 125-136.

Meybeck, M. 1979. Concentrations des eaux fluviales en éléments majeurs et apports en solution aux océans. *Revue de Géologie Dynamique et Géographie Physique*, 21, 215–246.

Meyer, H.-H. and Venzke, J.-F. 1985. Der Klængshóll-Kargletscher in Nordisland. *Natur und Museum*, 115, 29-46.

Miller, D. J. 1989. *The 10th Century Eruption of Eldgjá, Southern Iceland*. Reykjavík, Nordic Volcanological Institute, University of Iceland.

Milliman, J.D. and Syvitski, J.P.M. 1992. Geomorphic/tectonic control of sediment discharge to the ocean: the importance of small mountainous rivers. *Journal of Geology*, 100, 525–544.

Ministry of the Environment, Iceland 1992. *Release of Greenhouse Gases from Iceland in the Year 1990*. Reykjavík, Ministry of the Environment, May 1992.

Molewski, P. 1996. Nowe fakty dotyczace genezy zaglebien wytopiskowych na sandrach przedpola lodowca Skeiðará (Islandia). (New facts relating genesis of the kettle holes on the sandurs of the Skeiðará glacier (Iceland) foreland.) *Przeglad Geograficzny*, 68, 405-426.

Molewski, P. and Olszenski, A. 2000. Sedimentology of the deposits and their significance in the Gígjukvísl river gap, marginal zone of the Skeiðarárjökull, Iceland. *Polish Polar Studies, XXXVII Polar Symposium*, 235-258.

Moore, J.G. and Calk, L.C. 1991. Degassing and differentation in subglacial volcanoes, Iceland. *Journal of Volcanology and Geothermal Research*, 46, 157-180.

Mohr, N. 1786. *Forsøg til en Islandsk Naturhistorie med adskillige oekonomiske samt andre Anmerkninger.* Copenhagen, Christian Friderik Holm.

Moriwaki, H. 1990. Late- and postglacial shoreline displacement and glaciation in and around the Skagi peninsula, northern Iceland. *Geographical Reports of Tokyo Metropolitan University,* 25, 81-97.

Morris, W. 1911. Journals of Travel in Iceland 1871, 1873. London. In, *The Collected Works of William Morris,* Vol. 8, (London 1910-15).

Moulton, K.L., West, J. and Berner, R A. 2000. Solute flux and mineral mass balance approaches to the quantification of plant effects on silicate weathering. *American Journal of Science,* 300, 539–570.

Munster, S. I. 1598. *Cosmographey.* Basel.

Nanzyo, M., Shoji, S. and Dahlgren, R. 1993. Physical characteristics of volcanic ash soils. In, Shoji, S., Nanzyo, M. and Dahlgren, R.A. (eds.), *Volcanic Ash Soils. Genesis, Properties and Utilization.* Developments in Soil Science, Amsterdam, Elsevier, 189–207.

Näslund, J.-O. and Hassinen, S. 1996. Supraglacial sediment accumulations and large englacial water conduits at high elevations in Mýrdalsjökull, Iceland, *Journal of Glaciology.* 42, 190-192.

NASP members, 1994. Climatic changes in areas adjacent to the North Atlantic during the last glacial-interglacial transition (14-9 ka BP): a contribution to IGCP-253. *Journal of Quaternary Science,* 9, 185-198.

Náttúrufræðistofnun Íslands, 1998. *Vegetation Map of Iceland.* Compiled by Guðmundur Guðjónsson and Einar Gíslason.

Nelson, P.H.H. 1975. The James Ross Island volcanic group of north-east Graham Land. *British Antarctic Survey, Scientific Reports,* 54, 62 pp.

Nesbitt, H.W. and Wilson, R.E. 1992. Recent chemical weathering of basalts. *American Journal of Science,* 292, 740–777.

Nicholls, N., Gruza, G.V., Jouzel, J., Karl, T.R., Ogallo, L.A., and Parker, D.E. 1996. Observed climate variability and change. In, Houghton, J.T., Meira Filho, L.G., Callander, B.A., Harris, N., Kattenberg, A. and Maskell, K. (eds), *The Science of Climate Change, Contribution of Working Group 1 to the Second Assessment Report of the Intergovernmental Panel on Climatic Change.* New York, Cambridge University Press, 133-192.

Nielsen, N. 1928. Landskabet syd-öst for Hofsjökull i det indre Island. (Topography southeast of Hofsjökull, central Iceland.) *Geografisk Tidskrift,* Bd 42. Copenhagen.

Nielsen, N. 1937. *Vatnajökull. Kampen mellem ild og is* (Vatnajökull. The fight between ice and fire). H. Hagerup. København.

Noe-Nygaard, A. 1940. Sub-glacial volcanic activity in ancient and recent times. Studies in the palagonite-system of Iceland No. 1. *Folia Geographica Danica,* 1, 67pp.

Nordal, S. 1957. *The Historical Element in the Icelandic Family Sagas.* The fifteenth W. P. Ker Memorial Lecture, delivered in the University of Glasgow, 19 May, 1954. Glasgow, Jackson, Son and Co.

Norðdahl, H. 1981. A prediction of minimum age for the Weichselian maximum glaciation in North Iceland. *Boreas,* 10, 471-476.

Norðdahl, H. 1983. Late Quaternary stratigraphy of Fnjóskadalur central North Iceland: A study of sediments, ice-lake strandlines, glacial isostasy and ice-free areas. *LUNDQUA Thesis*, 12, Lund University, Lund.

Norðdahl, H. 1990. The Weichselian and Early Holocene deglaciation history of Iceland. *Jökull*, 40, 27-50.

Norðdahl, H. 1991. A review of the glaciation maximum concept and the deglaciation of Eyjafjörður, North Iceland. In, Maizels, J. K. and Caseldine, C. (eds.), *Environmental Change in Iceland: Past and Present*. Dordrecht, Kluwer Academic Publishers, 31-47.

Norðdahl, H. and Ásbjörnsdóttir, L. 1995. Ísaldarlok í Hvammsfirði. In, Hróarsson, B., Jónsson, D. and Jónsson, S. S. (eds.), *Eyjar í Eldhafi*. Reykjavík, Gott mál, 117-131.

Norðdahl, H. and Einarsson, Þ. 1988. Hörfun jökla og sjávarstöðubreytingar í ísaldarlok á Austfjörðum. (English summary: Late Weichselian deglaciation and sea-level changes in East and Sortheast Iceland.) *Náttúrufræðingurinn*, 58, 59-80.

Norðdahl, H. and Einarsson, Þ. 2001. Concurrent changes of relative sea-level and glacier extent at the Weichselian – Holocene boundary in Berufjörður, eastern Iceland. *Quaternary Science Reviews*, 20, 1607-1622.

Norðdahl, H. and Hafliðason, H. 1992. The Skógar Tephra, a Younger Dryas marker in North Iceland. *Boreas*, 21, 23-41.

Norðdahl, H. and Hjort, C. 1987. Aldur jökulhörfunar í Vopnafirði. *Geoscience Society of Iceland, Spring Meeting 1987*, 18-19.

Norðdahl, H. and Hjort, C. 1995. Lateglacial raised beaches and glacier recession in the Þistilfjördur – Bakkaflói area, north-eastern Iceland. *Jökull*, 43, 32-44.

Norðdahl, H. and Pétursson, H. G. 1994. Aldur jökulhörfunar og sjávarstöðubreytingar við Skjálfanda. *Geoscience Society of Iceland, Spring Meeting 1994*, 19.

Norðdahl, H. and Pétursson, H. G. 1998. High raised Bølling shorelines in Iceland and the mode of deglaciation. In, Wilson, J.R. (ed.), *Abstract volume: 23rd Nordic Geological Winter Meeting, 13-16 January, Århus 1998*. Department of Physics and Astronomy, Århus University, Århus, 219.

Norðdahl, H. and Pétursson, H. G. 2000. Termination of the last glaciation in Eyjafjörður, Central North Iceland, - an example of rapid deterioration of an outlet glacier. In, Russell, A.J. and Marren, P.M. (eds.), *Iceland 2000, Modern Processes and Past Environments – Abstracts*. Department of Geography Occasional Papers Series No. 21, Keele University, Department of Geography, Keele, 76.

Norðdahl, H. and Pétursson, H. G. 2002. Marine limit shorelines and LGM glacier extent in SW Vestfírðir, Iceland. In, Jónsson, S. S. (ed.), *Abstract volume: 25th Nordic Geological Winter Meeting, 6-9 January, Reykjavík 2002*. Reykjavík, Geoscience Society of Iceland, 143.

Norðdahl, H. and Sæmundsson, Þ. 1999. Jarðsaga Rauðamels og nágrennis. *Geoscience Society of Iceland, Spring Meeting 1999*, 34-35.

Norðdahl, H., Pétursson, H. and Sæmundsson, Þ. 2000. Late Weichselian deglaciation of Eastern and North-eastern Iceland and the configuration of the inland ice-sheet. In, Russell, A.J. and Marren, P.M. (eds.), *Iceland 2000, Modern Processes and Past*

Environments – Abstracts. Department of Geography Occasional Papers Series No. 21, Keele University, Department of Geography, 77.

Nummedal, D., Hine, A.C. and Boothroyd, J.C. 1987. Holocene evolution of the south-central coast of Iceland. In, Fitzgerald, D.M. and Rosen, P.S. (eds.), *Glaciated coasts.* London, Academic Press, 115-150.

Nye, J.F. 1976. Water flow in glacier - jökulhlaups, tunnels and veins. *Journal of Glaciology,* 17, 181-188.

Oelkers, E.H. 2001a. A general kinetic description of multi-oxide silicate mineral and glass dissolution. *Geochimica et Cosmochimica Acta,* 65, 3703–3719.

Oelkers, E.H. 2001b. An experimental study of forsterite dissolution rates as a function of temperature and aqueous Mg and Si concentrations. *Chemical Geology,* 175, 485–494.

Oelkers, E. H. and Gíslason, S.R. 2001. The mechanism, rates, and consequences of basaltic glass dissolution: I. An experimental study of the dissolution rates of basaltic glass as a function of aqueous Al, Si, and oxalic acid concentration at 25°C and pH = 3 and 11. *Geochimica et Cosmochimica Acta,* 65, 3671–3681.

Oelkers, E.H. and Schott J. 1998. Does organic adsorption affect alkalifeldspar dissolution rates? *Chemical Geology,* 151, 235–245.

Oelkers, E H. and Schott J. 2001. An experimental study of enstatite dissolution rates as a function of pH, temperature, and aqueous Mg and Si concentration, and the mechanism of pyronene/pyroxenoid dissolution. *Geochimica et Cosmochimica Acta,* 65, 1219–1231.

Oelkers, E.H., Schott, J. and Devidal, J.L. 1994. The effect of aluminium, pH, and chemical affinity on the rates of aluminosilicate dissolution reactions. *Geochimica et Cosmochimica Acta,* 58, 2011–2024.

Ogilvie, A.E.J. 1982. *Climate and Society in Iceland from the Medieval Period to the Late Eighteenth Century.* Unpublished Ph.D. Thesis, University of East Anglia, Norwich.

Ogilvie, A.E.J. 1984. The past climate and sea-ice record from Iceland. Part 1: Data to A. D. 1780. *Climatic Change,* 6, 131-152.

Ogilvie, A. E. J. 1986. The climate of Iceland 1701-1784. *Jökull,* 36, 57-73.

Ogilvie, A. E. J. 1991. Climatic change in Iceland AD c. 865 to 1598. *Acta Archaeologica,* 61, 233-251.

Ogilvie, A. E. J. 1992a. Documentary evidence for changes in the climate of Iceland, AD 1500 to 1800. In, Bradley, R. S. and Jones, P. D. (eds.), *Climate since AD 1500.* London, Routledge, 93-117.

Ogilvie, A.E.J. 1992b. 1816 – A year without a summer in Iceland? In, Harington, C.R. (ed.), *The Year Without a Summer? World Climate in 1816.* Ottawa, Canadian Museum of Nature, 331-354.

Ogilvie, A. E. J. 1997. Fisheries, climate and sea ice in Iceland: an historical perspective. In, Vickers, D. (ed.), *Marine Resources and Human Societies in the North Atlantic Since 1500.* Institute of Social and Economic Research, Memorial University, St Johns, 69-87.

Ogilvie, A.E.J. 2001. Climate and farming in northern Iceland, 1700-1850. In, Sigurðsson, I. and Skaptason, J. (eds.), *Aspects of Arctic and Sub-Arctic History.* Reykjavík, University of Iceland Press, 289-299.

Ogilvie, A.E.J. and Jónsdóttir, I. 2000. Sea ice, climate and Icelandic fisheries in historical times. *Arctic*, 53, 383-394.

Ogilvie, A.E.J. and Jónsson, T. 2001. 'Little Ice Age' research: a perspective from Iceland. In, Ogilvie, A.E.J. and Jónsson, T. (eds.), *The Iceberg in the Mist: Northern Research in Pursuit of a 'Little Ice Age'.* Dordrecht, Kluwer Academic Publishers, 9-52.

Ogilvie, A.E.J. and McGovern T.H. 2000. Sagas and science: climate and human impacts in the North Atlantic. In, Fitzhugh, W.W. and Ward, E.I. (eds.), *Vikings: The North Atlantic Saga.* Washington D.C., Smithsonian Instititution Press, 385-393.

Ogilvie, A.E.J. and Pálsson, G. 2004. Mood, magic and metaphor: Allusions to weather and climate in the *Sagas of Icelanders.* In, Strauss, S. and Orlove, B.S. (eds.), *Weather, Climate, Culture.* Oxford and New York, Berg Publishers.

Ogilvie, A.E.J., Barlow, L.K. and Jennings, A.E. 2000. North Atlantic climate ca. AD 1000: millennial reflections on the Viking discoveries of Iceland, Greenland and North America. *Weather*, 55, 34-45.

Okko, V. 1956. Glacial drift in Iceland. *Acta Geographica Fennica*, 15, 1-133.

Ólafsdóttir, R., Schlyter, P. and Haraldsson, H.V. 2001. Simulating Icelandic vegetation cover during the Holocene. Implications for long-term land degradation. *Geografiska Annaler*, 83, 203-215.

Ólafsdóttir, Þ. 1975. Jökulgarður á sjávarbotni út af Breiðafirði. (English summary: A moraine ridge on the Iceland shelf, west of Breiðafjördur.) *Náttúrufræðingurinn*, 45, 247-271.

Ólafsson, E. 1772. *Vice-Lavmand Eggert Olafsens og Land-Physici Biarne Povelsens Reise igiennem Island.* Sorøe, Videnskabernes Sielskab.

Ólafsson, E. 1981. *Ferðabók Eggerts Ólafssonar og Bjarna Pálssonar um ferðir þeirra á Íslandi árin 1752-1757.* (Travel book of Eggert Ólafsson and Bjarni Pálsson on their travels in Iceland in the years 1752-1757.) Reykjavík, Bókaútgáfan Örn og Örlygur.

Ólafsson, E. and Pálsson, B. 1975. *Travels in Iceland 1752-1757.* Revised English edition. Reykjavík, Bókaútgáfán Örn og Örlygur.

Ólafsson, J. 1979. The chemistry of Lake Mývatn and the River Laxá. *Oikos*, 32, 82-112.

Olafsson, J. 1999. Connections between oceanic conditions off N-Iceland, Lake Mývatn temperature, regional wind direction variability and the North Atlantic Oscillation. *Rit Fiskideildar*, 16, 41-57.

Ólafsson, Ó. 1784. *Beskrivelse over Fuglefangsten ved Drangøe udi Island.* Copenhagen.

Ólason, P. E. 1948. *Íslenzkar æviskár frá landnámstímum til ársloka 1940.* Reykjavík, Hið íslenzka bókmenntafélag.

Ólason, V. 1998. *Dialogues with the Viking Age: Narration and Representation.* Translated by Andrew Wawn. Reykjavík, Mál og menning.

Olavius, Ó. 1780. *Oeconomisk Reise igiennem de nordvestlige, nordlige, og nordostlige Kanter af Island, ved Olaus Olavius ... tilligemed Ole Henchels Underretning om de islandske Svovel-Miiner og Svovel-Raffinering, samt Christian Zieners Beskrivelse over nogle Surterbrands-Fielde i Island.* Copenhagen, Gyldendal.

Olsen, B.M. 1910. Um kornyrkju á Íslandi að fornu. *Búnaðarrit*, 24, 81-167.

Olszewski, A. and Weckwerth, P. 1999. The morphogenesis of kettles in the Höfðabrekkujökull forefield, Mýrdalssandur, Iceland. *Jökull*, 47, 71-88.

Orradóttir, B. 2002. *The Influence of Vegetation on Frost Dynamics, Infiltration Rate and Surface Stability in Icelandic Rangelands.* Unpublished M.Sc. Thesis, Texas A and M Univeristy, College Station, Texas.

Óskarsson, H., Arnalds, Ó., Guðmundsson, J. and Guðbergsson, G. 2003. Organic carbon in Icelandic Andosols. In, Arnalds, Ó. and Stahr, K. (eds.), *Volcanic Soil Resources, Occurrence, Development and Properties.* Catena Special Issue, Elsevier, in press.

Owen, L.A. and Derbyshire, E. 1989. The Karakoram glacial depositional system. *Zeitschrift für Geomorphologie*, 76, 33-73.

Owen, L.A. and Derbyshire, E. 1993. Quaternary and Holocene intermontane basin sedimentation in the Karakoram Mountains. In, Shroder, J.F. (ed.), *Himalaya to the Sea.* London, Routledge, 108-131.

Påhlsson, I. 1981. A pollen analytical study on a peat deposit at Lágafell, southern Iceland. In, Haraldsson, H. (ed.), The Markarfljót sandur area, southern Iceland: sedimentological, petrographical and stratigraphical studies. *Striae*, 15, 60-64.

Pálsson, G. 1995. *The Textual Life of Savants: Ethnography, Iceland and the Linguistic Turn.* Chur, Harwood Academic Publishers.

Pálsson, G. and Helgason, A. 1998. Schooling and skipperhood: the development of dexterity. *American Anthropologist*, 100 (4).

Pálsson, H. 1965. *Eftir þjóðveldið. Heimildir annála um íslenzka sögu 1263-98.* Reykjavík, Heimskringla.

Pálsson, S. 1945. *Ferðabók Sveins Pálssonar. Dagbækur og ritgerðir 1791-1797.* (Travel book of Sveinn Pálsson. Diaries and essays 1791-1797.) Reykjavík, Snælandsútgáfan h.f.

Pálsson, S. and Vigfússon G.H. 1996. *Gagnasafn aurburðarmælinga 1963–1995.* Orkustofnun Report OS-96032/VOD-05 B.

Pálsson, S., Zóphóníasson, S., Sigurðsson, O., Kristmannsdóttir, H. and Aðalsteinsson, H. 1992. *Skeiðarárhlaup og framhlaup Skeiðarárjökuls 1991.* Orkustofnun, OS-92035/VOD-09B.

Parfitt, R.L. 1990. Allophane in New Zealand – a review. *Australian Journal of Soil Research*, 28, 343–360.

Parfitt, R.L. and Childs, C.W. 1988. Estimation of forms of Fe and Al: a review, and analysis of contrasting soils by dissolution and Moessbauer methods. *Australian Journal of Soil Research*, 26, 121–144.

Parfitt, R.L. and Kimble, J.M. 1989. Conditions for formation of allophane in soils. *Soil Science Society of America Journal*, 53, 971–977.

Park, C.C. 1987. *Acid Rain: rhetoric and reality.* New York, Methuen.

Parry, M.L. 1978. *Climatic Change, Agriculture and Settlement.* Studies in Historical Geography. Folkestone, Dawson.

Paterson, W.S.B. 1994. *The Physics of Glaciers*. Oxford, Pergamon/Elsevier.

Peacock, M.A. 1926. The vulcano-glacial palagonite formation of Iceland. *Geological Magazine*, 63, 385-399.

Pétursson, H. and Larsen, G. 1992. An Early Holocene basaltic tephra bed in North Iceland, a possible equivalent to the Saksunarvatn Ash bed. In, Geirsdóttir, Á., Norðdahl, H. and Helgadóttir, G. (eds.), *Abstracts: 20th Nordic Geological Winter Meeting, 7 – 10 January, Reykjavík 1992*. The Icelandic Geoscience Society and the Faculty of Science, University of Iceland, Reykjavík, 133.

Pétursson, H. G. 1986. *Kvartergeologiske Undersökelser på Vest-Melrakkaslétta, Nordöst Island*. Unpublished Cand. Real. Thesis, University of Tromsö, Tromsö.

Pétursson, H. G. 1988. Eldvirkni á hlýindakafla á síðasta jökulskeiði. *Geoscience Society of Iceland, Spring Meeting 1999*, 25.

Pétursson, H. G. 1991. The Weichselian glacial history of west Melrakkaslétta, Northeastern Iceland. In, Maizels, J. K. and Caseldine, C. (eds.), *Environmental Change in Iceland: Past and Present*. Dordrecht, Kluwer Academic Publishers, 49-65.

Pétursson, H. G. and Norðdahl, H. 1994. Ísaldarlok á Langanesi. *Geoscience Society of Iceland, Spring Meeting 1994*, 11.

Pétursson, H. G. and Norðdahl, H. 1999. Forn fjörumörk og jöklar í Eyjafirði. *Geoscience Society of Iceland, Spring Meeting 1999*, 28.

Pjeturs, H. 1900. The glacial palagonite formation of Iceland. *Scottish Geographical Magazine*, 16, 265-293.

Pokrovsky, O. and Schott, J. 2000. Kinetics and mechanism of forsterite dissolution at $25°$ C and pH from 1 to 12. *Geochimica et Cosmochimica Acta*, 64, 3313–3326.

Pollock, J. B., Toon, O. B., Summers, A., Baldwin, B. and van Camp, W. 1976. Volcanic explosions and climate change: A theoretical assessment. *Journal of Geophysical Research*, 81, 1071-1083.

Poole, R. L. 1926. *Chronicles and Annals. A Brief Outline of Their Origin and Growth*. Oxford, Clarendon Press.

Poulain, P.-M., Warn-Warnas, A. and Niiler, P.P. 1996. Near surface circulation of the Nordic Seas as measured by Lagrangian drifters. *Journal of Geophysical Research*, 101(C8), 18237-18258.

Poulain, P.-M., Valdimarsson, H., Niiler, P.P. and Malmberg, S.-A. 2001. Drifter observations of surface circulation in the northern North Atlantic and in the Nordic Seas: statistical description and energetics. Pers. comm.

Price, R.J. 1966. Eskers near the Casement Glacier, Alaska. *Geografiska Annaler*, 48, 111-125.

Price, R.J. 1968. The University of Glasgow Breiðamerkurjökull Project (1964-67). *Jökull*, 18, 389-394.

Price, R.J. 1969. Moraines, sandar, kames and eskers near Breiðamerkurjökull, Iceland. *Transactions of the Institute of British Geographers*, 46, 17-43.

Price, R.J. 1970. Moraines at Fjallsjökull, Iceland. *Arctic and Alpine Research*, 2, 27-42.

Price, R.J. 1971. The development and destruction of a sandur, Breiðamerkurjökull, Iceland. *Arctic and Alpine Research*, 3, 225-237.

Price, R.J. 1973. *Glacial and Fluvioglacial Landforms*. Edinburgh, Oliver and Boyd.

Price, R.J. 1980. Rates of geomorphological changes in proglacial areas. In, Cullingford, R.A., Davidson, D.A. and Lewin, J. (eds.), *Timescales in Geomorphology*. Chichester, Wiley, 79-93.

Price, R.J. 1982. Changes in the proglacial area of Breiðamerkurjökull, southeastern Iceland: 1890-1980. *Jökull*, 32, 29-35.

Price, R.J. and Howarth, P.J. 1970. The evolution of the drainage system (1904-1965) in front of Breiðamerkurjökull, Iceland. *Jökull*, 20, 27-37.

Principato, S. M. 2000. Glacial geology of the Reykjarfjörður, NW Iceland. *Geological Society of America Abstracts* 32, A18-A19.

Raiswell, R. and Thomas, A.G. 1984. Solute acquisition in glacial melt waters. I. Fjallsjökull (southeast Iceland): bulk meltwater with closed-system characteristics. *Journal of Glaciology*, 30, 35–43.

Rampino, M. R. and Caldeira, K. 1992. Episodes of terrestrial geologic activity during the past 260 million years: a quantitative approach. *Celestial Mechanisms and Dynamics in Astronomy*, 51, 1-13.

Rampino, M. R. and Self, S. 1984. Sulphur-rich volcanic eruptions and stratospheric aerosols. *Nature*, 310, 677-679.

Rampino, M.R., Self, S. and Stothers, R. B. 1988. Volcanic winters. *Annual Review of Earth and Planetary Science*, 16, 73-99.

Rasmussen, T. L., van Weering, T. C. E. and Labeyrie, L. 1996. High resolution stratigraphy of the Faeroe-Shetland Channel and its relation to North Atlantic paleoceanography: the last 87kyr. *Marine Geology*, 131, 75-88.

Raymond, C.F. 1987. How do glaciers surge? A review. *Journal of Geophysical Research*, 92, 9121-9134.

Raymond, C.F., Jóhannesson, T., Pfeffer, T. and Sharp, M. 1987. Propagation of a glacier surge into stagnant ice. *Journal of Geophysical Research*, 92, 9037-9049.

Rea, B.R. and Whalley, W.B. 1996. The role of bedrock topography, structure, ice dynamics and pre-glacial weathering in controlling subglacial erosion beneath a high latitude, maritime icefield. *Annals of Glaciology*, 22, 121-125.

Rea, B.R., Whalley, W.B., Evans, D.J.A., Gordon, J.E. and McDougall, D.A. 1998. Plateau icefields: geomorphology and dynamics. *Quaternary Proceedings*, 6, 35-54.

Rea, B.R., Whalley, W.B., Rainey, M. and Gordon, J.E. 1996. Blockfields: old or new? Evidence and implications from some plateau blockfields in northern Norway. *Geomorphology*, 15, 109-121.

Reiss, N. M., Groveman, B. S. and Scott, C. M. 1980. Construction of a long time-series of seasonal mean temperature for New Brunswick, New Jersey. *Bulletin of the New Jersey Academy of Science*, 25, 1-11.

Richardson, S. 1997. *Deglaciation and Shoreline Displacement Adjacent to a Spreading Ridge, N. E. Iceland*. Unpublished Ph.D. Thesis, Royal Holloway, University of London.

Rist, S. 1956. *Icelandic Fresh Waters.* National Energy Authority Report, Reykjavík, 127 pp.

Rist, S. 1967. Jökulhlaups from the ice cover of Mýrdalsjökull on June 25, 1955 and January 20, 1956. *Jökull,* 17, 243-248.

Rist, S. 1983. Floods and flood danger in Iceland. *Jökull,* 33, 119-132.

Roberts, M.J. 2002. *Controls on Supraglacial Outlet Development during Glacial Outburst Floods.* Unpublished Ph.D. Thesis, Staffordshire University.

Roberts, M.J, Russell, A.J., Tweed, F.S, Harris, T.D. and Fay, H. 2002c. The causes and characteristics of July 2002 Skaftárhlaup, Tungnaárjökull, Iceland. *Eos,* 83, 6-10 December 2002, San Francisco, Fall Meeting Supplement, Abstract C62A-0918.

Roberts, M.J., Russell, A.J., Tweed, F.S. and Knudsen, Ó. 2000a. Rapid sediment entrainment and englacial deposition during jökulhlaups. *Journal of Glaciology,* 153, 349-351.

Roberts, M.J., Russell, A.J., Tweed, F.S. and Knudsen, Ó. 2000b. Ice fracturing during jökulhlaups: implications for englacial floodwater routing and outlet development. *Earth Surface Processes and Landforms,* 25, 1429-1446.

Roberts, M.J., Russell, A.J., Tweed, F.S. and Knudsen, Ó. 2001. Controls on englacial sediment deposition during the November 1996 jökulhlaup, Skeiðarárjökull, Iceland. *Earth Surface Processes and Landforms,* 26, 935-952.

Roberts, M.J., Russell, A.J, Tweed, F.S. and Knudsen, Ó. 2002b. Controls on the development of supraglacial floodwater outlets during jökulhlaups. In, Snorasson, Á., Finnsdóttir, H.P. and Moss, M. (eds.), *The Extremes of the Extremes: Extraordinary Floods.* IAHS Publication, 271, 71-76.

Roberts, M.J., Tweed, F.S., Russell, A.J., Knudsen, Ó. and Harris, T.D. 2003. Hydrologic and geomorphic effects of temporary ice-dammed lake formation during jökulhlaups. *Earth Surface Processes and Landforms,* 28, 723-737.

Roberts, M.J., Tweed, F.S., Russell, A.J., Knudsen, Ó., Lawson, D.E., Larson, G.J., Evenson, E.B. and Björnsson, H., 2002a. Glaciohydraulic supercooling in Iceland. *Geology,* 30, 439–442.

Róbertsdóttir, B. 1992. Gjóskulagatímatal fyrir Suðurlandsundirlendi. Forsöguleg gjóskulög frá Kötlu, áður nefnd 'Katla 5000'. In, *Veggspjaldaráðstefna Jarðfræðafélags Íslands, 28. apríl 1992.* Reykjavík, Jarðfræðafélag Íslands, 8-9.

Robinson, S. G., Maslin, M. A. and McCave, N. 1995. Magnetic susceptibility variations in Upper Pleistocene deep-sea sediments of the N.E. Atlantic: implications for ice rafting and palaeocirculation at the Last Glacial Maximum. *Paleoceanography,* 10, 221-250.

Robock, A. D. 1991. The volcanic contribution to climate change of the past 100 years. In, Schesinger, M.E. (ed.), *Greenhouse-Gas-Induced Climate Change: A Critical Appraisal of Simulations and Observations.* Amsterdam, Elsevier, 429-444.

Robock, A. 2000. Volcanic eruptions and climate. *Review of Geophysics,* 38, 191- 219.

Ruddiman, W. F. and Glover, L. K. 1972. Vertical mixing of ice-rafted volcanic ash in North Atlantic sediments. *Geological Society of America Bulletin,* 83, 2817-2836.

Ruddiman, W. F. and McIntyre, A. 1981a. The North Atlantic Ocean during the last deglaciation. *Palaeogeography, Palaeoclimatology, Palaeoecology,* 35, 145-214.

Ruddiman, W. F. and McIntyre, A. 1981b. The mode and mechanism of the last deglaciation: oceanic evidence. *Quaternary Research,* 16, 125-134.

Ruddiman, W. F., and others, 1994. Reconstructing the last glacial and deglacial ice sheets. *Eos,* 82-84.

Rundgren, M. 1995. Biostratigraphic evidence of the Alleröd – Younger Dryas – Preboreal oscillation in northern Iceland. *Quaternary Research,* 44, 405-416.

Rundgren, M. 1998. Early-Holocene vegetation of northern Iceland: pollen and plant macrofossil evidence from the Skagi peninsula. *The Holocene,* 8, 553-564.

Rundgren, M. 1999. A summary of the environmental history of the Skagi peninsula, northern Iceland, 11,300-7800 BP. *Jökull,* 4, 1-19.

Rundgren, M. and Ingólfsson, Ó. 1999. Plant survival in Iceland during periods of glaciation? *Journal of Biogeography,* 26, 387-396.

Rundgren, M., Ingólfsson, Ó., Björck, S., Jiang, H. and Hafliðason, H. 1997. Dynamic sea-level change during the last deglaciation of northern Iceland. *Boreas,* 26, 201-215.

Rushmer, E.L., Russell, A.J., Tweed, F.S., Knudsen, Ó. and Marren, P.M. 2002. The role of hydrograph shape in controlling glacier outburst flood (jökulhlaup) sedimentation: justification of field prototypes for flume modelling. In, Dyer, M.C., Thoms, M.C. and Olley, J.C. (eds.), *The Structure, Function and Management Implications of Fluvial Sedimentary Systems.* IAHS Publication 276, 305-313.

Russell, A.J. 1993. Obstacle marks produced by flow around stranded ice blocks during a glacier outburst flood (jökulhlaup) in west Greenland. *Sedimentology,* 40, 1091-1111.

Russell, A. J. and Knudsen, Ó. 1999a. Controls on the sedimentology of November 1996 jökulhlaup deposits, Skeiðarársandur, Iceland. In, Smith, N.D. and Rogers, J. (eds.), *Fluvial Sedimentology* VI. IAS. Special Publication, 28, 315-329.

Russell, A. J. and Knudsen, Ó. 1999b. An ice-contact rhythmite (turbidite) succession deposited during the November 1996 catastrophic outburst flood (jökulhlaup), Skeiðarárjökull, Iceland. *Sedimentary Geology,* 127, 1-10.

Russell, A.J. and Knudsen, Ó. 2002a. The effects of glacier-outburst flood flow dynamics in ice-contact deposits: November 1996 jökulhlaup, Skeiðarársandur, Iceland. In, Martini, I.P., Baker, V.R. and Garzón, G. (eds.), *Flood and Megaflood Processes and Deposits: Recent and Ancient Examples.* IAS Special Publication, 32, 67-83.

Russell, A. J. and Knudsen, Ó. 2002b. The influence of channel flood history on the impact of the November 1996 jökulhlaup, Skeiðarárjökull, Iceland. In, Snorasson, Á., Finnsdóttir, H. P., Moss, M. (eds.), *The Extremes of The Extremes: Extraordinary Floods.* IAHS Publication 271, 243-247.

Russell, A. J. and Marren, P. M. 1999. Proglacial fluvial sedimentary sequences in Greenland and Iceland: a case study from active proglacial environments subject to jökulhlaups. In, Jones, A.P., Tucker, M.E. and Hart, J.K. (eds.), *The Description and Analysis of Quaternary Stratigraphic Field Sections.* Technical Guide 7, Quaternary Research Association, London, 171-208.

Russell, A. J., Tweed, F. S. and Knudsen, Ó. 2000. Flash flood at Sólheimajökull heralds the reawakening of an Icelandic subglacial volcano. *Geology Today,* May-June, 103-107.

Russell, A.J., Fay, H., Harris, T.D., Tweed, F. and Roberts, M.J. 2003b. Preservation of jökulhlaups within subglacial sediments. *The XVI INQUA Congress*, July 23 - 30, 2003 Reno, Nevada USA.

Russell, A. J., Knight, P., G., and van Dijk, T. A. G. P. 2001a. Glacier surging as a control on the development of proglacial fluvial landforms and deposits, Skeiðarárjökull, Iceland. *Global and Planetary Change*, 28, 173-184.

Russell, A.J., Knudsen, Ó., Maizels, J.K. and Marren, P.M. 1999a. Channel cross-sectional area changes and peak discharge calculations on the Gígjukvísl during the November 1996 jökulhlaup, Skeiðarársandur, Iceland. *Jökull*, 47, 1-14.

Russell, A. J., Knudsen, Ó., Fay, H., Marren, P. M., Heinz, J. and Tronicke, J. 2001b. Morphology and sedimentology of a giant supraglacial, ice-walled, jökulhlaup channel, Skeiðarársandur, Iceland. *Global and Planetary Change*, 28, 203-226.

Russell, A.J., Knudsen, Ó., Tweed, F.S., Marren, P.M., Rice, J.W., Roberts, M.J., Waitt, R.B. and Rushmer, L. 2000. Giant jökulhlaups from the northern margin of Vatnajökull ice cap, Iceland. *American Geophysical Union Fall meeting*, San Francisco, December 15-19.

Russell, A.J. Roberts, M.J. Tweed, F.S. Knudsen, Ó. Harris, T.D. Rushmer, E.L. Marren, P.M. and Carrivick, J.L. 2003a. Rapid downstream discharge attenuation during recent Icelandic jökulhlaups. *EGS-AGU-EUG Joint Assembly Nice, France*, 06 - 11 April 2003.

Russell, A.J., Tweed, F. S., Knudsen, Ó., Roberts, M.J., Harris, T.D. and Marren, P.M. 2002. The geomorphic impact and sedimentary characteristics of the July 1999 jökulhlaup on the Jökulsá á Sólheimasandi, Mýrdalsjökull, southern Iceland. In, Snorasson, Á., Finnsdóttir, H.P. and Moss, M. (eds.), *The Extremes of the Extremes: Extraordinary Floods*. IAHS Publication 271, 249-254.

Russell, A.J., Tweed, F.S., Knudsen, Ó., Roberts, M.J. Harris, T. Marren, P.M., Guðmundsson M.T. and Rice, J. submitted. The causes, characteristics and impacts of the volcanically induced subglacial outburst, July 1999, Sólheimajökull, Iceland. *GSA Bulletin*.

Russell, A. J., Tweed, F. S., Knudsen, Ó., Roberts, M. J. and Waller, R. I. 1999b. Ice fracturing and glacier sediment entrainment during two recent Icelandic jökulhlaups. *Eos*, (Fall Meeting Supplement) 80, 400.

Sæmundsson, K. 1967. Vulkanismus und Tektonik das Hengill-Gebietes in südwest-Island. *Acta Naturalia Islandica*, II(7), 1-103.

Sæmundsson, K. 1972. Jarðfræðiglefsur um Torfajökulssvæðið. (English summary: Notes on the geology of Torfajökull.) *Náttúrufræðingurinn*, 42, 81-99.

Sæmundsson, K. 1973. Straumrákaðar klappir í kringum Ásbyrgi. (English summary: Grooving on lava surfaces at Ásbyrgi NE-Iceland.) *Náttúrufræðingurinn*, 43, 52-60.

Sæmundsson, K. 1977. *Geological Map of Iceland, Sheet 7, NE-Iceland*. Reykjavík, Icelandic Museum of Natural History and Iceland Geodetic Survey.

Saemundsson, K. 1979. Outline of the geology of Iceland. *Jökull*, 29, 7-28.

Sæmundsson, K. 1991. Jarðfræði Kröflukerfisins. In, Garðarsson, A. and Einarsson, Á. (eds.), *Náttúra Mývatns*. Reykjavík, Hið íslenska náttúrufræðifélag, 25-95.

Sæmundsson, K. 1992. Geology of the Þingvallavatn area. *Oikos*, 64, 40-68.

Sæmundsson, K. 1995. Um aldur stóru dyngjanna á utanverðum Reykjanesskaga. In, Hróarsson, B., Jónsson, D. and Jónsson, S. S. (eds.), *Eyjar í Eldhafi*. Reykjavík, Gott mál, 165-178.

Sæmundsson, K., Pringle, M. S. and Harðarson, B. S. 2000. Um aldur berglaga í Kröflukerfinu. *Geoscience Society of Iceland, Spring Meeting 2000*, 26-27.

Sæmundsson, Þ. 1988. *Setmyndun í Rauðamel á Vestanverðum Reykjanesskaga*. Unpublished 4th year Thesis, University of Iceland, Reykjavík.

Sæmundsson, Þ. 1994. The deglaciation history of the Hofsárdalur valley, Northeast Iceland. In, Warren, W. P. and Croot, D. (eds.), *Formation and Deformation of Glacial Deposits*. Rotterdam, Balkema, 173-187.

Sæmundsson, Þ. 1995. Deglaciation and shoreline displacement in Vopnafjörður, north-eastern Iceland. *Lundqua Thesis*, 33, 1-106.

Sæmundsson, Þ. and Norðdahl, H. 2002. Rauðamelur, a Weichelian Interstadial on the Reykjanes peninsula, South-western Iceland. In, Jónsson, S. S. (ed.), *Abstract volume: 25th Nordic Geological Winter Meeting, 6-9 January, Reykjavík 2002*. Geoscience Society of Iceland, Reykjavík, 211.

Safn til Sögu Íslands IV. 1907-1915. Copenhagen and Reykjavík, 186-294.

Schove, D. J. 1954. Summer temperatures and tree-rings in North-Scandinavia AD 1461-1950. *Geografiska Annaler*, 36, 40-80.

Schunke, E. and Zoltai, S.C. 1988. Earth hummocks (thufur). In, Clark M.J. (ed.), *Advances in Periglacial Geomorphology*. New York, John Wiley, 231-245.

Schwar, J. 1978. Moorkundliche Untersuchungen am Laugarvatn (Südwest-Island). *Berichte aus der forschungsstelle Neðri-Ás*, Hveragerði, 29, 1-19.

Schythe, J. C. 1840. En Fieldreise i Island i Sommeren 1840. (A mountain trip in Iceland in the summer of 1840.) *Krøyers Naturhistorisk* Tidskrift, 4, 331-394, 499-500.

Self, S., Keszthelyi, L. and Þórðarson, Þ. 1998. The importance of pahoehoe. *Annual Review of Earth and Planetary Sciences*, 26, 81-110.

Self, S., Rampino, M. R. and Barbera, J. J. 1981. The possible effects of large 19th and 20th Century volcanic eruptions on zonal and hemispheric surface temperatures. *Journal of Volcanological and Geothermal Research*, 11, 41-60.

Self, S., Þórðarson, Þ. and Keszthelyi, L. 1997. Emplacement of continental flood basalt lava flows. In, Mahoney, J. J. and Coffin, M. F. (eds.), *Large Igneous Provinces: Continental, Oceanic, and Planetary Flood Volcanism*. Washington D. C., AGU, 100, 381-410.

Self, S., Þórðarson, Þ., Keszethelyi, L., Walker, G. P. L., Hon, K., Murphy, M.T., Long, P. and Finnemore, S. 1996. A new model for the emplacement of the Columbia River Basalt as large, inflated pahoehoe sheet lava flow fields. *Geophysical Research Letters*, 23, 2689-2692.

Sejrup, H. P., Sjoholm, J., Furnes, H., Beyer, I., Eide, L., Jansen, E. and Mangerud, J. 1989. Quaternary tephrochronology on the Iceland Plateau, north of Iceland. *Journal of Quaternary Science*, 4, 109-114.

Shakesby, R.A. 1989. Variability in Neoglacial moraine morphology and composition, Storbreen, Jotunheimen, Norway: within-moraine patterns and their implications. *Geografiska Annaler*, 71A, 17-29.

Sharp, M.J. 1984. Annual moraine ridges at Skálafellsjökull, south-east Iceland. *Journal of Glaciology*, 30, 82-93.

Sharp, M.J. 1985a. 'Crevasse-fill' ridges: a landform type characteristic of surging glaciers? *Geografiska Annaler*, 67A, 213-220.

Sharp, M.J. 1985b. Sedimentation and stratigraphy at Eyjabakkajökull - an Icelandic surging glacier. *Quaternary Research*, 24, 268-284.

Sharp, M.J. 1988. Surging glaciers: geomorphic effects. *Progress in Physical Geography*, 12, 533-559.

Sharp, M. and Dugmore, A.J. 1985. Holocene glacier fluctuations in East Iceland. *Zeitschrift für Gletscherkunde und Glazialgeologie*, 21, 341-349.

Sharpe, R.P. 1953. Glacial features of Cook County, Minnesota. *American Journal of Science*, 251, 855-883.

Shoji, S., Nanzyo, M. and Dahlgren, R.A. 1993. Volcanic ash soils. Genesis, properties and utilization. *Developments in Soil Science*, 21, Amsterdam,Elsevier.

Shoji, S., Nanzyo, M., Dahlgren, R.A. and Quantin, P. 1996. Evaluation and proposed revisions of criteria for Andosols in the World Reference Base for Soil Resources. *Soil Science*, 161, 604-615.

Shreve, R.L. 1985. Esker characteristics in terms of glacier physics. Katahdin esker system, Maine. *Bulletin of the Geological Society of America*, 96, 639-646.

Sigbjarnarson, G. 1969. The loessial soil formation and the soil erosion on Haukadalsheiði. (In Icelandic. Extended English summary.) *Náttúrufræðingurinn*, 39, 68–118.

Sigbjarnarson, G. 1973. Katla and Askja. *Jökull*, 23, 45-51.

Sigbjarnarson, G. 1983. The Quaternary alpine glaciation and marine erosion in Iceland. *Jökull*, 33, 87-98.

Sigbjarnarson, G. 1990. Hlaup og hlaupfarvegir. In, Sigbjarnarson, G. (ed.), *Vatnið og Landið. Vatnafræðin*. Orkustofnun, Reykjavík, 129-141.

Sigfúsdóttir, A. B. 1969. Hitabreytingar á Íslandi 1846-1968. (Temperature variations in Iceland 1846-1968.) In, Einarsson, M.Á, (ed.), *Hafísinn* ('The Sea Ice'). Reykjavík, Almenna bókafélagið, 70-79.

Sigfússon, B. 1956. *KL = Kulturhistorisk Leksikon for nordisk middelalder fra vikingetid reformationstid* I- 1956: Danish edition. Copenhagen, Rosenkilde og Bagger.

Sigurðsson, F. and Einarsson, K. 1988. Groundwater resources of Iceland—availability and demand. *Jökull*, 38, 35–53.

Sigurðsson, F. and Ingimarsson, J. 1990. The hydraulic conductivity of Icelandic rocks. In, Sigbjarnarson, G. (ed.), *Vatnið og Landið*. Reykjavík, Orkustofnun, 121-128 (in Icelandic).

Sigurðsson, G. 2002. *Túlkun íslendingasagna í ljósi munnlegar hefðar*. ('Íslendingasögur and oral tradition: a discourse on method'). Reykjavík, Stofnun Árna Magnússonar á Íslandi.

Sigurðsson, H. 1990. Evidence of volcanic aerosol loading of the atmosphere and climate response. *Paleo*, 89, 227-289.

Sigurðsson, O. 1998. Glacier variations in Iceland 1930-1995 - From the database of the Iceland Glaciological Society. *Jökull*, 45, 3-25.

Sigurðsson, O. 1999. Jökulhlaup úr Sólheimajökuli 17-18 júlí 1999. *Jöklarannsóknafélag Íslands*, 74, 6-7.

Sigurðsson, O. 2002a. Jökulhlaup úr Kverkfjöllum. *Jöklannsóknafélag Íslands*, 87, 4-7.

Sigurðsson, O. 2002b. Skaftárhlaup í júlí 2002. *Jökularannsóknafélag Íslands*, 89, 6-8.

Sigurðsson, O. and Jónsson, T. 1995. Relation of glacier variations to climate change in Iceland. *Annals of Glaciology*, 21, 263-270.

Sigurðsson, O., Snorrason, Á. and Zóphóníasson, S. 1992. Jökulhlaupaannáll 1984-1988. *Jökull*, 42, 73-80.

Sigurðsson, O., Zóphóníasson, S. and Ísleifsson, E. 2000. Jökulhlaup úr Sólheimajökuli 18 júlí 1999, *Jökull*, 49, 75-80.

Sigurðsson, S. 1977. Birki á Íslandi (útbreiðsla og ástand). In, sex vinir Hákonar Bjarnasonar (eds.), *Skógarmál: þættir um gróður og skóga tileinkaðir Hákoni Bjarnasyni sjötugum*, Reykjavík, 146-177.

Sigurgeirsson, M.Á. and Leósson, M.A. 1993. Gjóskulög í Sogamýri: tvö gjóskulög frá upphafi nútíma. *Náttúrufræðingurinn*, 62, 129-137.

Sigurvinsson, J. R. 1982. *Kvarter Landmótun Útnessins Milli Dýrafjarðar og Önundarfjarðar*. Unpublished 4th year Thesis, University of Iceland, Reykjavík.

Sigvaldason, G.E. 1959. Mineralogische Untersuchungen über Gesteinszersetzug durch postvulkanische Aktivität in Island. *Contribution to Mineralogy and Petrology*, 6, 405-426.

Sigvaldason, G.E. 1968. Structure and products of subaquatic volcanoes in Iceland. *Contributions to Mineralogy and Petrology*, 18, 1-16.

Sigvaldason, G.E., Annertz, K. and Nilsson, M. 1992. Effect of glacier loading/unloading on volcanism: Postglacial volcanic production rate of the Dyngjufjöll area, central Iceland. *Bulletin of Volcanology*, 54, 385-392.

Sigþórsdóttir, M. 1976. *Frjógreining úr mó undan Búrfellshrauni. Rannsóknarverkefni við Verkfræði- og raunvísindadeild H.Í.* (mimeograph).

Símonarson, L. A. and Leifsdóttir, Ó. E. 2002. Late-Holocene sea-level changes in South and Southwest Iceland reconstructed from littoral molluscan stratigraphy. *The Holocene, 12*, 149-158.

Skaftadóttir, Þ. 1974. *Um Tvö Frjólínurit af Rosmhvalanesi og Ágrip af Jarðsögu Þess*. B.Sc. Thesis, Reykjavík, Verkfræði- og raunvísindadeild H.Í. (mimeograph).

Skilling, I.P. 1994. Evolution of an englacial volcano: Brown Bluff, Antarctica. *Bulletin of Volcanology*, 56, 573-591.

Skilling, I.P. 2002. Basaltic pahoehoe lava-fed deltas: large-scale characteristics, clast generation, emplacement processes and environmental discrimination. In, Smellie, J.L. and Chapman, M. (eds.), Ice-Volcano Interaction on Earth and Mars. *Geological Society of London Special Publication*, 202, 91-113.

Smellie, J.L. 2000. Subglacial eruptions. In, Sigurðsson, H. (ed.), *Encyclopedia of Volcanoes*. San Diego, Academic Press, 403-418.

Smellie, J.L. 2002. The 1969 subglacial eruption on Deception Island (Antarctica): events and processes during an eruption beneath a thin glacier and implications for volcanic hazards. In, Smellie, J.L., and Chapman, M. (eds.) Ice-volcano interaction on Earth and Mars. *Geological Society of London Special Publication*, 202, 59-79.

Smellie, J.L. and Chapman, M. 2002 (eds.). *Ice-volcano interaction on Earth and Mars*. Geological Society of London Special Publication, 202, 431 pp.

Smellie, J.L., Hole, M.J and Nell, P.A.R. 1993. Late Miocene valley-confined subglacial volcanism in northern Alexander Island, Antarctic Peninsula. *Bulletin of Volcanology*, 55, 273-288.

Smith, K. 2001. Holocene jökulhlaups, north western Mýrdalsjökull, Iceland – new evidence and wider significance. *Abstracts Earth System Processes*, 24-28 June, Edinburgh, Scotland. The Geological Society of America and The Geological Society of London.

Smith, L. C., Alsdorf, D. E., Magilligan, F. J., Gomez, B., Mertes, L. A. K., Smith, N. D. and Garvin, J. B. 2000. Estimation of erosion, deposition, and net volumetric change caused by the 1996 Skeiðarársandur jökulhlaup, Iceland, from synthetic aperture radar interferometry. *Water Resources Research*, 36, 1583-1594.

Smith, L. M. 2001. *Holocene Paleoenvironmental Reconstruction of the Continental Shelves Adjacent to the Denmark Strait*. Unpublished Ph.D. Thesis, University of Colorado.

Smith, L. M. and Licht, K. J. 2000. Radiocarbon Date List IX: Antarctica, Arctic Ocean, and the Northern North Atlantic. *INSTAAR Occasional Papers*, No. 54, University of Colorado, 138 pp.

Snorrason, S.P. and Vilmundardóttir, E.G. 2000. Pillow lava sheets: Origins and flow patterns. *Volcano/Ice Interaction on Earth and Mars, Reykjavík, Abstract Volume*, 45.

Snorrason, Á., Jónsson, P., Pálsson, S., Árnason, S., Sigurðsson, O., Víkingsson, S., Sigurðsson, Á. and Zóphóníasson, S. 1997. Hlaupið á Skeiðarársandi haustið 1996: útbreiðsla, rennsli og aurburður. In, Haraldsson, H. (ed.), *Vatnajökull: Gos og hlaup*. Vegagerðin, 79-137.

Snorrason, Á., Jónsson, P., Pálsson, S., Árnason, S., Víkingsson, S. and Kaldal, I. 2002. November 1996 jökulhlaup on Skeiðarársandur outwash plain, Iceland. In, Martini, I.P., Baker, V.R. and Garzón, G. (eds.), *Flood and Megaflood Processes and Deposits: Recent and Ancient Examples*. IAS Special Publication, 32, 55-65.

Soil Survey Staff, 1998. *Keys to Soil Taxonomy*. 8th edition. USDA-NRCS, Washington, D.C.

Spedding, N. and Evans, D.J.A. 2002. Sediments and landforms at Kvíárjökull, southeast Iceland: a reappraisal of the glaciated valley landsystem. *Sedimentary Geology*, 149, 21-42.

Spring, U. and Hutter, K. 1981. Numerical studies of jökulhlaups. *Cold Regions Science and Technology*, 4, 227-244.

Stefánsson, A. and Gíslason, S.R. 2001. Chemical weathering of basalts, SW Iceland: effect of rock crystallinity and secondary minerals on chemical fluxes to the ocean. *American Journal of Science*, 301, 513–556.

Stefánsson, A., Gíslason, S.R. and Arnórsson, S.A. 2001. Dissolution of primary minerals in natural waters II. Mineral saturation state. *Chemical Geology*, 172, 251–276.

Stefánsson, U. 1962. North Icelandic Waters. *Rit Fiskideildar* III. Bind, Vol 3.

Stefánsson, U. 1969. Near Surface Temperature in the Icelandic Coastal Waters. *Jökull*, 19, 29.

Stein, R., Nam, S.-I., Grobe, H. and Hubberten, H. 1996. Late Quaternary glacial history and short-term ice-rafted debris fluctuations along the East Greenland continental margin. In, Andrews, J. T., Austen, W. A., Bergetsen, H. and Jennings, A. E. (eds.), *Late Quaternary Paleoceanography of North Atlantic Margins*. Geological Society, London, 135-151.

Steindórsson, S. 1937. Jurtagróðurinn og jökultíminn. *Náttúrufræðingurinn*, 7, 93-100.

Steindórsson, S. 1954. Um aldur og innflutning íslenzku flórunnar. *Ársrit Ræktunarfélags Norðurlands*, 54, 53-72.

Steindórsson, S. 1962. On the age and immigration of the Icelandic flora. *Societas Scientiarum Islandica*, 35, 1-157.

Steinþórsson, S., Harðarson, B.S., Ellam, R.M. and Larsen, G. 2000. Petrochemistry of the Gjálp 1996 subglacial eruption, Vatnajökull, SE Iceland. *Journal of Volcanology and Geothermal Research*, 98, 79-90.

Stephensen, M. 1806. *Eftirmæli Átjándu Aldar eptir Krists híngadburd, frá Ey-konunni Íslandi*. Leirárgørdum, Islands opinbera Vísinda-Stiptun.

Stephensen, M. 1808. *Island i det Attende Aarhundrede, Historiskpolitisk Skildret*. Copenhagen, Gyldendalske Boghandling.

Stokes, C. R. and Clark, C. D. 2001. Palaeo-ice streams. *Quaternary Science Reviews*, 20, 1437-1457.

Stoner, J. S. and Andrews, J. T. 1999. The North Atlantic as a Quaternary magnetic archive. In, Maher, B. and Thompson, R. (eds.), *Quaternary Climates, Environments and Magnetism*. Cambridge, Cambridge University Press.

Storm, G. (ed.) 1888. *Islandske Annaler Intil 1578*. Det norske historiske kilderskriftfond. Christiania. Reprinted Oslo, 1977.

Stothers, R. B. 1989. Turbulent atmospheric plumes above line sources with an application to volcanic fissure eruptions on the terrestrial planets. *Journal of Atmospheric Science*, 46, 2662-2670.

Stothers, R. B. 1993. Flood basalts and extinction events. *Geophysical Research Letters*, 20, 1399-1402.

Stothers, R. B. 1996. The great dry fog of 1783. *Climate Change*, 32, 79-89.

Stothers, R. B. 1998. Far reach of the tenth century Eldgjá eruption, Iceland. *Climate Change*, 39, 715- 726.

Stothers, R. B., Wolff, J. A., Self, S. and Rampino, M. R. 1986. Basaltic fissure eruptions, plume height and atmospheric aerosols. *Geophysical Research Letters*, 13, 725-728.

Stötter, J. 1990. Neue Beobachtungen und Überlegungen zur postglazialen Landschaftsgeschichte Islands am Beispiel des Svarfaðar - Skíðadals. *Münchener Geographische Abhandlungen*, B8, 83-104.

Stötter, J. 1991a: Geomorphologische und landschaftsgeschichtliche Untersuchungen im Svarfaðardalur-Skíðadalur, Tröllaskagi, N-Island. *Münchener Geographische Abhandlungen*, B9.

Stötter, J. 1991b. New observations on the postglacial glacial history of Tröllaskagi, northern Iceland. In, Maizels, J.K. and Caseldine, C. (eds.), *Environmental Change in Iceland: Past and Present*, Kluwer, Dordrecht, 181-192.

Stötter, J. and Wastl, M. 1999. Landschafts- und Klimageschichte Nordislands im Postglazial. *Geographischer Jahresbericht aus Österreich*, 56, 49-68.

Stötter, J., Wastl, M., Caseldine, C. and Häberle, T. 1999. Holocene palaeoclimatic reconstructions in Northern Iceland: approaches and results. *Quaternary Science Reviews*, 18, 457-474.

Straka, H. 1956. Pollenanalytische Untersuchungen eines Moorprofiles aus Nord-Island. *Neues Jahrbuch für Geologie und Palaeontologie*, 6, 262-272.

Stuiver, M. and Reimer, P.J. 1993. Extended ^{14}C data base and revised CALIB 3.0 ^{14}C age calibration program. *Radiocarbon*, 35, 215-230.

Stuiver, M., Reimer, P. J., Bard, E., Beck, J. W., Burr, G. S., Hughen, K. A., Kromer, B., McCormac, F. G., v. d. Plicht, J. and Spurk, M. 1998. INTCAL98 Radiocarbon age calibration 24,000-0 cal BP. *Radiocarbon*, 40, 1041-1083.

Sturluson, S. 1979. *Heimskringla* I. *Íslenzk Fornrit* XXVI. Reykjavík, Hið íslenzka fornritafélag.

Sturluson, S. 1999. *Heimskringla. History of the Kings of Norway.* Translated with Introduction and Notes by Lee M. Hollander. Austin, The American Scandinavian Foundation, University of Texas Press.

Sveinbjörnsdóttir, Á. E. and Johnsen, S. J. 1991. The late glacial history of Iceland. Comparison with isotopic data from Greenland and Europe, and deep sea sediments. *Jökull*, 40, 83-96.

Sveinbjörnsdóttir, Á. E., Eiríksson, J., Geirsdóttir, Á., Heinemeier, J. and Rud, N. 1993. The Fossvogur marine sediments in SW Iceland – confined to the Alleröd/Younger Dryas transition by AMS ^{14}C dating. *Boreas*, 22, 147-157.

Sveinsson, G. 1919. *Kötlugosið 1918 og afleiðingar þess.* (The 1918 Katla eruption and its impact.) Reykjavík, Gutenberg.

Sveinsson, P. 1992. Kötluför, 2. september 1919. (A trip to Katla, 2 September 1919.) *Jökull*, 42, 89-93.

Swift, J.H. 1980. *Seasonal Processes in the Iceland Sea with Especial Reference to Their Relationship to the Denmark Strait Overflow.* Unpublished Ph.D. Thesis, University of Washington, Seattle.

Swift, J.H. 1984. The circulation of the Denmark Strait and Iceland-Scotland overflow waters in the North Atlantic. *Deep-Sea Research*, 31, 1339-1355.

Syverson, K.M., Gaffield, S.J. and Mickelson, D.M. 1994. Comparison of esker morphology and sedimentology with former ice-surface topography, Burroughs Glacier, Alaska. *Bulletin of the Geological Society of America*, 106, 1130-1142.

Syvitski, J. P., Andrews, J. T. and Dowdeswell, J. A. 1996a. Sediment deposition in an iceberg-dominated glacimarine environment, East Greenland: basin fill implications. *Global and Planetary Change*, 12, 251-270.

Syvitski, J. P. M., Lewis, C. F. M. and Piper, D. J. W. 1996b. Paleoceanographic information derived from acoustic surveys of glaciated continental margins: examples from eastern Canada. In, Andrews, J. T., Austin, W., Bergsten, H. and Jennings, A. E. (eds.), *Late Quaternary Paleoceanography of the North Atlantic Margins*. Geological Society, London, 51-76.

Syvitski, J. P. M., Jennings, A. E. and Andrews, J. T. 1999. High-resolution seismic evidence for multiple glaciation across the southwest Iceland shelf. *Arctic and Alpine Research*, 31, 50-57.

The Statistical Bureau of Iceland, 1984. *Handbook of Statistics 1984*. The Statistical Bureau of Iceland, Reykjavík. (In Icelandic.)

Thomas, G.P. and Connell, R.J. 1985. Iceberg drop, dump and grounding structures from Pleistocene glacio-lacustrine sediments, Scotland. *Journal of Sedimentary Petrology*, 55, 243-249.

Thompson, A. P. and Jones, A. 1986. Rates and causes of proglacial river terrace formation in southeast Iceland: an application of lichenometric dating techniques. *Boreas*, 15, 231-246.

Thorarensen, S. 1792. *Tanker til høiere Eftertanke, om Uaar og dets Virkninger, samt om Føde eller Korn-Magaziners Oprettelse i Haarde Aar; Med tilføide Specielle Beregninger, til de Handlendes Underretning, over Exporterne fra Handelstederne i Nord- og Øster Amtet, i Aarene 1789 og 1790*. ('Thoughts for Greater Consideration regarding Dearth Years and their Effects, in addition to the setting up of Food or Grain Reserves in Severe Years'.) Copenhagen.

Thorarensen, P. M. 1997 (written in 1839). Sandfells- og Hofssóknir í Öræfum. (Description of the parishes of Sandfell and Hof.) In, Jónsson, J.A. and Sigmundsson, S. (eds.), *Skaftafellssýsla. Sýslu- og sóknalýsingar Hins íslenska bókmenntafélags*. Reykjavík, Sögufélag, 145-153.

Þórarinsson, S. 1939. The ice-dammed lakes of Iceland, with particular reference to their values as indicators of glacier oscillations. *Geografiska Annaler*, 21, 216-242.

Þórarinsson, S. 1943. Vatnajökull. The scientific results of the Swedish-Icelandic investigations 1936-37-38. Chapter XI. Oscillations of the Icelandic glaciers in the last 250 years. *Geografiska Annaler*, 25, 1-56.

Þórarinsson, S. 1944. *Tefrokronologiska studier på Island*. Munksgaard, Copenhagen.

Þórarinsson, S. 1950. Glacier outburst floods in the river Jökulsá á Fjöllum. *Náttúrufræðingurinn*, 20, 113-133.

Þórarinsson, S. 1953a. Oscillations of Iceland glaciers during the last 250 years. *Geografiska Annaler*, 25, 1-54.

Þórarinsson, S. 1953b. Some aspects of the Grímsvötn problem. *Journal of Glaciology*, 2, 267-274.

Þórarinsson, S. 1955. Nákuðungslögin við Húnaflóa. *Náttúrufræðingurinn*, 25, 172-186.

Þórarinsson, S. 1956a. On the variations of Svínafellsjökull, Skaftafellsjökull and Kvíárjökull in Öræfi. *Jökull*, 6, 1-15.

Þórarinsson, S. 1956b. *The Thousand Years Struggle against Ice and Fire*. Reykjavík, Menningarsjóður.

Þórarinsson, S. 1956c. Mórinn í Seltjörn. (English summary. The submerged peat in Seltjörn.) *Náttúrufræðingurinn*, 26, 179-193.

Þórarinsson, S. 1957. The jökulhlaup from the Katla area in 1955 compared with other jökulhlaups in Iceland. *Jökull*, 7, 21-25.

Þórarinsson, S. 1958. The Öræfajökull eruption 1362. *Acta Naturalia Islandica*, II, 2, 100 pp.

Þórarinsson, S. 1959a. Um möguleika á Því að segja fyrir um næsta Kötlugos. (On the possibilities of predicting the next eruption in Katla.) *Jökull*, 9, 6-18.

Þórarinsson, S. 1959b. Some geological problems involved in the hydro-electric development of the Jökulsá á Fjöllum. *Report to the State Electricity Authority*. 35pp.

Þórarinsson, S. 1961. Uppblástur á Íslandi í ljósi öskulagarannsókna (Wind erosion in Iceland. A tephrochronological study). *Icelandic Forestry Society Yearbook*, 1961, 17–54 .

Þórarinsson, S. 1964. Sudden advances of Vatnajökull outlet glaciers 1930-1964. *Jökull*, 14, 76-89.

Þórarinsson, S. 1969. Glacier surges in Iceland, with special reference to the surges of Brúarjökull. *Canadian Journal of Earth Sciences*, 6, 875-882.

Þórarinsson, S. 1970. Tephrochronology and medieval Iceland. In, Berger, R. (ed.), *Scientific Methods in Medieval Archaeology*. Los Angeles, Berkeley, London, University of California Press.

Þórarinsson, S. 1974a. Sambúð lands og lýðs í ellefu aldir. (Relation of land and people in eleven centuries.) In, Líndal, S. (ed.), *Saga Íslands* I. Reykjavík, Hið íslenzka bókmenntafélag, Sögufélagið, 29-97.

Þórarinsson, S. 1974b. *Vötnin Stríð. Saga Skeiðarárhlaupa og Grímsvatnagosa*. (The swift flowing rivers. The history of Grímsvötn jökulhlaups and eruptions.) Reykjavík, Menningarsjóður.

Þórarinsson, S. 1975. Katla og annáll Kötlugosa. (Katla and an annal of Katla eruptions.) In, Jónsson, P. (ed.), *Árbók Ferðafélags Íslands 1975*. Ferðafélag Íslands, Reykjavík, 125-149.

Þórarinsson, S. 1979. On the damage caused by volcanic eruptions with special reference to tephra and gases. In, Sheets, P. D. and Grayson, D. K. (eds.), *Volcanic Activity and Human Geology*. New York, Academic Press, 125-159.

Þórarinsson, S. 1981. Greetings from Iceland: ash-falls and volcanic aerosols in Scandinavia. *Geografiska Annaler*, 63, 109-118.

Þórarinsson, S. and Rist, S. 1955. Rannsókn á Kötlu og Kötluhlaupi sumarið 1955. (Investigations on Katla and a Katla jokulhlaup in the summer of 1955.) *Jökull*, 5, 37-40.

Þórarinsson, S., Einarsson, Þ., Sigvaldason, G. and Elísson, G. 1964. The submarine eruption off the Vestmann Islands 1963-64. *Bulletin of Volcanology*, XXVII, 1-12.

Þórarinsson, S. and Sæmundsson, K. 1979. Volcanic activity in historical time. *Jökull*, 29, 29-32.

Þórarinsson, Þ. 1974. Þjóðin lifði en skógurinn dó. *Ársrit Skógræktarfélags Íslands*, 25, 16-29.

Þórðardóttir, Þ. 1977. Primary production in North Icelandic waters in relation to recent climatic change. *Polar Oceans: Proceedings of the Oceanographic Congress*, 655-665.

Þórðardóttir, Þ. 1984. Primary production North of Iceland in relation to water masses in May-June 1970-1980. *Council for the Exploration of the Sea, C.M.*, 1984/L20, 1-17.

Þórðarson, Þ. 1995. *Volatile Release and Atmospheric Effects of Basaltic Fissure Eruptions*. Department of Geology and Geophysics. Honolulu, University of Hawaii.

Þórðarson, Þ. and Self, S. 1993. The Laki (Skaftár Fires) and Grímsvötn eruptions in 1783-1785. *Bulletin of Volcanology*, 55, 233-263.

Þórðarson, Þ. and Self, S. 1996. Sulphur, chlorine and fluorine degassing and atmospheric loading by the Roza eruption, Columbia River Basalt group, Washington, USA. *Journal of Volcanological and Geothermal Research*, 74, 49-73.

Þórðarson, Þ. and Self, S. 1998. The Roza Member, Columbia River Basalt Group: A gigantic pahoehoe lava flow field formed by endogenous processes? *Journal of Geophysical Research*, 103, 27,411-27,445.

Þórðarson, Þ. and Self, S. 2001. Real-time observations of the Laki sulphuric aerosol cloud in Europe 1783 as documented by Professor S. P. van Swinden at Franeker, Holland. *Jökull*, 50, 65-72.

Þórðarson, Þ. and Self, S. 2003. Atmospheric and environmental effects of the 1783-84 Laki eruption, Iceland: a review and reassessment. *Journal of Geophysical Research*, 108, 10.1029/2001JD002042.

Þórðarson, Þ., Miller, D. J., Larsen, G., Self, S. and Sigurðsson, H. 2001. New estimates of sulphur degassing and atmospheric mass-loading by the 934 AD Eldgjá eruption, Iceland. *Journal of Volcanological and Geothermal Research*, 108, 33-54.

Þórðarson, Þ., Self, S., Miller, D. J., Larsen, G. and Vilmundardóttir, E. G. 2003. Sulphur release from flood lava eruptions in the Veiðivötn, Grímsvötn and Katla volcanic systems, Iceland. In, Oppenheimer, C., Pyle, D.M. and Barclay, J. (eds.), *Volcanic Degassing*. Geological Society of London Special Publication, 213, 103-121.

Þórðarson, Þ., Self, S., Óskarsson, N. and Hulsebosch, T. 1996. Sulphur, chlorine, and fluorine degassing and atmospheric loading by the 1783-1784 AD Laki (Skaftár Fires) eruption in Iceland. *Bulletin of Volcanology*, 58, 205-225.

Þórhallsdóttir, Þ. E. 1997. Tundra ecosystems of Iceland. In, Wiegolaski, F.E. (ed.), Polar and Alpine Tundra. *Ecosystems of the World. 3*, New York, Elsevier, 85–96.

Þorkelsson, J. 1887. Þáttur af Birni Jónssyni á Skarðsá. *Tímarit hins íslenzka bókmenntafjelags*, 8, 34-96.

Þorkelsson, Þ. 1923. Eldgosin 1922. (The eruptions 1922.) *Tímarit V.F.Í.*, 1923, 29-40.

Þorláksson, H. 1991. *Vaðmál og verðlag: vaðmál í utanlandsviðskiptum og bískap Íslendings á 13. og 14. öld*. Reykjavík.

Þorláksson, H. 1992. *Vaðmál og verðlag: vaðmál í utanlandsviðskiptum og búskap Íslendinga á 13. og 14. öld*. ('Wadmal and value: wadmal in foreign trade and the Icelandic economy'.) Unpublished Ph.D. Thesis, University of Iceland, Reykjavík.

Thoroddsen, Þ. 1892. Postglaciale marine aflejringer, kystterrasser og strandlinjer i Island. *Geografisk Tidskrift*, 11, 209-225.

Thoroddsen, Þ. 1892-1904. *Landfræðissaga Íslands. Hugmyndir manna um Ísland, Náttúruskoðun og rannsóknir fyrr og síðar* I-IV. Reykjavík 1892-96, Copenhagen, 1898-1904. Hið íslenzka bókmenntafélag.

Thoroddsen, Þ. 1906. *Island. Grundriss der Geographie und Geologie.* Petermanns Mitteilungen, Ergänzungshefte No. 152 und 153. Gotha, Justus Perthes.

Thoroddsen, Þ. 1908-22. *Lýsing Íslands* I-IV. Copenhagen, Hið íslenzka bókmenntafélag.

Thoroddsen, Þ. 1916-17. *Árferði á Íslandi í þúsund ár.* Copenhagen. Hið íslenzka fræðafélag.

Thoroddsen, Þ. 1933. *Lýsing Íslands.* Jöklar (Description of Iceland. Glaciers). Vol. II. Reykjavík, Sjóður Þorvaldar Thoroddsen.

Thoroddsen, Þ. 1959. *Ferðabók*, III. bindi. Reykjavík, Snæbjörn Jónsson.

Thors, K. 1974. *I. Sediments of the Vestfirðir Shelf, NW Iceland, and II. Geology of the Úlfarsfell Area, SW Iceland.* Ph.D. Thesis, University of Manchester, 167 pp.

Thors, K. 1978. The sea-bed of the southern part of Faxaflói, Iceland. *Jökull*, 28, 42-52.

Thors, K. and Boulton, G. S. 1991. Deltas, spits and littoral terraces associated with rising sea level: Late Quaternary examples from northern Iceland. *Marine Geology*, 98, 99-112.

Thors, K. and Helgadóttir, G. 1991. Evidence from south west Iceland of low sea level in early Flandrian times. In, Maizels, J. K. and Caseldine, C. (eds.), *Environmental Change in Iceland: Past and Present.* Dordrecht, Kluwer Academic Publishers, 93-104.

Thors, K. and Helgadóttir, G. 1999. Seismic reflection profiles from Ísafjarðardjúp, Jökulfirðir, and Djúpáll, NW Iceland. *Abstracts Geological Society of America*, 31, A74.

Thorseth, I. H., Furnes, H. and Tumyr, O. 1995. Textural and chemical effects of bacterial activity on basaltic glass: an experimental approach. *Chemical Geology*, 119, 139–160.

Thorson, G. 1957. Bottom communities. In: Hedgpeth, J. W. (ed.), Treatise on marine ecology and paleoecology: I. Ecology. *Memoirs of the Geological Society of America*, 67, 461-534.

Þorsteinsson, I. 1980. Environmental data, botanical composition and production of plant communities and the plant preference of sheep. *Journal of Agricultural Research in Iceland*, 12, 85–99.

Thwaites, F.T. 1926. The origin and significance of pitted outwash. *Journal of Geology*, 34, 308-319.

Thwaites, F.T. 1935. *Outline of Glacial Geology.* Edwards Bros.

Tómasson, H. 1973. Hamfarahlaup í Jökulsá á Fjöllum. (English summary: Catastrophic floods in Jökulsá á Fjöllum.) *Náttúrufræðingurinn*, 43, 12-34.

Tómasson, H. 1974. Grímsvatnahlaup 1972, mechanism and sediment discharge. *Jökull*, 24, 27-38.

Tómasson, H. 1986. Glacial and volcanic shore interactions. Part I: On land. In, Sigbjarnarson, G. (ed), *Iceland Coastal and River Symposium Proceedings*, Reykjavík, University of Iceland, 7–16.

Tómasson, H. 1993. Jökulstífluð vötn á Kili og hamfarahlaup í Hvítá í Árnessýslu. *Náttúrufræðingurinn*, 62, 77-98.

Tómasson, H. 1996. The jökulhlaup from Katla in 1918. *Annals of Glaciology*, 22, 249-254.

Tómasson, H. 2002. Catastrophic floods in Iceland. In, Snorasson, Á., Finnsdóttir, H.P. and Moss, M. (eds.), *The Extremes of the Extremes: Extraordinary Floods*. IAHS Publication 271, 121-126.

Tómasson, H. and Vilmundardóttir, E.G. 1967. The lakes Stórisjór and Langisjór. *Jökull*, 17, 280-299.

Tómasson, H., Ingólfsson, P. and Pálsson, S. 1980. Comparison of sediment load transport in the Skeiðará jökulhlaups in 1972 and 1976. *Jökull*, 30, 21-33.

Torfason, H. 1974. *Skorradalur – Andakíll, Landmótun og Laus Jarðlög*. Unpublished B.S. Thesis, University of Iceland, Reykjavík.

Tryggvason, E. 1960. Earthquakes, jökulhlaups and subglacial eruptions. *Jökull*, 10, 18-22.

Tuffen, H., Gilbert, J. and McGarvie, D. 2001. Products of an effusive subglacial rhyolite eruption: Bláhnúkur, Torfajökull, Iceland. *Bulletin of Volcanology*, 63, 179-190.

Tuffen, H, McGarvie, D.W., Gilbert, J.S. and Pinkerton, H. 2002. Physical volcanology of a subglacial-to-emergent rhyolitic tuya at Rauðufossafjöll, Torfajökull, Iceland. In, Smellie, J.L. and Chapman, M. (eds.), *Ice-volcano interaction on Earth and Mars*. Geological Societyof London Special Publication, 202, 213-236.

Turney, C. S. M., Harkness, D. D. and Lowe, J. J. 1997. The use of microtephra horizons to correlate Late-glacial lake sediment successions in Scotland. *Journal of Quaternary Science*, 12, 525-531.

Turville-Petre, G. 1953. *Origins of Icelandic Literature*. Oxford, Clarendon Press.

Turville-Petre, G. and Olszewska, E.S. 1942. *The Life of Gudmund The Good, Bishop of Hólar*. The Viking Society for Northern Research, Coventry, Curtis and Beamish, Ltd. Coventry.

Tushingham, A. M. and Peltier, W. R. 1991. Ice-3G: a new global model of Late Pleistocene deglaciation based upon geophysical prediction of post-glacial relative sea level change. *Journal of Geophysical Research*, 96, 4497-4523.

Tweed, F.S. 2000a. Jökulhlaup initiation by ice-dam flotation: The significance of glacier debris content. *Earth Surface Processes and Landforms*, 25, 105-108.

Tweed, F.S. 2000b. An ice-dammed lake in Jökulsárgil: predictive modelling and geomorphological evidence. *Jökull*, 48, 1-11.

Tweed, F. S. and Russell, A. J. 1999. Controls on the formation and sudden drainage of glacier-impounded lakes: implications for jökulhlaup characteristics. *Progress in Physical Geography*, 23, 79-110.

Tweed, F.S., Roberts, M.J., Finnegan, D.C., Russell, A.J., Knudsen, Ó. and Gomez, B., 2000. Englacial flood routes during the November 1996 jökulhlaup as revealed by aerial photography 1996-2000. *Eos*, 81, F503–504.

Urey, H.C. 1952. *The Planets: Their Origin and Development*. New Haven, Yale University Press.

Valdimarsson, H. and Malmberg, S.-A. 1999. Near-surface circulation in Icelandic waters derived from satellite tracked drifters. *Rit Fiskideildar*, 16, 23-39.

van Dijk, T.A.G.P. 2002. *Glacier Surges as a Control on the Development of Proglacial Fluvial Landforms and Deposits.* Unpublished Ph.D. Thesis, University of Keele.

van Dijk, T.A.G.P. and Sigurðsson, O. 2002. Surge-related floods at Skeiðarárjökull Glacier, Iceland: implications for ice-marginal outwash deposits. In, Snorasson, Á., Finnsdóttir, H.P. and Moss, M. (eds.), *The Extremes of the Extremes: Extraordinary Floods.* IAHS Publication 271, 193-198.

van der Meer, J.J.M., Kjaer, K.H. and Krüger, J. 1999. Subglacial water-escape structures and till structures, Sléttjökull, Iceland. *Journal of Quaternary Science*, 14, 191-205.

van Kreveld, S., Sarnthein, M., Erlenkeuser, H., Grootes, P., Jung, S., Nadeau, M. J., Pflaumann, U. and Voelker, A. 2000. Potential links between surging ice sheets, circulation changes, and the Dansgaard-Oeschger cycles in the Irminger Sea, 60-18 ka. *Paleoceanography*, 15, 425-442.

van Vliet-Lanoe, B., Bourgeois, O. and Dauteuil, O. 1998. Thufur formation in northern Iceland and its relation to Holocene climate change. *Permafrost and Periglacial Processes*, 9, 347–365.

Vasari, Y. 1972. The history of the vegetation of Iceland during the Holocene. In, Vasari, Y., Hyvärinen, H. and Hicks, S. (eds.), *Climatic Changes in Arctic Areas During the Last Ten-Thousand Years.* Acta Universitatis Ouluensis, A3 Geologica, 239-252.

Vasari, Y. 1973. Post-glacial plant succession in Iceland before the period of human interference. In, Khotinsky, N.A. (ed.) *Palynology of Holocene, Proceedings of the III. International Palynological Conference.* Moscow, Publishing House Nauka, 7-14.

Vasari, Y. and Vasari, A. 1990. L'histoire Holocène des lacs Islandais. In, Devers, S. (ed.), *Études Pour Jean Malaurie*, Éditions Plon, Paris, 277-293.

Vasey, D.E. 2001. A quantitative assessment of buffers among temperature variations, livestock, and the human population of Iceland, 1784-1900. *Climatic Change*, 48, 243-263.

Vésteinsson, O. 1998. Patterns of settlement in Iceland: a study in prehistory. *Saga-Book of the Viking Society*, 25, 1-29.

Vésteinsson, O. 2000. *The Christianization of Iceland: Priests, Power, and Social Change 1000-1300.* Oxford and New York, Oxford University Press.

Vídalín, Þ. Þ. 1754. Dissertatiuncula de Montibus Islandiae Chrystalinis. (Dissertation on the glaciers in Iceland.) (Written in 1695.) *Hamburgisches Magazin*, Band XIII. Hamburg und Leipzig.

Vigfússon, G., Sigurðsson, J. and others, (eds.) 1856-78: *Biskupa Sögur* I-II. Copenhagen, Hið íslenzka bókmenntafélag.

Vigfússon, G. (ed.) 1878. *Sturlunga Saga* I-II. Oxford, Clarendon Press.

Vigfússon, G. and York Powell, F. (eds. and transl.) 1905. *Íslendingabók* (in) *Origines Islandicæ. A Collection of the more important sagas and other native writings relating to the settlement and early history of Iceland* Vol. I. Oxford, Clarendon Press, 288-305.

Vilhjálmsson, H. 1997. Climatic variations and some examples of their effects on the marine ecology of Icelandic and Greenland waters, in particular during the present century. *Rit Fiskideildar*, 15, 1-29.

Vilmundardóttir, E. G. 1977. Tungnaárhraun. (Tugnaá lavas.) Reykjavík, Orkustofnun, OS-ROD-7702.

Vilmundardóttir, E.G., Snorrason, S.P. and Larsen, G. 2000. *Geological Map of Subglacial Volcanic Area Southwest of Vatnajökull Icecap, Iceland, 1:50,000.* Reykjavík, Orkustofnun.

Vilmundarson, Þ. 1969. Heimildir um hafís á síðari öldum. In, Einarsson, M.Á. (ed.), *Hafísinn.* Reykjavík, Almenna bókafélagið, 313-332.

Voelker, A.H.L. 1999. Zur Deutung Der Dansgaard-Oeschger Ereignisse in ultra-hochauflosenden Sedimentprofilen aus dem Europäischen Nordmeer. *Christian-Albrechts-Universität zur Kiel, Reports Institüt für Geowissenschaften* Nr. 9, 271 pp.

Voelker, A. H. L., Sarnthein, M., Grootes, P. M., Erlenkeuser, H., Laj, C., Mazaud, A., Nadeau, M.-J. and Schleicher, M. 1998. Correlation of marine ^{14}C ages from the Nordic Seas with the GISP2 isotope mecord: Implications for ^{14}C calibration beyond 25 ka BP. *Radiocarbon,* 40, 517-534.

Von Troil, U. 1777. *Bref rörande en resa til Island MDCCLXXII.* Uppsala, Magnus Swederus.

Von Troil, U. 1808. Letters on Iceland: containing observations on the civil, literary ecclesiastical and natural history ... made during a voyage ... in 1772, by J. Banks assisted by Dr. Solander Extr. fr. *A general collection of the best and most interesting voyages and travels* ... By John Pinkerton, Vol. I, London.

Von Troil, U. 1961. *Bréf frá Íslandi.* Haraldur Sigurðsson translated. Reykjavík, Bókaútgáfa Menningarsjóðs.

Vorren, T.O. and Labert, J.S. 1997. Trough mouth fans – paleoclimate and ice-sheet monitors. *Quaternary Science Reviews,* 16, 865-881.

Wada, K. 1985. The distinctive properties of Andosols. *Advances in Soil Science,* 2, 173-229.

Wada, K., Arnalds, Ó., Kakuto, Y., Wilding, L.P. and Hallmark, C.T. 1992. Clay minerals in four soils formed in eolian and tephra materials in Iceland. *Geoderma,* 52, 351–365.

Waitt, R.B. 1998. Cataclysmic flood along Jökulsá á Fjöllum, north Iceland, compared to repeated colossal jökulhlaups of Washington's channelled scabland. *15th International Sedimentological Congress, Alicante, Abstracts,* 811-12.

Waitt, R.B., 2002. Great Holocene floods along Jökulsá á Fjöllum, north Iceland. In, Martini, P.I., Baker, V.R. and Garzon, G. (eds.), *Flood and Megaflood Processes and Deposits: Recent and Ancient Examples.* IAS Special Publication, 31, 37-51.

Walden, J., Oldfield, F. and Smith, J. (eds.) 1999. *Environmental Magnetism. A Practical Guide.* Quaternary Research Association, London.

Walder, J.S. and Costa, J.E. 1996. Outburst floods from glacier-dammed lakes: the effect of mode of drainage on flood magnitude. *Earth Surface Processes and Landforms,* 21, 701-723.

Walker, G.P.L. 1965. Some aspects of Quaternary volcanism. *Transactions of the Leicester Literary and Philosophical Society,* 59, 25-40.

Walker, G.P.L. and Blake, D.H. 1966. The formation of a palagonite breccia mass beneath a valley glacier in Iceland. *Quarterly Journal of the Geological Society of London,* 122, 45-61.

Walker, J.C.G., Hays, P. B. and Kasting, J.F. 1981. A negative feedback mechanism for the long-term stabilization of Earth´s surface temperature. *Journal of Geophysical Research*, 86, 9776–9782.

Waller, R.I, Russell, A.J., van Dijk, T.A.G.P. and Knudsen, Ó. 2001. Jökulhlaup related ice fracture and the supraglacial routing of water and sediment, Skeiðarárjökull, Iceland. *Geografiska Annaler*, 83A, 29-38.

Waltershausen, S. von. 1847. *Physisch-Geographische Skizze von Island*. Göttingen.

Walther, J. V. and Wood, V.J. 1986. Mineral-fluid reactions. In Walther, J. V. and Wood, V. J. (eds), Fluid-Rock Interactions During Metamorphism. New York, Springer-Verlag, 194-212.

Wang, S.-W. and Zhao, Z.-C. 1981. Droughts and floods in China, 1470-1979. In, Wigley, T. M. L., Ingram, M. J. and Farmer, G. (eds.), *Climate and History: Studies in Past Climates and Their Impact on Man*. Cambridge, Cambridge University Press, 271-288.

Warren, W.P. and Ashley, G.M. 1994. Origins of the ice-contact stratified ridges (eskers) of Ireland. *Journal of Sedimentary Research*, 64, 433-449.

Wastl, M. 2000. *Reconstruction of Holocene Palaeoclimatic Conditions in Northern Iceland Based on Investigations of Glacier and Vegetation history*. Unpublished Ph.D. Thesis, University of Innsbruck.

Wastl, M., Stötter, J. and Caseldine, C. 2001. Reconstruction of Holocene variations of the upper limit of tree or shrub birch growth in northern Iceland based on evidence from Vesturárdalur – Skíðadalur, Tröllaskagi. *Arctic, Antarctic, and Alpine Research*, 33, 191-203.

Wastl, M., Stötter, J. and Venzke, J.-F. 2001. Gletschergeschichtliche Untersuchungen zum Übergang Spätglazial/Postglazial in Nordisland. *Norden*, 14, 127-144.

Wastl, M., Stötter, J. and Venzke, J.-F. 2003. Neue Beiträge zur spätglazialen und holozänen Gletschergeschichte in Nordisland. *Norden*, 15, 137-158.

Wawn, A. (ed.) 1987. *The Iceland Journal of Henry Holland 1810*. London, The Hakluyt Society, Second Series, No. 168.

Webb, F. M., Marshall, S. J., Björnsson, H. and Clarke, G. K. C. 1999. Modelling the variations in glacial coverage of Iceland from 120,000 BP to Present. *EOS Abstract Vol*, 80(40), F332.

Welch, R. 1967. *The Application of Aerial Photography to the Study of a Glacial Area. Breiðamerkur, Iceland*. Unpublished Ph.D. Thesis, University of Glasgow.

Werner, R. and Schmincke, H.-U. 1999. Englacial vs lacustrine origin of volcanic table mountains: evidence from Iceland. *Bulletin of Volcanology*, 60, 335-354.

West, J.F. (ed.) 1970-76. *The Journals of the Stanley Expedition to the Faroe Islands and Iceland in 1789*, I-III. Tórshavn, Føroya Fróðskaparfélag.

Whalley, W.B. and Martin, H.E. 1994. Rock glaciers in Tröllaskagi, their origin and climatic significance. *Münchener Geographische Abhandlungen*, B12, 289-308.

Whalley, W.B., Palmer, C.F., Hamilton, S.J., Gordon, J.E. and Martin, H.E. 1995a. The dynamics of rock glaciers: data from Tröllaskagi, north Iceland. In, Slaymaker O. (ed.), *Steepland Geomorphology*. Chichester, Wiley, 129-145.

Whalley, W.B., Palmer, C.F., Hamilton, S.J. and Martin, H.E. 1995b. An assessment of rock glacier sliding using seventeen years of velocity data: Nautardalur rock glacier, north Iceland. *Arctic and Alpine Research*, 27, 345-351.

Whiting, P. 1996. Sediment sorting over bed topography. In, Carling, P.A. and Dawson, M.A. (eds.), *Advances in Fluvial Dynamics and Stratigraphy*. London, Wiley, 203-228.

Wilson, L. and Head, J.W. 2002. Heat transfer and melting in subglacial basaltic volcanic eruptions: implications for volcanic deposit morphology and meltwater volumes. In, Smellie, J.L. and Chapman, M. (eds.), *Ice-Volcano Interaction on Earth and Mars*. Geological Society of London Special Publication, 202, 2-26.

Wisniewski, E., Andrzejewski, L. and Olszewski, A. 1999. Relief of the Höfðabrekkujökull forefield, South Iceland, in light of geomorphological mapping. *Jökull*, 47, 59-70.

Wohletz, K.H. 1983. Mechanisms of hydrovolcanic pyroclast formation: grain size, scanning electron microscopy, and experimental studies. *Journal of Volcanology and Geothermal Research*, 17, 31-64.

Wolff-Boenisch, D., Gíslason, S.R. and Oelkers, E.H. 2002. Dissolution rates of volcanic glasses of different chemical composition. *Geochimica et Cosmochimica Acta Special Supplement*. Goldschmidt Conference Abstracts 2002, A843.

Wood, C. A. 1984. The amazing and portentous summer of 1783. *Eos*, 65, 410.

Wood, C. A. 1992. Climatic effects of the 1783 Laki eruption. In, Harington, C. R. (ed.), *The Year Without a Summer?* Ottawa, Canadian Museum of Nature, 58-77.

Woods, A. W. 1993. A model of the plumes above basaltic fissure eruptions. *Geophysical Research Letters*, 20, 1115-1118.

Yoon, S. H., Chough, S. K., Thiede, J. and Werner, F. 1991. Late Pleistocene sedimentation on the Norwegian continental slope between 67 degrees and 71 degrees N. *Marine Geology*, 99, 187-207.

Zielinski, T. and van Loon, A. J. 2002. Present-day sandurs are not representative of the geological record. *Sedimentary Geology*, 152, 1-5.

Zielinski, G. A., Germani, M. S., Larsen, G., Ballie, M. G. L., Withlow, S., Twicker, M. S. and Taylor, K. 1995. Evidence of the Eldgjá (Iceland) eruption in the GISP2 Greenland ice core: Relationship to eruption processes and climatic conditions in the tenth Century. *The Holocene*, 5, 129-140.

Zimanowski, B. 1998. Phreatomagmatic explosions. In, Freundt, A. and Rosi, M. (eds.), *From Magma to Tephra*. Amsterdam, Elsevier, 25-54.

Zutter, C. 1997. *The Cultural Landscape of Iceland: A Millennium of Human Transformation and Environmental Change*. Unpublished Ph.D. Thesis, University of Edmonton.

Zutter, C. 2000. Wood and plant-use in 17th – 19th century Iceland: archaeobotanical analysis of Reykholt, western Iceland. *Environmental Archaeology*, 5, 73-82.

COLOUR SUPPLEMENT INDEX

Colour plates of the black and white figures as printed in chapters:

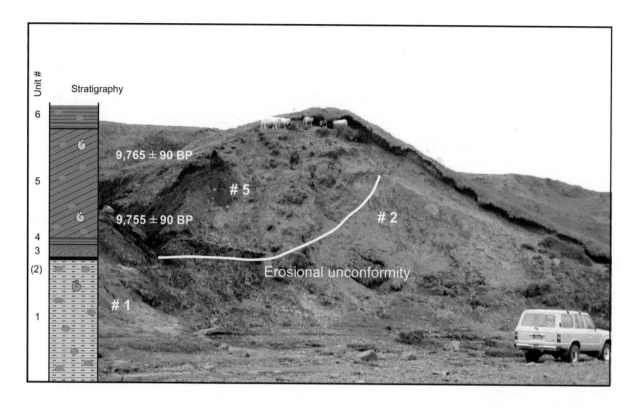

Fig. 3.10 The Mjóhylur locality in the Laxárdalur valley in the Dalir district in western Iceland (based on: Norðdahl and Ásbjörnsdóttir, 1995).

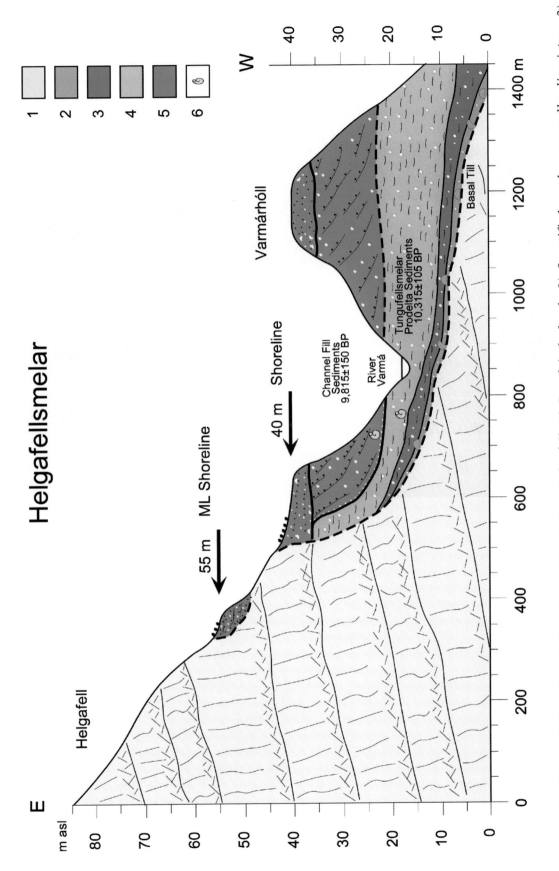

Fig. 3.14 The Varmárhóll – Helgafellsmelar section below Helgafell. 1) Basaltic bedrock. 2) Stratified sandy – gravelly diamicton. 3) Poorly stratified sand and gravel. 4) Fossiliferous sandy prodelta sediments. 5) Fossiliferous channel fill and foreset sediments. 6) Radiocarbon dated sample of marine shells.

Fig. 3.22   The raised marine limit shoreline at Sveltingur (Fig. 3.21) about 15 km south of Kópasker in northeastern Iceland.

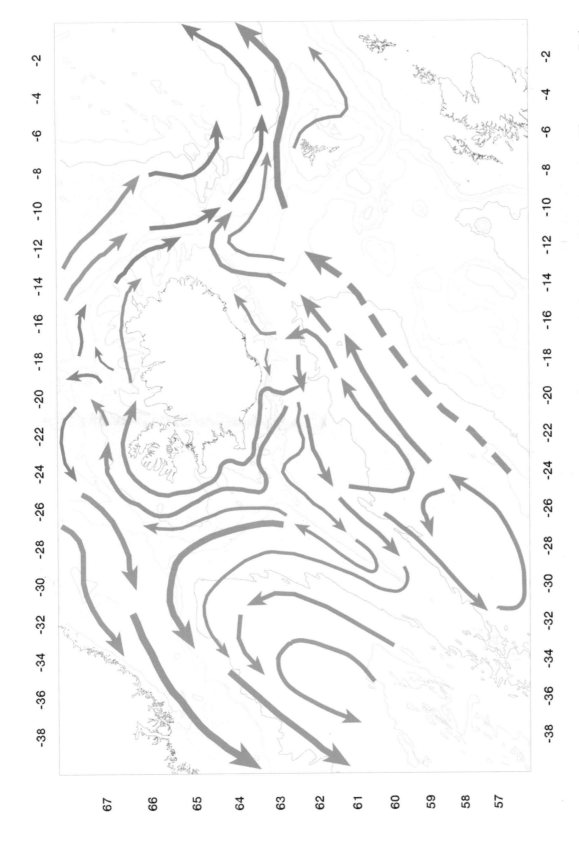

Fig. 4.1b Schematic presentation of the surface circulation in Icelandic waters. Red arrows: Atlantic water; Blue arrows: Polar water; Green arrows: mixed waters. Depth contours are 200, 500, 1000, 2000 and 3000 m (from Valdimarsson and Malmberg (1999). Values around the margin are in degrees of latitude and longitude (negative values are degrees west, positive are degrees north)

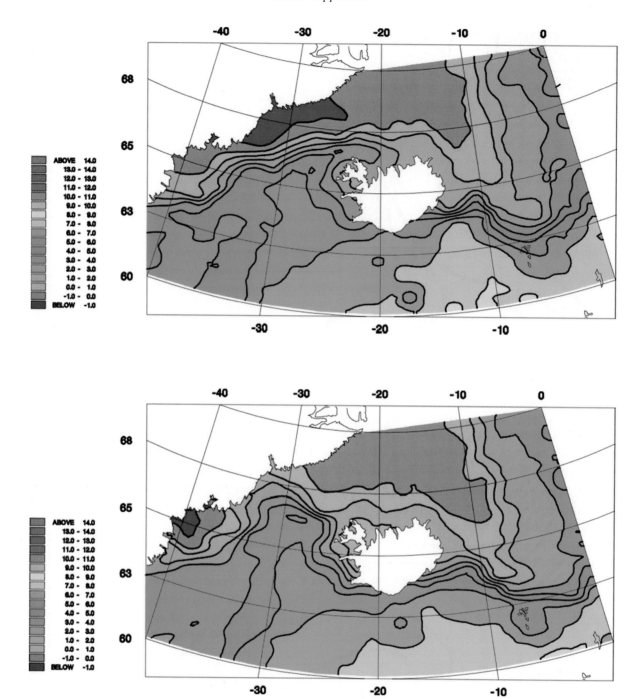

Fig. 4.2 Sea surface temperature in the ocean around Iceland in February 1994 (above) and February 1995 (below). Values around the margin are in degrees of latitude and longitude (negative values are degrees west, positive are degrees north)

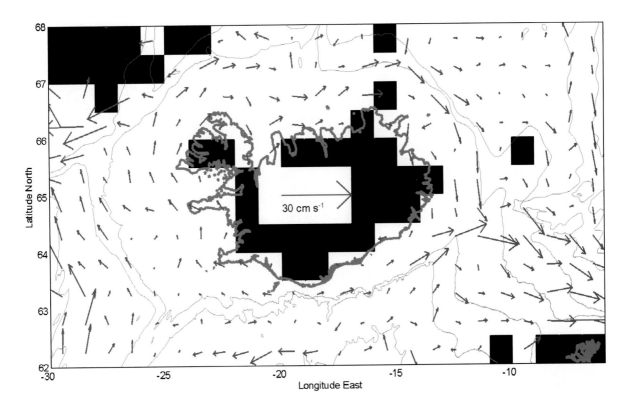

Fig. 4.4 Mean surface circulation binned from all drogued data 1990–1999 (from Poulain et al., 2001).

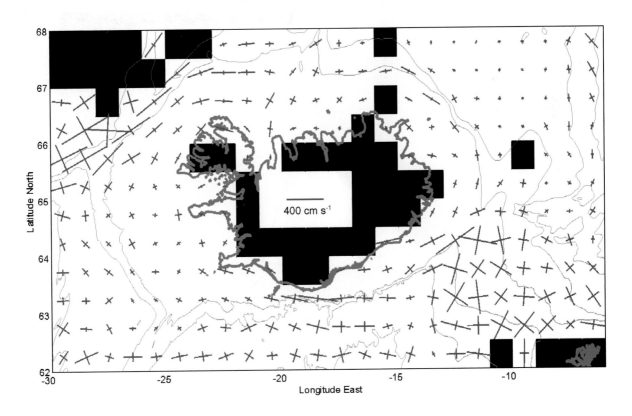

Fig. 4.5 Principal axes of velocity variance for the same data as in previous figure (from Poulain et al., 2001).

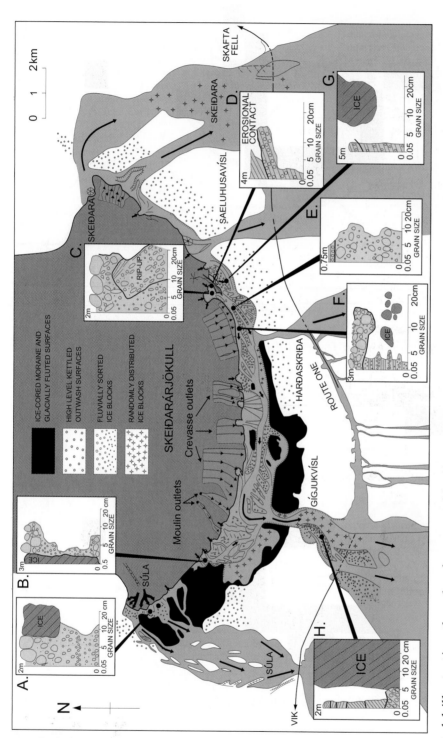

Fig. 7.31 Model illustrating the role of topographically generated backwater effects on the style and spatial distribution of proglacial jökulhlaup deposition (after Russell & Knudsen, 1999a). The model shows how vertical sedimentary characteristics might vary depending on whether jökulhlaup waters were confined by moraine ridges (A & B) or were free to spread out from the ice-margin (C–F). Note increased spatial variability of expected sedimentary successions associated with the moraine-confined conditions where backwater conditions act as a major control on sedimentation at different times during the flood. Unconfined jökulhlaup successions are divided into locations where waning-stage sediment flux decreases (A) and where sediment flux remains high even on the waning flow stage (B). In the moraine-confined scenario, backwater conditions result in much finer grained deposits downstream of the backwater zone (D). Sedimentary successions associated with ice-contact proglacial fans may also vary depending upon degree of backwater ponding or whether deposition is into shallow fast flows (C) or into sluggish deep flows (F).

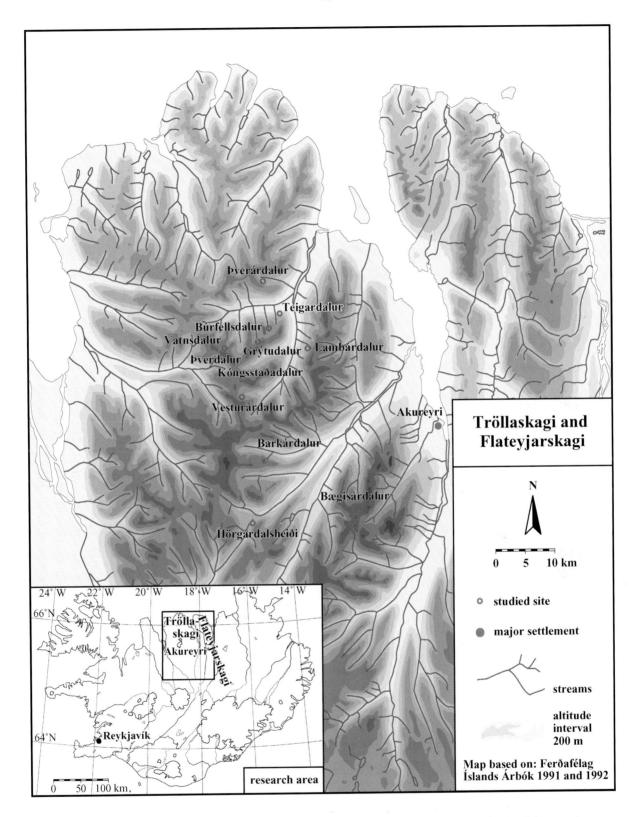

Fig. 9.1 Location of the sites studied for the reconstruction of Holocene glacier history in Northern Iceland.

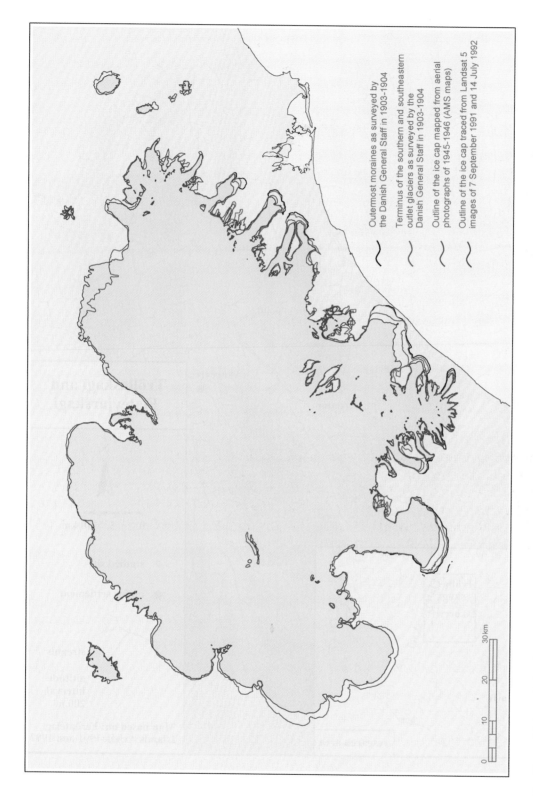

Outermost moraines as surveyed by
the Danish General Staff in 1903-1904

Terminus of the southern and southeastern
outlet glaciers as surveyed by the
Danish General Staff in 1903-1904

Outline of the ice cap mapped from aerial
photographs of 1945-1946 (AMS maps)

Outline of the ice cap traced from Landsat 5
images of 7 September 1991 and 14 July 1992

0   10   20   30 km

Fig. 10.6 Outline of Vatnajökull at four different times. The outermost one is based on the outermost moraines mapped by the Danish General Staff in 1903-1904 (only at the termini of the southern part). The second is the margin of Vatnajökull surveyed by the Danish General Staff in 1903-1904 (only the southern and southeastern part). The third is from U. S. Army Map Service (AMS/ Series C762, 1:50,000 scale) maps compiled from aerial photographs of 1945-1946. The fourth is traced from Landsat 5 images taken on 7 September 1991 and 14 July 1992.

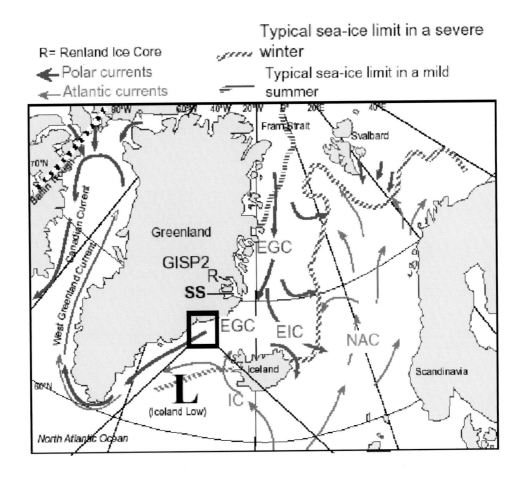

Fig 11.1. The surface currents around Iceland and also the position of the ice edge in a severe winter and a mild summer. 'R' is the site of the Renland ice core. The site of the GISP2 ice-record is shown and also the study site (Jennings and Weiner, 1996) of the Nansen fjord area (indicated by a square). SS - Scoresby Sund; EGC - East Greenland Current; IC - Irminger Current; EIC - East Irminger Current; NAC - North Atlantic Current; NAC - Norwegian Atlantic Current. The diagram has been modified after Hurdle, 1986.